Methods in Molecular Biology™

For further volumes:
http://www.springer.com/series/7651

Metabolomics Tools for Natural Product Discovery

Methods and Protocols

Edited by

Ute Roessner and Daniel Anthony Dias

School of Botany, The University of Melbourne, Parkville, Victoria, Australia

Humana Press

Editors
Ute Roessner
School of Botany
The University of Melbourne
Parkville, Victoria, Australia

Daniel Anthony Dias
School of Botany
The University of Melbourne
Parkville, Victoria, Australia

ISSN 1064-3745 ISSN 1940-6029 (electronic)
ISBN 978-1-4939-6301-0 ISBN 978-1-62703-577-4 (eBook)
DOI 10.1007/978-1-62703-577-4
Springer New York Heidelberg Dordrecht London

Printed on acid-free paper

Humana Press is a brand of Springer
Springer is part of Springer Science+Business Media (www.springer.com)

Preface

The inspiration behind this volume of *Methods in Molecular Biology: Metabolomics Tools for Natural Product Discoveries* is to unite the diverse methodologies and protocols that have developed from classical natural product chemistry now transitioning to modern day metabolomics to identify bioactive secondary metabolites for the purpose of drug discovery. Natural product chemistry maybe regarded as an older science with the isolation of many natural products between the 1970s and 1990s mainly due to the advent of improved, highly sensitive separation methods, mass spectrometry, and nuclear magnetic resonance spectroscopy. Therefore, it is worthwhile to summarize recent advancements in current comprehensive analytical platforms and present how metabolomics is now being integrated into this classical field to dereplicate and profile natural product extracts.

The aims of this book are to discuss and in part review, how natural product chemistry is transitioning to metabolomics as a result of the advent of comprehensive analytical platforms. Applications for the extraction of selected natural products (secondary metabolites) from less common sources such as bryophytes and hard corals are presented. Various biological assays including anticancer, anti diabetic, antibacterial, and various metabolomic, biomarker, and dereplication strategies are discussed. Comprehensive applications and strategies for gas chromatography-mass spectrometry (GC-MS) (polar metabolite profiling and fatty acid analysis), liquid chromatography-mass spectrometry (LC-MS) (untargeted profiling and lipidomics), and nuclear magnetic resonance (NMR) spectroscopy (profiling) metabolomic-based studies are discussed. Finally, protocols and strategies for the structure elucidation of isolated natural products by NMR, determination of absolute configurations, and the discovery, biosynthesis, and engineering of novel enterocin and wailupemycin polyketide analogues and the synthesis of K_1 and K_2 melatonin metabolites are presented.

Methods in Molecular Biology: Metabolomics Tools for Natural Product Discoveries is targeted at chemists, biologists, pharmacologists, students, and researchers in related fields to appreciate the current available methodologies and protocols for natural product isolation, biomarker discovery, dereplication, biological assays, and comprehensive metabolomic platforms available for high-throughput analyses.

Parkville, Victoria, Australia

Ute Roessner
Daniel Anthony Dias

Contents

Contributors

HANI AL SALAMI • *Curtin Health Innovation Research Institute, School of Pharmacy, Curtin University, Perth, Australia*
ABIGAIL L. ALBRIGHT • *School of Chemistry, The University of Melbourne, Melbourne, Australia; Bio21 Institute (Molecular Science and Biotechnology Institute), The University of Melbourne, Melbourne, Australia*
YOSHINORI ASAKAWA • *Faculty of Pharmaceutical Sciences, Tokushima Bunri University, Tokushima, Japan*
RIMA CACCETTA • *School of Pharmacy, Curtin Health Innovation Research Institute, Curtin University, Perth, Australia*
DAMIEN L. CALLAHAN • *Metabolomics Australia, School of Botany, The University of Melbourne, Parkville, VIC, Australia*
YOUNG HAE CHOI • *Natural Products Laboratory, Institute of Biology, Leiden University, Leiden, The Netherlands*
ALYSHA M. DE LIVERA • *Metabolomics Australia, Bio21 Institute (Molecular Science and Biotechnology Institute), The University of Melbourne, Melbourne, Australia*
DAVID P. DE SOUZA • *Metabolomics Australia, Bio21 Institute (Molecular Science and Biotechnology Institute), The University of Melbourne, Melbourne, Australia*
DANIEL ANTHONY DIAS • *School of Botany, The University of Melbourne, Parkville, Victoria, Australia*
RUANGELIE EDRADA-EBEL • *Strathclyde Institute of Pharmacy and Biomedical Sciences, University of Strathclyde, Glasgow, UK*
CANDACE E. ELLIOTT • *School of Botany, The University of Melbourne, Melbourne, Australia*
PHILIPPE J. EUGSTER • *Phytochemistry and Bioactive Natural Products, School of Pharmaceutical Sciences, University of Geneva and University of Lausanna, Geneva, Switzerland*
GAËTAN GLAUSER • *Chemical Analytical Service of the Swiss Plant Science Web, Institute of Biology, University of Neuchâtel, Neuchâtel, Switzerland*
BENJAMIN R. GORDON • *Australian Research Council Centre of Excellence for Coral Reef Studies, AIMS@JCU, The Comparative Genomics Centre and School of Pharmacy and Molecular Sciences, James Cook University, Townsville, QLD, Australia*
ELISA J. HAYHOE • *Environmental and Biotechnology Centre, Swinburne University of Technology, Hawthorn, Australia*
HELMUT M. HÜGEL • *School of Applied Sciences, RMIT University, Melbourne, Australia*
NIRUPAMA SAMANMALIE JAYASINGHE • *Metabolomics Australia, School of Botany, The University of Melbourne, Melbourne, Australia*
OLIVER A.H. JONES • *School of Applied Sciences, RMIT University, Melbourne, Australia*
JOHN A. KALAITZIS • *School of Biotechnology and Biomolecular Sciences, The University of New South Wales, Sydney, Australia*
HYE KYONG KIM • *Natural Products Laboratory, Institute of Biology, Leiden University, Leiden, The Netherlands*

WILLIAM LEGGAT • *Australian Research Council Centre of Excellence for Coral Reef Studies, The Comparative Genomics Centre and School of Pharmacy and Molecular Sciences, James Cook University, Townsville, QLD, Australia*
AGNIESZKA LUDWICZUK • *Faculty of Pharmaceutical Sciences, Tokushima Bunri University, Tokushima, Japan; Department of Pharmacognosy with Medical Plant Unit, Medical University of Lublin, Lublin, Poland*
JANICE MCCAULEY • *School of Chemistry, University of Wollongong, Wollongong, Australia; Shoalhaven Marine and Freshwater Centre, School of Biological Sciences, University of Wollongong, Wollongong, Australia*
CHERIE A. MOTTI • *Australian Institute of Marine Science, Cape Cleveland, Australia*
MOSHE OLSHANSKY • *Bioinformatics Division, Walter and Eliza Hall Institute, Parkville, VIC, Australia*
ENZO A. PALOMBO • *Environmental and Biotechnology Centre, Swinburne University of Technology, Hawthorn, Australia*
LÚCIA P. SANTOS PIMENTA • *Departamento de Química, Instituto de CiênciasExatas, Universidade Federal de Minas Gerais, Belo Horizonte, Brazil; Natural Products Laboratory, Institute of Biology, Leiden University, Leiden, The Netherlands*
UTE ROESSNER • *School of Botany, The University of Melbourne, Parkville, Victoria, Australia*
THUSITHA WASNTHA THILAKA RUPASINGHE • *Metabolomics Australia, Bio21 Institute (Molecular Science and Biotechnology Institute), The University of Melbourne, Melbourne, Australia*
JOHN ROBERT SHEEDY • *Department of Zoology, The University of Melbourne, Parkville, VIC, Australia*
DANIELLE SKROPETA • *School of Chemistry, University of Wollongong, Wollongong, Australia; Centre for Medicinal Chemistry, University of Wollongong, Wollongong, Australia*
TERENCE P. SPEED • *Bioinformatics Division, Walter and Eliza Hall Institute, Parkville, VIC, Australia*
AHMED FARES TAWFIKE • *Strathclyde Institute of Pharmacy and Biomedical Sciences, University of Strathclyde, Glasgow, UK*
SYLVIA URBAN • *School of Applied Sciences, Health Innovations Research Institute (HIRi), RMIT University, Melbourne, Australia*
ROBERT VERPOORTE • *Natural Products Laboratory, Institute of Biology, Leiden University, Leiden, The Netherlands*
CHRISTINA VIEGELMANN • *Strathclyde Institute of Pharmacy and Biomedical Sciences, University of Strathclyde, Glasgow, UK*
JONATHAN M. WHITE • *School of Chemistry, The University of Melbourne, Melbourne, Australia; Bio21 Institute (Molecular Science and Biotechnology Institute), The University of Melbourne, Melbourne, Australia*
JEAN-LUC WOLFENDER • *Phytochemistry and Bioactive Natural Products, School of Pharmaceutical Sciences, University of Geneva and University of Lausanne, Geneva, Switzerland*
ANA ZIVANOVIC • *School of Chemistry, University of Wollongong, Wollongong, Australia; Centre for Medicinal Chemistry, University of Wollongong, Wollongong, Australia*

Chapter 1

Bryophytes: Liverworts, Mosses, and Hornworts: Extraction and Isolation Procedures

Yoshinori Asakawa and Agnieszka Ludwiczuk

Abstract

There are more than 20,000 species of bryophytes in the world. Among them, almost of liverworts (Marchantiophyta) possess beautiful blue, yellow colored or colorless cellular oil bodies from which over several hundred new terpenoids, acetogenins, and aromatic compounds including flavonoids with more than 40 new carbon skeletons have been isolated. Some of the isolated compounds from liverworts show antimicrobial, antifungal, antiviral, allergenic contact dermatitis, cytotoxicity, insect antifeedant and mortality, antioxidant, nitric oxide (NO) production and plant growth inhibitory, neurotrophic and piscicidal activity, tublin polymerization inhibitory, muscle relaxing, and liver X-receptor (LXR)α agonist and (LXR) β antagonist activities, among others. The bio- and chemical diversity, chemical analysis of bryophytes including extraction, distillation, purification, TLC, GC and GC-MS, and HPLC analysis of oil bodies of liverworts are surveyed.

Key words Bryophytes, Liverworts, Bio- and chemical diversity, Extraction, Isolation, Essential oils, Terpenoids, Bis-bibenzyls, Biological activity

1 Introduction

The bryophytes are found everywhere in the world except in the sea. They grow on the tree and soil, in lakes and rivers, even in Antarctic island [1]. The bryophytes are placed taxonomically between algae and pteridophytes; there are about 20,000 species in the world. They are further divided into three phyla: Bryophyta (mosses 14,000 species) (Fig. 1), Marchantiophyta (liverworts 6,000 species) (Fig. 2), and Anthocerotophyta (hornworts 300 species) (Fig. 3).

Liverworts are considered to be the oldest terrestrial green plants, although no strong scientific evidence for this has appeared in the literature. This hypothesis was mainly based on the resemblance of the present-day liverworts to the first land plant fossils, the spores of which date back almost 500 million years. Among the bryophytes almost all liverworts possess beautiful blue, yellow

Ute Roessner and Daniel Anthony Dias (eds.), *Metabolomics Tools for Natural Product Discovery: Methods and Protocols*, Methods in Molecular Biology, vol. 1055, DOI 10.1007/978-1-62703-577-4_1, © Springer Science+Business Media, LLC 2013

Fig. 1 *Polytrichum* species (Bryophyta)

Fig. 2 *Radula* species (Marchantiophyta)

Fig. 3 *Anthocerotae* species (Anthocerotophyta)

Fig. 4 Oil bodies of *Frullania vethii* (Frullaniaceae, Marchantiophyta)

colored or colorless cellular oil bodies (Fig. 4) which are peculiar, membrane-bound cell organelles that consist of ethereal terpenoids and aromatic oils suspended in carbohydrates- or protein-rich matrices, while mosses and hornworts do not. These oil bodies are very important biological markers for the classification in the Marchantiophyta [2–15].

Phytochemistry of bryophytes has been neglected for a long time because they are morphologically very small and difficult to collect a large amount as pure samples. Their identification is also very difficult even under the microscope. They are considered to be nutritionally useless to human diets. Generally, bryophytes are not damaged by bacteria and fungi, insect larvae and adults, snails, slugs, and other mammals. Furthermore, some liverworts cause intense allergenic contact dermatitis.

A number of bryophytes, in particular, mosses have been widely used as medicinal plants in China, to cure burns, bruises, external wounds, snake bite, pulmonary tuberculosis, neurasthenia, fractures, convulsions, scald, uropathy, pneumonia, neurasthenia, etc. [16–21]. Many species of liverworts have characteristic fragrant odors and with intense hot and bitter taste. Some *Fissidens* species belonging to the Bryophyta produce sweet tasting secondary metabolites. We have been interested in these biologically active substances found in bryophytes and have studied more than 1,000 species of bryophytes collected in the world with respect to their chemistry, pharmacology, and application as sources of cosmetics, and medicinal or agricultural drugs. The biological activities of liverworts are due to terpenoids, aromatic compounds, and acetogenins which constitute oil bodies in each species [3, 4, 9, 22–32].

1.1 Biodiversity of Bryophytes

As presented in Fig. 5, there are 54 endemic genera (black numbers) in southern hemispheric countries, such as New Zealand and Argentina [33]. In South-East Asia and Japan, relatively a large number of endemic genera has been recorded, however, South

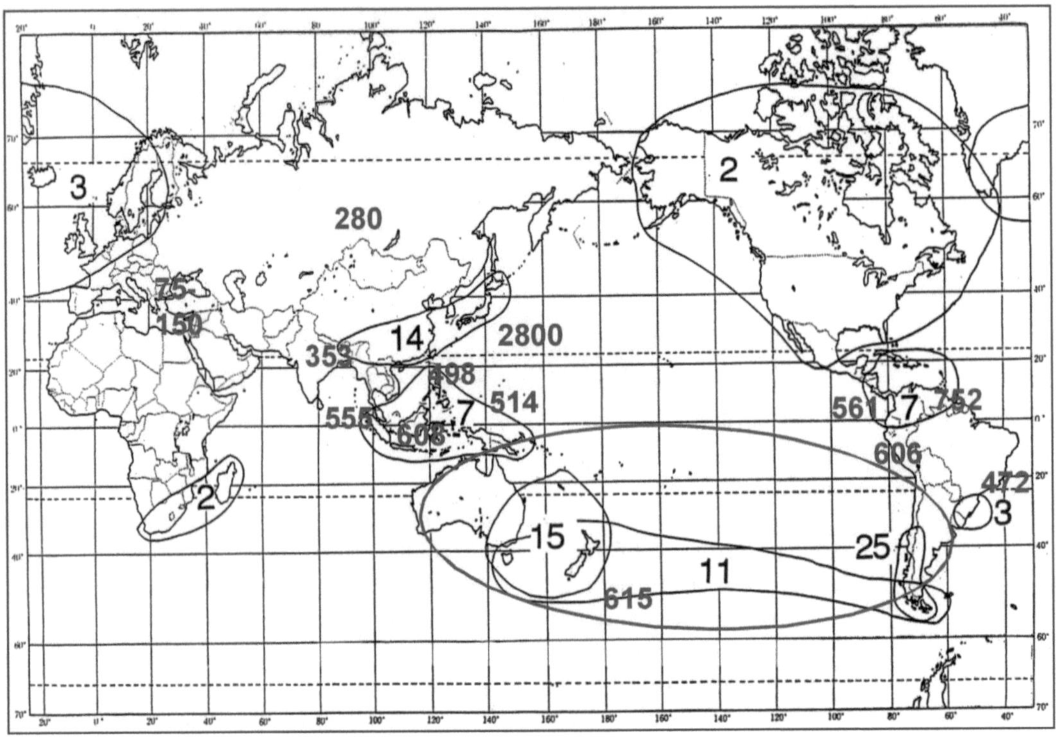

Fig. 5 Distribution of endemic genera (*black letters*) and species numbers (*red letters*) of bryophytes (Color figure online)

Africa, Africa, Madagascar, and both North America and Europe are very poor regions of endemic genera. The red numbers in Fig. 5 illustrates the species numbers of bryophytes [34].

The Marchantiophyta (liverworts) includes two subclasses: the Jungermanniidae and Marchantiidae, and six orders, 49 families, 130 genera, and 6,000 species. Still many new species have been recorded in the literatures. The richness of endemic genera of bryophytes in the southern hemisphere suggests that the bryophytes might originate from the past Antarctic islands since 350–400 million years ago and migrated to the northern hemisphere with a long range evolutionary process. In the southern hemisphere, New Zealand is one of the most charming countries to see many different species of the Marchantiophyta which are totally different from those found in the northern Asia, including Japan. In Japan, Yaku Island which has been protected by UNESCO is the most interesting place to observe many species of the Marchantiophyta.

In the tropical regions, such as south-east Asia, Borneo, Sumatra and Papua New Guinea and Colombia, Ecuador, and Venezuela, there are rainforests where numerous liverworts species have been found, but many different species like the Lejeuneaceae

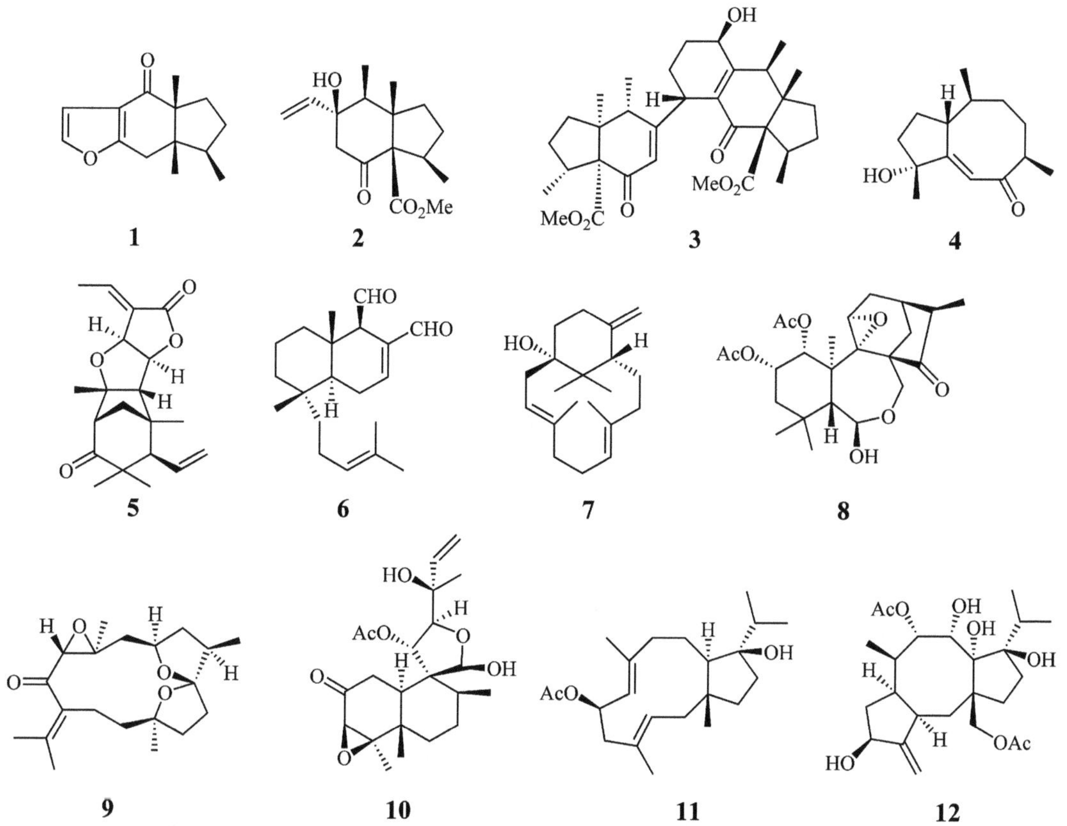

Fig. 6 Sesqui- (**1–4**) and diterpenoids (**5–12**) found in liverworts

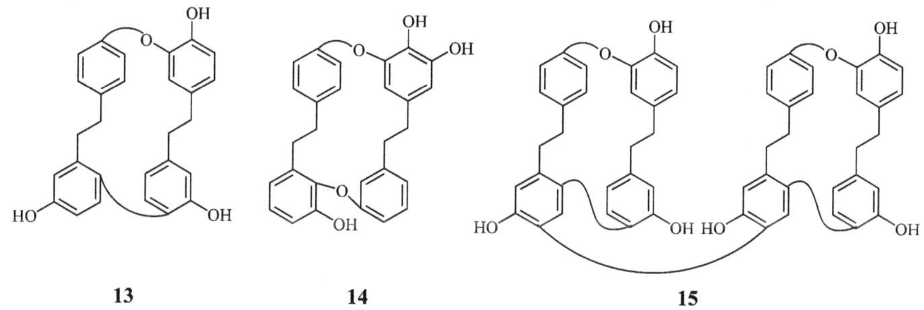

Fig. 7 Bis-bibenzyls found in liverworts

species intermingle with each other and it is very time consuming work to purify all of them. In Columbia and Ecuador, the Marchantiophyta species grow in the high mountains, over 2,000 m where people live, but not in the lower level of their lands.

1.2 Chemical Diversity of Bryophytes

Typical secondary metabolites isolated from liverworts are illustrated in Figs. 6, 7, 8, 9. Among bryophytes, liverworts are characterized by greatest chemical diversity. Almost all of liverworts produce a great number of and high amount of terpenoids (**1–12**),

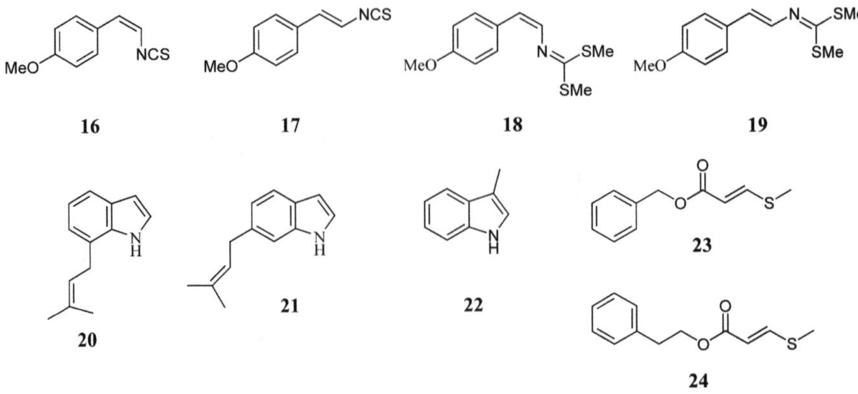

Fig. 8 Nitrogen and/or sulfur containing aromatic compounds found in liverworts

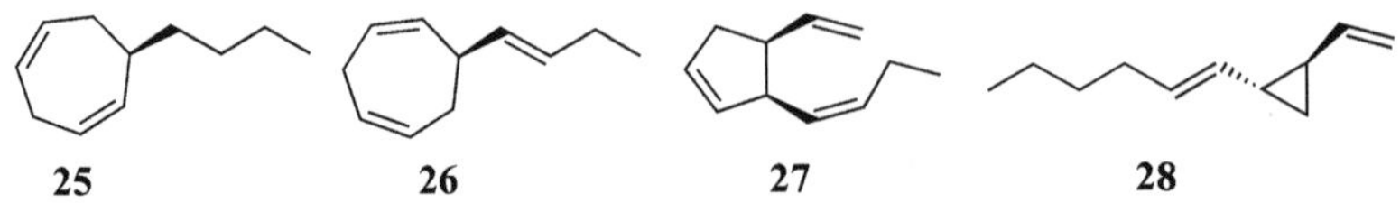

Fig. 9 Acetogenins (sex pheromones of brown algae) found in liverworts

aromatic compounds (**13–24**), and acetogenins (**25–28**). However, the presence of nitrogen, sulfur, or both nitrogen and sulfur containing compounds in bryophytes is very rare. The most characteristic chemical phenomenon of liverworts is that most of sesqui- and diterpenoids are enantiomers of those found in higher plants, although there are a few exceptions such as germacrane- and guaiane-type sesquiterpenoids. It is noteworthy that the different species of the same genera, such as *Frullania dilatata* and *F. tamarisci* (Frullaniaceae) each produces different sesquiterpene lactone enatiomers with some liverworts, such as *Lepidozia* species (Lepidoziaceae), biosynthesize both enantiomers.

One of the most significant compounds in liverworts is *bis-bibenzyl* derivatives (**13–15**). Among which marchantin A (**14**) has been isolated from *Marchantia polymorpha* and *M. paleacea* var. *diptera* in high yield (30.2 g from 8 kg of the dried *M. polymorpha* and 79.5 g from 6.67 kg of *M. paleacea* var. *diptera*). Flavonoids are ubiquitous components in bryophytes and have been isolated from or detected in both the Marchantiophyta and the Bryophyta.

Recently several nitrogen-containing compounds (**16–19**) have been isolated from the Mediterranean liverwort, *Corsinia coriandrina* belonging to the Corsiniaceae (Marchantiales) [35], two prenyl indole derivatives (**20, 21**) from the *Riccardia* species belonging to the Dilaniaceae [4], skatole (**22**) from the *Asterella* or *Mannia* belonging to the Aytoniaceae [4] and the Tahitian *Cyathodium foetidissimum* belonging to the Cyathodiaceae [13], and benzyl- and β-phenethyl β-methylthioacrylates (**23, 24**) from Isotachidaceae [4].

Typical acetogenins are sex pheromones, dictyotene (**25**), ectocarpene (**26**), multifidene (**27**), and dictyopterene (**28**) from liverworts, which have been isolated from brown algae [12]. Highly evolved liverworts, belonging to the Marchantiaceae, produce phytosterols, such as campesterol, stigmasterol, and sitosterol which are ubiquitous components in higher plants. Almost all liverworts elaborate α-tocopherol and squalene. The characteristic components of the Bryophyta are highly unsaturated fatty acids and alkanones, such as 5,8,11,14,17-eicosapentaenoic acid, 7,10,13,16,19-docosapentaenoic acid, and 10,13,16-nonadecatrien-7-*yn*-2-one; hopane-type triterpenoids; flavonoids; benzonaphthoxanthenones; and some nitrogen-containing substances. The neolignan is one of the most important chemical markers of the Anthocerotophyta [4]. The presence of hydrophobic terpenoids is very rare in the Marchantiophyta. A few bitter kaurene glycosides have been found in the *Jungermannia* species, however a number of flavonoid glycosides have been detected both in Marchantiophyta and in Bryophyta [2, 4].

In this chapter, bio- and chemical diversity, chemical analysis of secondary metabolites in bryophytes are discussed.

2 Materials

2.1 Collecting Equipment

For collecting bryophytes you are necessary the following:

1. Paper bags, gardening gloves, pencils, tags.
2. Knife, trowel, or large spatula for prying or scarping up the specimen from the substrate.
3. Field notebook to record habitat and location information.

2.2 Thin-Layer Chromatography

1. Stationary phases: aluminum foils or glass plates covered by silica gel Si60 F_{254} (0.2 mm thickness of silica gel).
2. Solvents: analytical grade *n*-hexane, ethyl acetate, chloroform, methanol, and distilled water.
3. Mobile phases: (a) *n*-hexane:ethyl acetate, 4:1 (*v*/*v*); (b) *n*-hexane:ethylacetate, 1:1 (*v*/*v*); (c) chloroform:methanol:water, 65:35:5 (*v*/*v*/*v*) or the same solvent system in different ratios of solvents.

2.3 TLC Spray Reagents

1. Godin reagent I: Prepare two solutions: (a) 1 % solution of vanillin in ethanol mix in 1:1 ratio with 3 % solution of terchloric acid in water; (b) 10 % solution of sulfuric acid in ethanol (*see* **Note 1**) [36].
2. Godin reagent II: Prepare two solutions: (a) 1 % vanillin in ethanol and (b) 5 % sulfuric acid in ethanol (*see* **Note 2**) [37].
3. 30 % sulfuric acid: 30 % solution of sulfuric acid in ethanol.

2.4 Gas Chromatography

Capillary columns: The most popular capillary columns used for GC and GC-MS analysis of volatiles from bryophytes are

1. Capillary column with 100 % dimethyl polysiloxane phase composition (e.g., DB-1, HP-1, CPSil-5 CB).
2. Capillary column with 95 % dimethyl/5 % diphenyl polysiloxane phase composition (e.g., DB-5, HP-5MS).
3. Capillary column with 50 % dimethyl/50 % diphenyl polysiloxane phase composition (e.g., DB-17HT).
4. Capillary column with 14 % cyanopropylphenyl/86 % dimethyl polysiloxane phase composition (e.g., DB-1701; CPSil-19 CB).
5. For chiral analysis of essential oils received from bryophytes permethylated *β*-cyclodextrin capillary column (e.g., Cyclodex-B) or fused silica capillaries with octakis(2,6-di-*O*-methyl-3-*O*-pentyl)-γ-cylodextrin, heptakis(2,6-di-*O*-methyl-3-*O*-pentyl)-β-cylodextrin,orheptakis(6-*O*-*tert*-butyldimethylsilyl-2,3-di-*O*-mehyl)-β-cyclodextrin in OV 1701 (50 %,w/w) are used.
6. For preparative GC, a stainless steel column (1.85 m×4.3 mm) with 10 % polydimethylsiloxane SE-30 on Chromosorb W-HP or with 2.5 % octakis(2,6-di-*O*-methyl-3-*O*-pentyl)-γ-cylodextrin in OV-1701 (50 % w/w) on Chromosorb W-HP can be used [38, 39].

2.5 Open Column Chromatography

1. Columns: glass columns with different length and diameters (e.g., 30 cm×1.5 cm or 46 cm×5 cm).
2. Sorbents: silica gel 60 (70–230 mesh), Sephadex LH-20.
3. Solvents: analytical grade *n*-hexane, ethyl acetate, chloroform, dichloromethane, and methanol.

2.6 Preparative High-Performance Liquid Chromatography

1. Chromatography columns: Cosmosil 5SL-II (10 mm×250 mm) for normal phase HPLC and Cosmosil 5C18-AR (10 mm×250 mm) for reverse phase HPLC.
2. Solvents: HPLC-grade *n*-hexane, ethyl acetate, methanol, and acetonitrile. Glass-distilled deionized water passed through a high-purity organic remover for HPLC.
3. Mobile phases: For normal phase HPLC the mixtures of *n*-hexane/EtOAc are used as mobile phases, while for RP-HPLC mixtures of acetonitrile and water.

3 Methods

Bryophytes, like other plants, are complex matrices, producing a range of secondary metabolites with different functional groups and polarities. Prior to choosing an extraction method, it is necessary to establish the target of the extraction. The typical extraction process should start from proper preparation of plant material.

3.1 Collection of Bryophytes and Purification of Plant Samples

1. Select specimens in good condition and collect plants by use of knife, trowel, or large spatula. If the liverwort is growing on a tree, collect it with the bark (*see* **Note 3**).
2. Place all specimens of a single species from one locality into one collection bag (paper bag) (*see* **Note 4**).
3. As each specimen is collected, assign a unique collection number, date of collection, name(s) of collector(s), and location information.
4. After returning all collected specimens should be purified. Remove soil and other accompanying materials very carefully (*see* **Note 5**). Don't be afraid to wash purified material thoroughly, even it may take more than one washing. Excess moisture after washing can be removed leaving the plant between paper towels.

3.2 Voucher Specimens and Drying Procedure

1. Each collected specimens need to be correctly identified. If you have any doubt, send a duplicate specimen to a taxonomic expert for confirmation. A specialized taxonomist should give you detailed data concerning collected plant (i.e., classification into its species, genus, family, order, and class).
2. After identification, the names of the collected specimens, the place and date of collection, condition of collection site, collector name(s), and one who identified your samples should be recorded as part of a voucher deposited in a herbarium for future reference.
3. The remaining samples should be gently dried under shade and placed in cool place with no exposure to sunlight. Air dried plant material should be stored at room temperature (*see* **Note 6**). On the other hand, instead of drying, purified plant material can be deep-frozen (−20 °C) (*see* ref. [40]).

3.3 Extraction of Nonpolar Secondary Metabolites with Organic Solvents

1. Air-dried plant material should be grinded before extraction. In case of small samples (1–2 g of dried plant materials), these can be crushed by use of mortar and pestle.
2. Put your plant material to mortar, add 3 mL of solvent, like *n*-hexane, diethyl ether, or methylene chloride, and homogenize sample by pestle. If necessary add more solvent.
3. Transfer the homogenized sample to a proper flask and leave for maceration (*see* **Note 7**) for 1 or 2 weeks. After this time, solvent should be filtered from the plant material and fresh solvent should be added. Continue maceration for 1 of 2 weeks. Both extracts received from two maceration steps should be combined and evaporated at temperature less than 40 °C (*see* **Note 8**).

4. In case of larger samples (more than 2 g) plant materials should be mechanically ground (*see* **Note 9**) before extraction. This step is especially necessary for mosses and some thallus liverworts, e.g., *Marchantia*, *Conocephalum*, and *Wiesnerellaceae*, which are very hard. In order to obtain large amounts of their crude extracts, these samples should be mechanically ground before extraction.
5. Mix plant material with organic solvents. Put the flask with extraction mixture to the ultrasonic bath for a few minutes (*see* **Note 10**). After ultrasonic extraction leave the flask with plant material and solvent for maceration for 1 or 2 weeks. Repeat maceration as described in **step 3**.

3.4 Extraction of Polar Secondary Metabolites with Organic Solvents

Methanol is often used for the extraction to obtain hydrophilic compounds from bryophytes (*see* **Note 11**). There are two ways to extract polar secondary metabolites from bryophytes: continuation of extraction of the plant material remaining after procedures described in Subheading 3.3, or start new extraction from fresh air-dried sample(s).

1. If you want to continue procedures described in Subheading 3.3, dry the remaining plant materials, which were already extracted by the nonpolar solvent (*see* **Note 12**) mix with methanol. Put the flask with extraction mixture to the ultrasonic bath for 30 min. After ultrasonic extraction leave the flask with plant material and solvent for maceration for 1 or 2 weeks. Repeat maceration as described in Subheading 3.3, **step 3**.
2. If you are interested in the extraction of only polar secondary metabolites from the bryophytes, mix air-dried and mechanically ground plant material with methanol and follows procedure described in **step 3**. After evaporation of the solvent you can use crude methanol extract for further studies or redissolve this crude extract in water and participate it with diethyl ether, ethyl acetate, and *n*-butane successively by the use of liquid–liquid extraction.

3.5 Hydro- and Steam Distillation of Volatile Components

In order to obtain essential oils from bryophytes, hydrodistillation or steam distillation techniques can be used (*see* **Note 13**).

1. The essential oil can be isolated from fresh or air-dried plant material using Likens-Nickerson-type apparatus. Distill samples for 3 h with distillation speed 3 mL/min. Use *n*-pentane (50 mL) as the collection solvent. The oil recovered in *n*-pentane should be concentrated at room temperature under reduced pressure, collected in a vial and finally concentrated to minimum volume under nitrogen flux. Essential oil can be stored at −20 °C in the dark until analysis [40, 41].
2. The essential oil from the aqueous homogenates of fresh or air-dried plants can be prepared by a 2 h hydrodistillation in

Clevenger-type apparatus using *n*-hexane (1 mL) as collection solvent [38, 39].

3.6 TLC Analysis of Crude Extracts

The TLC analysis is the simplest, fastest, and cheapest method and is routinely used for first checking of crude extracts. Analytical TLC plates can be used to get an idea about the degree of polarity of different extract components. They are also widely applied in the detection of compounds through several separation steps. Moreover, they can be used to predict the separation pattern on column chromatography, and thus help in selecting the best column chromatographic systems.

1. For thin-layer chromatography, aluminium foil or glass plates covered by silica gel Si60 F_{254} (0.2 mm thickness of silica gel) can be used.
2. Two solvent systems are used for checking the crude extracts received by extraction by nonpolar solvents: (a) *n*-hexane + ethyl acetate, 4:1 (*v*/*v*), and (b) *n*-hexane + ethyl acetate, 1:1 (*v*/*v*). Mixtures of chloroform, methanol, and water [e.g., $CHCl_3 + MeOH + H_2O$, 65:35:5 (*v*/*v*/*v*)] or the same solvent system in different ratios are recommended as mobile phases for extracts received by use of polar solvents.
3. The presence of the spots is visualized by UV lights (254 or 365 nm) and by spraying with Godin reagent or 30 % sulfuric acid. After spraying, plates should be heating at 105 °C for 3 min. In Fig. 10a, b, TLC analysis of several ether extracts of the New Zealand and Japanese liverworts are shown. Very beautiful clear spots have been seen in the plates. A major compound is easily purified on the TLC plate using UV light or iodide vapor in the glass vessel.

3.7 GC and GC-MS Analysis of Essential Oils and Volatiles Present in Extracts

The most powerful method to identify the volatile chemical constituents of the crude extracts of the liverworts as well as essential oils is GC and GC-MS.

1. The sample should be dissolved in a dry solvent with a concentration <1 mg/mL to avoid overloading the mass spectrometer. Received sample should appear clear with no debris. If your sample is not clear filter the solution because debris can clog the injection needle (*see* **Note 14**).
2. Inject 1 μL of sample manually or by use of auto sampler. You can use one of the following conditions for your analysis:
   (a) *GC-FID analysis conditions*: Oven temperature programmed to rise from 45 to 175 °C at 3 °C/min, subsequently at 15 °C/min to 300 °C, and then hold isothermal for 10 min; injector and detector temperatures 280 and 300 °C, respectively; carrier gas, hydrogen, adjusted to a linear velocity of 30 cm/s. Sample injection by using a split sampling technique, ratio 1:50 [40, 41].

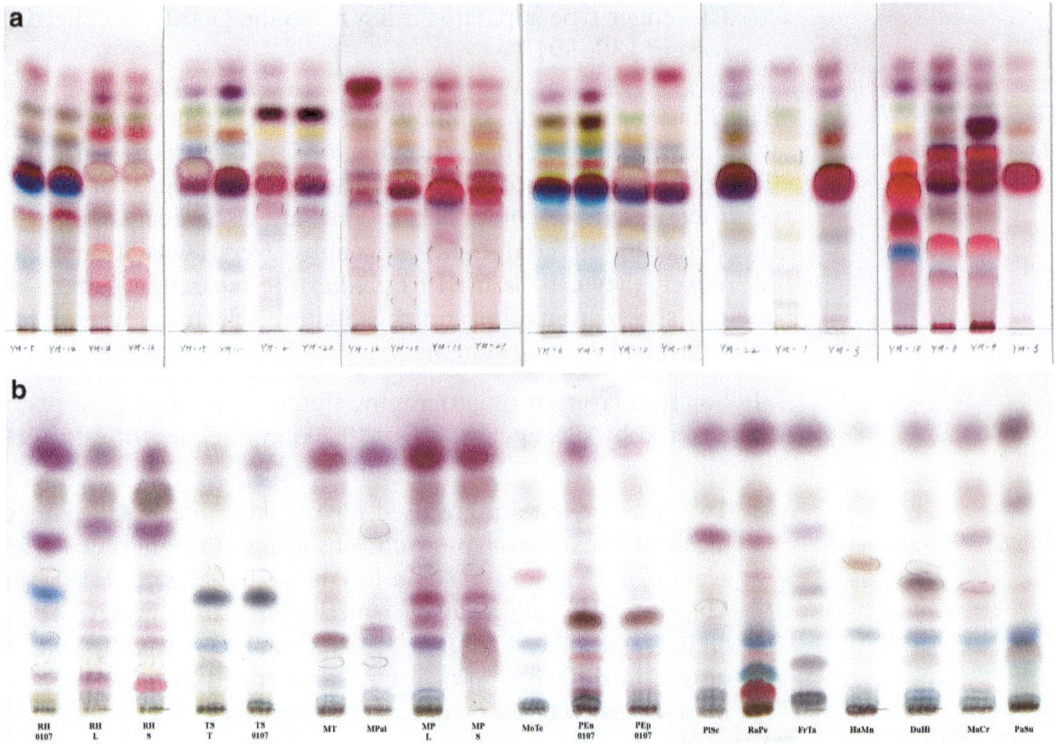

Fig. 10 (**a**) TLC of the crude extracts of several New Zealand liverworts (solvent system: *n*-hexane/ethyl acetate 4:1; spraying Godin reagent, heating at 105 °C). (**b**) TLC of the crude extracts of several Japanese liverworts (solvent system: *n*-hexane/ethyl acetate 4:1; spraying Godin reagent, heating at 105 °C)

(b) *GC/MS analysis conditions 1*: Oven temperature programmed to rise from 45 °C to 175 °C at 3 °C/min, subsequently at 15 °C/min to 280 °C, and then hold isothermal for 10 min; injection temperature: 280 °C. Carrier gas: adjusted to a linear velocity of 30 cm/s. Sample injection by using a split sampling technique, ratio 1:40. Ion trap temperature, 220 °C; ionization energy, 70 eV; scan-range, *m/z* 40–300; scan time, 1 s. These conditions were used for DB-1 capillary column (30 m × 0.25 mm, 0.25 μm film thickness) [40, 41].

GC/MS analysis conditions 2: Oven temperature: 50 °C with 3 min initial hold, and then to 50 °C, temperature programmed at 5 °C/min, and 15 min at 250 °C. Injection temperature: 280 °C. Carrier gas: Helium at 1 mL/min. The detector operated in electron impact mode (70 eV with 3 scans/s and mass range *m/z* 40–500) at 230 °C. These conditions were used for DB-1 and HP-5MS capillary columns (30 m × 0.25 mm, 0.25 μm film thickness) [13].

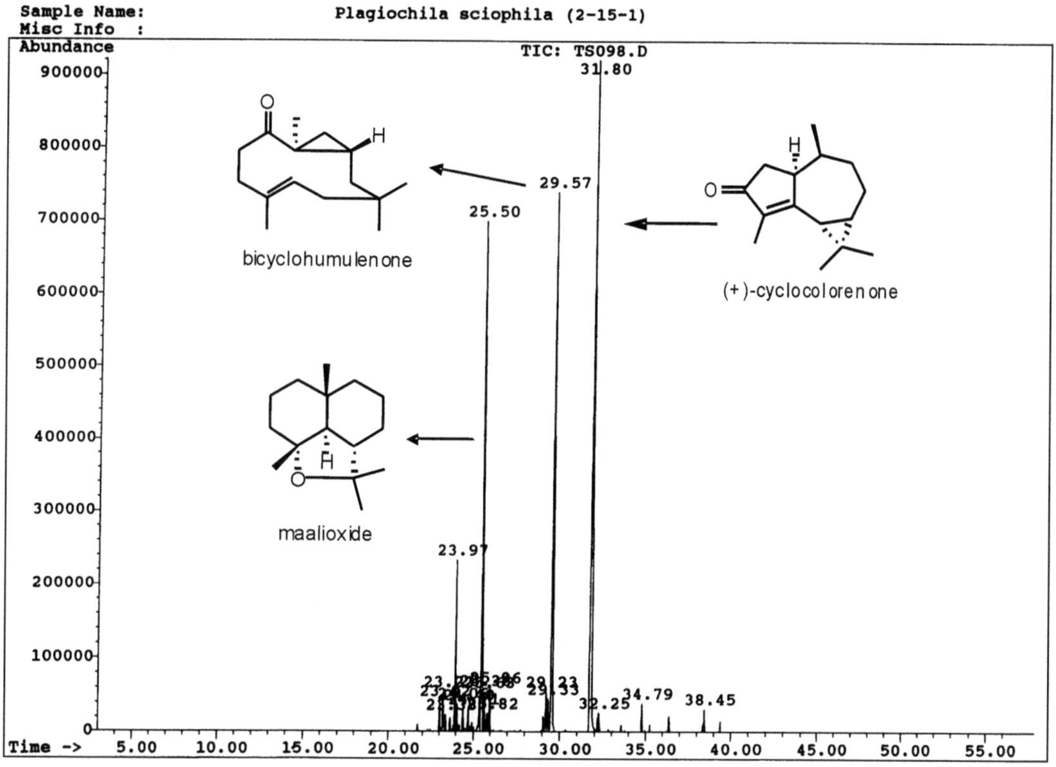

Fig. 11 GC/MS of essential oil of the liverwort *Plagiochila sciophila*

(c) *GC/MS conditions for chiral analysis*: Oven temperature: 70 °C with 8 min initial hold, and then to 230 °C temperature programmed at 10 °C/min, and 18 min at 230 °C. Injection temperature: 250 °C. Carrier gas: Helium at 1 mL/min. The detector is operated in electron impact mode (70 eV with 3 scans/s and mass range m/z 40–500) at 230 °C. These conditions were used for Cyclodex-B capillary column (30 m×0.25 mm, 0.25 μm film thickness).

(d) *Preparative GC/FID*: Helium gas as carrier at a flow rate of 240 mL/min; injector and detector temperatures are 200 and 250 °C, respectively [38, 39].

3. *Analysis of GC/MS data*: The retention indices can be calculated relative to C_8–C_{27} *n*-alkanes. For identification of compounds computer supported spectral library [42], mass spectra of reference compounds, as well as MS data from references [43] and our own library databases are used. Then the identities are confirmed by comparison of their retention indices with those of reference compounds and published data (*see* Fig. 11 and **Note 15**).

3.8 Purification of the Crude Extracts by Chromatographic Techniques

The objectives at this stage are to simplify the extract composition by dividing it into groups of compounds sharing similar physicochemical characteristics and/or to remove the bulk of unwanted materials and thus enrich the extract with respect to the target compounds. The most simple and widely used is open column chromatography on silica gel 60 (70–230 mesh) and/or on Sephadex LH-20 (*see* **Note 16**).

3.8.1 Sample Application, Elution, and Fractions Collection

1. Dissolve the sample in a minimum of the mobile phase. Using a Pasteur pipette carefully apply the sample solution, taking care not to disturb the bed (*see* **Note 17**).
2. Open the outlet valve and allow the sample to flow into the bed. When the sample has been adsorbed completely, the glass column sides can be washed carefully with a minimal amount of the mobile phase and then allowed to flow again. Care must be taken to ensure the bed is not disturbed and is not allowed to become dry.
3. Fill the column with the initial mobile phase required for the separation and allow the mobile phase to flow.
4. Collect the fractions. For complex separations, small fractions can be collected (for a 100-mL column, fractions of 5–10 mL are ideal). For crude separations, fractions of larger volumes can be collected.
5. Repeat as required for solvents of different polarity. For a nonpolar extract, e.g., *n*-hexane extract, and silica gel column a typical step gradient sequence might be as follows: initial composition of 100 % *n*-hexane, followed by at least 1 column volume each of 5 % EtOAc in *n*-hexane, 10–100 % EtOAc in *n*-hexane (increment of 10 % in each step), and 5–20 % MeOH in EtOAc (increment of 5 % in each step). For Sephadex LH-20 column the isocratic elution using CH_2Cl_2–MeOH (1:1), $CHCl_3$–MeOH (1:1), or pure MeOH as solvent is usually applied. For a MeOH extract, the reverse phase silica gel column can also be used and a typical step gradient sequence might be as follows: 10, 30, 60, 80, and 100 % MeOH in water.

3.8.2 HPLC Analysis

For further purification of compounds occurring in fractions preparative HPLC can be employed. Prior to HPLC analysis sample should be dissolved in the mobile phase, or another suitable solvent and filtered through a 0.45-mm filter. A precolumn is essential for protecting the more expensive HPLC column, and the precolumn should be changed regularly.

In our laboratory, preparative HPLC is usually done by using either Cosmosil 5SL-II (10 mm×250 mm) or Cosmosil 5C18-AR (10 mm×250 mm) columns. For normal phase HPLC the mixtures of *n*-hexane/EtOAc are used as mobile phases, while for RP-HPLC mixtures of acetonitrile and water. Ultraviolet (UV)

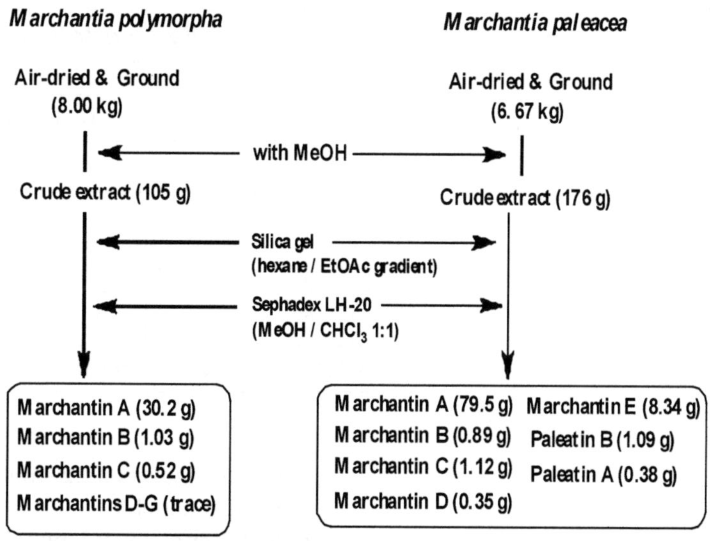

Fig. 12 Isolation procedure of methanol extracts from liverworts *Marchantia polymorpha* and *Marchantia paleacea*

detection is the preferred detection technique for HPLC analysis of natural products because of its ease of use and high sensitivity. However, one shortcoming of UV detectors is the inability to detect compounds that lack UV chromophores. Moreover, some of the solvents used in normal-phase chromatography are themselves strong absorbers of UV light, which means that low detector wavelength settings are not possible. The refractive index (RI) detector was, until recently, the only available alternative to UV detection. However, RI detectors are much less sensitive and are limited to isocratic elution. Gradient elution involves the mixing of solvents of differing RI, thus giving rise to large baseline drifts. These drawbacks make the detection of low concentrations of non-chromophoric metabolite notoriously difficult. As illustrated in Figs. 12 and 13, the examples of isolation procedures for methanol and diethyl ether extracts are presented.

4 Notes

1. Spray solution A onto previously dried TLC plate, and then continue with solution B. This is followed by heating TLC plate at 105 °C for 3 min.
2. Spray the TLC plate firstly with vanillin (solution A) and then with H_2SO_4 (solution B), or just before derivatization process mix solutions A and B in 1:1 (*v*/*v*) ratio and spray TLC plate using this mixture. Stock both solutions separately in dark place to avoid vanillin oxygenation.

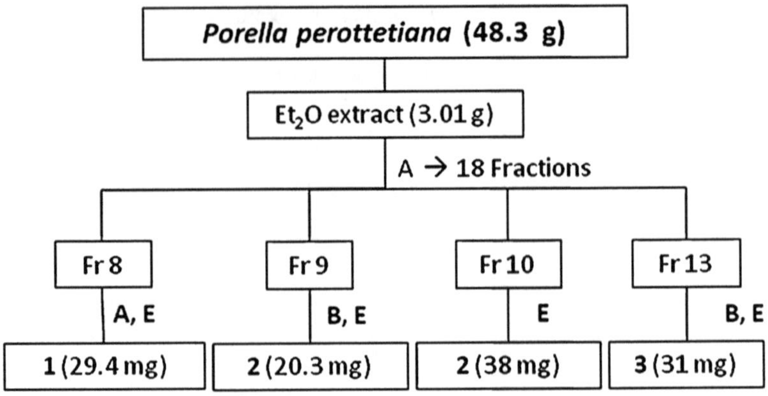

(1) 7-Oxopinguisenol-12-methyl ester
(2) Perrottetianal A
(3) 4α,5β-epoxy-8-epi-inunolide

A : CC on silica gel (*n*-hexane-EtOAc gradient)
B : CC on Sephadex LH-20 (CH_2Cl_2-MeOH 1:1)
C : Prep. HPLC (*n*-hexane-EtOAc 9:1)
D : Prep. HPLC (*n*-hexane-EtOAc 4:1)
E : Prep.HPLC (*n*-hexane-EtOAc 3:2)

Fig. 13 Isolation procedure of diethyl ether extract from liverwort *Porella perrottetiana* [44]

3. Thalloid liverworts belonging to the Marchantiales and hornworts grow on humid soil, while stem-leafy liverworts (Jungermanniales) such as *Frullania*, *Lejeunea*, *Porella*, *Radula*, *Plagiochila* species are found in rocks covered by humid soil or on living or dead trunk of trees (like oak, birch, etc.). The species of *Riccardia* (Metzgeriales), and *Chilosyphus* and *Gymnocolea* (Jungermanniales) grow often inside in river, lake, and ponds. On contrary, many of mosses like *Mnium*, *Fissidens*, *Pogonatum*, etc. habitats on concrete walls, dried or wet rocks. Most of *Sphagnum* species grow in lakes and ponds, and *Funallia* species in the stream of small and narrow rivers. The large moss species, *Polytrichum* grow on wet soil.
4. Plastic trays or boxes including plasticizer, phthalate esters (di-2-ethylhexyl phthalate) will not be recommended to carry liverwort samples.
5. When one collects certain bryophytes species, so many different species and/or dead trees' leaves, twigs, and living roots, even small animals intermingle into the species which one wishes study chemically. Thus, elimination of all such accompanying materials should be very carefully carried out. Such time consuming treatment is one of the reasons why bryophytes

chemistry has been neglected for one century. Purification of bryophyte samples is normally carried out by hands or sharp pinsetter; however, even purified samples still include very similar species when one treats very miniature species, such as the *Lejeunea*.

6. The authors recommend to use fresh dried materials directly to extract, if not, so many volatile components will evaporate or many unstable compounds will oxidize during air exposure.
7. Maceration should be done in room temperature in a dark place, to avoid the decomposition of unstable components.
8. Solvent removal should be done immediately after extraction to minimize the loss of compounds unstable in solution. The evaporation of solvents is also very carefully carried out. Too much vacuum in the evaporator and high temperatures in the water bath, are so dangerous, since many mono- and sesquiterpene hydrocarbons and volatile aromatic compound occurring in liverworts will degrade. For organic solvents, the extract is concentrated by evaporation under reduced pressure (using a rotary evaporator) at a temperature below 40 °C to minimize the degradation of thermally labile compounds.
9. The crushing machine is heated sometimes till more than 100 °C. The authors recommend the machine equipped with cool-water circulation system.
10. The cellular oil bodies present in the liverworts are easily extracted for a few minutes by using ultrasonic apparatus.
11. Soxhlet apparatus using ethanol are not recommended to extract the samples, because they damage many unstable terpenoids including dialdehydes, hemiacetals and many highly unsaturated fatty acids and aromatic compounds during a long extraction period. Ethyl acetate is also not a good extraction medium, because it is hydrolyzed by organic acids included in the natural products to give acetic acid and ethanol, which react with certain compounds in liverwort extracts during a long extraction period and many different artifacts are created.
12. Evaporation of the nonpolar solvent from the plant material before extraction with the methanol, allow the better penetration of the plant tissues by this solvent, and better extraction efficiency.
13. High temperature, exposure of air and hot water or hot steam during a long distillation time decompose many unstable terpenoids, aromatics, or acetogenins to form a number of artifacts. Due to small plant samples, which we can usually collect, both methods are not so frequently used. So comparison of chemical components obtained by both hydrodistillation and steam distillation and those from solvent extracts is absolutely necessary to recognize whether they are natural or nonnatural products.

14. Because extracts often contain water originating from the plant sample, thus, it is necessary to dehydrate the extracts. Dehydration is most commonly achieved by using anhydrous sodium sulfate or magnesium sulfate. Samples are dehydrated by adding the anhydrous sodium or magnesium sulfate directly to the extract and then filtering the dry solvent solution, or passing the samples through a column or funnel packed with anhydrous sodium or magnesium sulfate.
15. These GC/MS fingerprints are very important tool for identification of each liverwort.
16. Before preparation of a Sephadex LH-20 column, swell gel overnight by suspending the stationary phase in mobile phase (e.g., methanol). Use a sufficient mobile phase to get slurry with a pourable consistency. Allow enough time to ensure the gel is completely swollen.
17. If your extract is not soluble in the initial mobile phase or is very viscous, dissolve it with a suitable solvent, and put silica gel and mix equally in mortar and pestle. After complete evaporation of the solvent, the silica gel impregnated the extract put on the bed of the column which was filled with silica gel or Sephadex LH-20 was filled, then power the solvent as indicated in Subheading 3.8.1, **step 3**.

Acknowledgments

This work was supported in part by a Grant-in-Aid for the Scientific Research (A) (No. 11309012) from Ministry of Education, Culture, Sports, Science and Technology.

References

1. Schofield WB (1985) Introduction to bryophytes. Macmillan Publishing Company, New York
2. Asakawa Y, Tokunaga N, Toyota M et al (1979) Chemosystematics of bryophytes I. The distribution of terpenoids in bryophytes. J Hattori Bot Lab 45:395–407
3. Asakawa Y (1982) Chemical constituents of the Hepaticae. In: Herz W, Grisebach H, Kirby GW (eds) Progress in the chemistry of organic natural products, vol 42. Springer, Vienna, pp 1–285
4. Asakawa Y (1995) Chemical constituents of the bryophytes. In: Herz W, Kirby WB, Moore RE, Steglich W, Tamm C (eds) Progress in the chemistry of organic natural products, vol 65. Springer, Vienna, pp 1–618
5. Asakawa Y (2001) Recent advances in Phytochemistry of bryophytes—acetogenins, terpenoids and bis(bibenzyl)s from selected Japanese, Taiwanese, New Zealand, Argentinean and European liverworts. Phytochemistry 56:297–312
6. Asakawa Y (2004) Chemosystematics of the Hepaticae. Phytochemistry 65:623–669
7. Asakawa Y, Ludwiczuk A (2008) Bryophytes–chemical diversity, bioactivity and chemosystematics. Part I. Chemical diversity and bioactivity. Rośliny Lecznicze w Polsce i na Świecie (Medical Plants in Poland and in the World) 2:33–53
8. Asakawa Y, Ludwiczuk A (2008) Bryophytes–chemical diversity, bioactivity and chemosystematics. Part II. Chemosystematics. Rośliny Lecznicze w Polsce i na Świecie (Medical Plants in Poland and in the World) 3(4):43–54
9. Asakawa Y, Ludwiczuk A, Nagashima F (2013) Chemical constituents of bryophytes: bio- and chemical diversity, biological activity and

chemosystematics. In: Kinghorn DA, Falk H, Kobayashi J (eds) Progress in the chemistry of organic natural products, vol 95. Springer, Vienna, pp 1–796
10. Ludwiczuk A, Asakawa Y (2008) Distribution of terpenoids and aromatic compounds in selected southern hemispheric liverworts. Fieldiana, Botany 47:37–58
11. Ludwiczuk A, Asakawa Y (2010) Chemosystematics of the liverworts collected in Borneo. Tropical Bryol 31:33–42
12. Ludwiczuk A, Nagashima F, Gradstein SR et al (2008) Volatile components from the selected Mexican, Ecuadorian, Greek, German and Japanese liverworts. Nat Prod Commun 3: 133–140
13. Ludwiczuk A, Komala I, Pham A et al (2009) Volatile components from selected Tahitian liverworts. Nat Prod Commun 4:1387–1392
14. Ludwiczuk A, Gradstein SR, Nagashima F, Asakawa Y (2011) Chemosystematics of *Porella* (Marchantiophyta, Porellaceae). Nat Prod Commun 6:315–321
15. Sukkharak P, Ludwiczuk A, Asakawa Y et al (2011) Studies on the genus *Thysananthus* (Marchantiophyta, Lejeuneaceae) 3. Terpenoid chemistry and chemotaxonomy of selected species of *Thysananthus* and *Dendrolejeunea fruticosa*. Cryptogamie Bryol 32:199–209
16. Garnier G, Bezaniger-Beauquesne L, Debraux G (1969) Ressources médicinales de la flore francaise, vol 1. Vigot Frères Éditeurs, Paris, pp 78–81
17. Suire C (1975) Chimie des bryophytes. Rev Bryol et Lichenol 41:105–256
18. Ding H (1982) Zhong guo Yao Yun Bao zi Zhi Wu. Kexue Jishu Chuban She, Shanghai, pp 1–409
19. Wu PC (1982) Some uses of mosses in China. Bryol Times 13:5
20. Ando H, Matsuo A (1984) Applied bryology. In: Schultze-Motel W (ed) Advances in bryology, vol 2. J Cramer, Vaduz, pp 133–224
21. Asakawa Y (1999) Phytochemistry of bryophytes: biologically active terpenoids and aromatic compounds from liverworts. In: Romeo J (ed) Phytochemicals in human health protection, nutrition, and plant defense. Kluwer Academic/ Plenum Publishers, New York, pp 319–342
22. Asakawa Y (1990) Biologically active substances from bryophytes. In: Chopra RN, Bhatla SC (eds) Bryophyte development: physiology and biochemistry. CRC Press, Boca Raton, pp 259–287
23. Asakawa Y (1990) Terpenoids and aromatic compounds with pharmacological activity from bryophytes. In: Zinsmeister DH, Mues R (eds) Bryophytes: their chemistry and chemical taxonomy. Oxford University Press, Oxford, pp 369–410
24. Asakawa Y (1993) Biologically active terpenoids and aromatic compounds from liverworts and inedible mushroom *Cryptoporus volvatus*. In: Colegate SM, Molyneux RJ (eds) Bioactive natural products: detection, isolation, and structural determination. CRC Press, Boca Raton, pp 319–347
25. Asakawa Y (2007) Biologically active compounds from bryophytes. Chenia 9:73–104
26. Asakawa Y (2008) Liverworts-potential source of medicinal compounds. Curr Pharm Des 14:3067–3088
27. Asakawa Y (2008) Recent advances of biologically active substances from Marchantiophyta. Nat Prod Commun 3:77–92
28. Asakawa Y, Ludwiczuk A (2009) Marchantiophyta: bio- and chemical diversity and bioactivity. Malaysian J Sci 28:229–257
29. Asakawa Y, Toyota M, Nagashima F et al (2008) Chemical constituents of selected Japanese and New Zealand liverworts. Nat Prod Commun 3:289–300
30. Asakawa Y, Ludwiczuk A, Nagashima F et al (2009) Bryophytes: bio- and chemical diversity, bioactivity and chemosystematics. Heterocycles 77:99–150
31. Harinantenaina L, Asakawa Y (2007) Malagasy liverworts, source of new and biologically active compounds. Nat Prod Commun 2:701–709
32. Xie CF, Lou HX (2009) Secondary metabolites in bryophytes: an ecological aspect. Chem Biodivers 6:303–312
33. Inoue H (1988) Bryophytes as an indicator of continental drift (Gondwana land). Kagakuasahi 8:116–121
34. Mohamed H, Baki BB, Nasrulhaq-Boyce A, Lee PKY (2008) Bryology in the new millenium. University of Malaya, Kuala Lumpur, pp 1–513
35. von Reuβ SH, König WA (2005) Olefinic isothiocyanates and iminodithiocabonates from the liverwort *Corsinia coriandrina*. Eur J Org Chem 1184–1188
36. Godin P (1954) A new spray reagent for paper chromatography of polyols and ketoses. Nature 174:134
37. Ludwiczuk A, Nyiredy S, Wolski T (2005) Separation of the ginsenosides fraction obtained from the roots of *Panax quinquefolium* L. cultivated in Poland. J Planar Chromatogr – Mod TLC 18:104–107
38. Paul C, Konig WA, Wu C-L (2001) Sesquiterpenoid constituents of the liverworts *Lepidozia fauriana* and *Lepidozia vitrea*. Phytochemistry 58:789–798
39. Adio AM, Paul C, König WA, Muhle H (2002) Vilatile components from European liverworts

Marsupella emarginata, *M. aquatica* and *M. alpina*. Phytochemistry 61:79–91
40. Figueiredo AC, Sim-Sim M, Costa MM et al (2005) Comparison of the essential oil composition of four *Plagiochila* species: *P. bifaria*, *P. maderensis*, *P. retrosa* and *P. stricta*. Flavour Fragr J 20:703–709
41. Figueiredo AC, Sim-Sim M, Barroso JG et al (2009) Liverwort *Radula* species from Portugal: chemotaxonomical evaluation of volatiles composition. Flavour Fragr J 24:316–325
42. Konig WA, Joulain D, Hochmuth DH (2008) Mass Finder 4.0, terpenoid and related constituents of essential oils. http://www.massfinder.com/
43. Joulain D, Konig WA (1998) The atlas of spectral data of sesquiterpene hydrocarbons. E. B.-Verlag, Hamburg, pp 1–658
44. Komala I (2011) Phytochemical studies on the selected Indonesian, Japanese and Tahitian Liverworts. Ph.D. thesis, Tokushima Bunri University, Japan, pp 1–151

Chapter 2

Plant Tissue Extraction for Metabolomics

Ute Roessner and Daniel Anthony Dias

Abstract

Plants are not only important producers of foods and energy storages (e.g., sugars, carbohydrates, proteins, and fats) in the form of grains, fruits, and vegetables, they also provide many valuable products to human existence including wood, fibers, oils, resins, pigments, antioxidants, and sources of medicine. Most importantly in light of this book, plants have been a source of therapeutic and health promoting compounds throughout history. This chapter describes several essential considerations for the extraction process when aiming to study plant metabolism or to characterize the chemical composition of plant originated samples using metabolomics technologies.

Key words Plants, Metabolomics, Tissue extraction, Metabolites, GC-MS, LC-MS

1 Introduction

Plants play an important role as the primary producers of food and the source of metabolic energy of which all other life forms sustain off. Photosynthesis in green plant tissues captures sunlight energy, water, and carbon dioxide to build the primary building blocks including, sugars, carbohydrates, proteins, and fats. Plants provide foods (e.g., grains, fruits, and vegetables) and there are many plant products which are fundamental to the sustainability to all life on earth. These include wood and wood products, vitamins, antioxidants, fibers, drugs, oils, latex, pigments, and resins. In addition, plants have played an important role in medicinal applications throughout generations. Most medicinal drugs (about 80 %) are originated from the plant kingdom. However to date, only 2 % of all species in the plant kingdom have been explored for their therapeutic potential which provides us endless opportunities to discover novel bioactive compounds with wide medicinal or health supporting applications.

Ute Roessner and Daniel Anthony Dias (eds.), *Metabolomics Tools for Natural Product Discovery: Methods and Protocols*, Methods in Molecular Biology, vol. 1055, DOI 10.1007/978-1-62703-577-4_2, © Springer Science+Business Media, LLC 2013

1.1 Plant-Specific Features to Be Considered for Metabolomics

The major differences between plants and many other organisms are that they contain chlorophyll, are immobile, and have no sensory organs or nervous system. Most plant cells contain plastids and large vacuoles and most importantly are surrounded by often rigid cell walls. Seed-producing plants are divided into three organs: leaves, stems, and roots. Leaves are the main photosynthesizing organs, therefore critical for the survival of the rest of the plant as well as for providing sufficient energy for reproduction. The roots are usually grown underground and have two main functions. Firstly, to anchor the plant in the soil and secondly to absorb water and essential nutrients (both macro- and micro-nutrients) from the soil to be transported to the upper sections of the plant system. Roots of some plant species (e.g., legumes) develop symbiotic relationships with nitrogen fixating bacteria to enhance their own amounts of nitrogen fixated for protein storage. The main function of the stems is to transport water, soluble carbon sources, nutrients, and hormones between roots, leaves, and the reproductive organs.

Apart from considerations about the complex plant tissue anatomy and physiology, there are specific challenges to be considered when aiming to do conduct a metabolomics or comprehensive chemical characterization of any plant tissue. Plant metabolism is highly light dependent and many metabolite levels are altered during the course of a day depending on photosynthetic or respiratory activities. During light conditions many synthesizing and storage processes are functional, whereas during dark processes storage products may be degraded to provide the plant cells with energy through respiration. This consideration has to be reflected when harvesting samples for a metabolomics experiment which needs to be at very similar time points of a day (*see* **Note 1**). Another important point to consider is the strength of light which leaves are exposed to, e.g., leaves in bright light and leaves in the shade will have different metabolite profiles (*see* **Note 2**).

1.2 Extraction of Plant Metabolites

A prerequisite for an efficient extraction process of plant tissues is that the often rigid cell walls are broken to release the metabolites in the extraction solvent. Therefore it is important to investigate an appropriate homogenization process. There are many techniques of homogenizing plant sections and mechanical shear seems to be the most efficient. This can be achieved via mortar and pestle but many laboratories are now using automatic grinders such as ball or cryomills. It may be important to observe under a microscope how well the material was ground and if all macromolecular structures were spatially homogenized to a fine powder. Especially important if working with fresh and frozen plant tissues is to make sure that the homogenization process is carried out under freezing conditions, ideally under liquid nitrogen. Since this is not always achievable, plant material can be freeze-dried prior to homogenization and extraction; however the grinding process needs to be carried out under very dry conditions (*see* **Note 3**).

The next step is to optimize and validate an appropriate extraction method for the plant tissue as well as for the metabolites of interest. If an untargeted metabolomics approach is utilized, the extraction process should be as crude and extract as many metabolites as possible from the tissues. In our laboratory, we use a two-step procedure using methanol in the first instance and water in the second since this covers both polar and apolar compounds simultaneously. Both supernatants after removing all insoluble components such as protein, starch, cell wall, and other high-molecular weight carbohydrates are then combined to allow analysis of a large coverage of the metabolome (*see* **Note 4**).

When extracting plant metabolites, one has to consider that the plant metabolome is very dynamic and concentrations of metabolites may range over several magnitudes. Some compounds may be found in molar ranges such as sucrose or glucose and others, often the ones of interest (secondary metabolites), are found only at very low abundances. Therefore, if metabolites of interest only exist in low concentrations, specific approaches may need to be employed removing high abundant compounds therefore enriching and concentrating low abundant metabolites. However, care has to be taken when removing certain compound classes with similar physiochemical properties since it may lead to unwanted losses of the metabolite(s) of interest.

This chapter aims to emphasize general points of consideration when aiming to extract plant metabolites (primary and secondary). Methods for crude tissue extractions to capture as many metabolites as possible allowing to analyze a "snap shot" of a large portion of the metabolome are presented within. In light of the topic of this book, these approaches also allow to detect and particularly identify as many metabolites as possible for a discovery strategy for novel natural products with a desired activity. For a detailed description of the challenges and approaches in plant-based metabolomics, the reader is referred to Villas-Boas et al. [1], Beckles and Roessner [2], and Roessner and Beckles [3].

2 Materials

There are a wide range of methodologies published for plant tissue extractions. It is important that for every tissue type, multiple extraction procedures should be optimized and validated before settling on the most appropriate method would then be carried out throughout an experiment. Important parameters for consideration for instance are how easy is it to obtain/harvest the tissue? (e.g., in glass house or on the field or if tissue of interest requires dissection from surrounding tissue types); sample size (is there sufficient material available or are samples sizes minimal?); tissue homogenization (availability of appropriate instrumentation?);

storage of frozen tissue or freeze dry tissue for ease of storage; solvent extraction (availability of clean solvents and incubators?); removal of insoluble material from the metabolite extract (availability of appropriate centrifuges); reproducibility of extraction, the stability of metabolites of interest throughout the extraction procedure; recoveries of metabolites throughout extraction process and how to best store extract prior to analysis are several factors to consider. This chapter describes two extraction procedure methods commonly used in our laboratory for untargeted profiling of plant metabolites using GC-MS (modified from ref. [4]) and LC-MS (modified from ref. [5]). The methods are applicable to all plant tissues for untargeted profiling since they are aimed to obtain a crude extract of the tissue without minimal losses of metabolites. It is important to mention that for any targeted approach for selected classes of metabolites, a specific extraction procedure may need to be employed to enhance enrichment of low abundant secondary metabolites.

2.1 Tissue Harvesting and Quenching of Metabolism

1. Reaction tubes (e.g., 2 mL safe lock Eppendorf tubes) labeled with identifiable names for each sample to be harvested.
2. Liquid nitrogen in an appropriate storage container.

2.2 Storage of Samples

1. –80 °C freezer if frozen.
2. Freeze drier.
3. Dark, dry, and cool storage if material is freeze dried.

2.3 Homogenization of Tissue Using Mortar and Pestle or Automatic Grinders

1. Mortar and pestle.
2. Automatic grinders such as ball mills, ultra turrax, or cryomills.

2.4 Weighing of Tissue

1. New labeled reaction tubes (e.g., 2 mL safe lock Eppendorf tubes).
2. Analytical balance (to at least four decimal places).
3. Liquid nitrogen if working with frozen fresh tissue.
4. Spatula.

2.5 Metabolite Extraction for GC-MS Profiling

1. Solvents: 100 % (v/v) methanol and distilled water.
2. Internal standards (e.g., $^{13}C_6$-Sorbitol and $^{13}C^{15}N$-Valine).
3. Vortexer.
4. Thermoshaker for reaction tubes (e.g., Eppendorf).
5. Bench top centrifuge for reaction tubes (up to 11, 337×g).
6. Speed vacuum.
7. Silica beads.

2.6 Metabolite Extraction for LC-MS Profiling

1. Solvents: 2:3:3 (v/v/v) mixture of water:acetonitrile: isopropanol.
2. Internal standards (e.g., $^{13}C_6$-Sorbitol and $^{13}C^{15}N$-Valine).
3. Vortexer.
4. Sonicator.
5. Bench top centrifuge for reaction tubes (up to 11, 337×*g*).

3 Methods

3.1 Tissue Harvesting and Quenching of Metabolism

Since metabolic reactions are extremely fast it is crucial to harvest tissue and quench metabolism as quickly as possible. This has to be considered when aiming to determine a "snap shot" of the steady-state level of the metabolome. Equally important is to avoid metabolic changes due to wounding of the tissue. Ideally, samples are frozen in liquid nitrogen as quickly as possible which is relatively easy when working with whole leaves or other easily accessible plant tissue types. However, if a particular tissue of interest requires to be dissected from surrounding tissues prior to freezing, it is important to keep the time frame for this process as similar as possible for all samples harvested. Below the protocols are described when working with easily accessible tissues such as leaves or roots.

1. Prepare liquid nitrogen in appropriate storage container.
2. Harvest leaves or roots, put in reaction tube.
3. Freeze in liquid nitrogen (*see* **Note 5**).

3.2 Storage of Samples

1. Place reaction tubes with frozen tissue in −80 °C freezer.
2. If no freezer available, dry tissue in freeze drier under freezing conditions.
3. Store freeze dried samples in a dark, dry, and cool environment (*see* **Note 6**).

3.3 Homogenization of Tissue

1. If working with frozen tissue, homogenize tissue using mortar and pestle under freezing conditions (*see* **Note 5**), if tissue is freeze dried, there is no need for freezing conditions as long the process is kept under dry conditions.
2. If automatic grinder devices are available such as ball or cryo mills, use as instructed by the manufacturer.

3.4 Weighing of Tissue

1. If working with frozen tissue, prepare liquid nitrogen in appropriate storage containers.
2. Prepare new labeled reaction tubes.
3. Pre-cool reaction tubes in liquid nitrogen.
4. Place tubes on balance and set balance to zero (tare).

5. Place tissue powder in cooled tubes using a pre-cooled spatula or spoon (*see* **Note 5**).
6. Read weigh and record (*see* **Note 7**).
7. Place tube with weighed tissue back into liquid nitrogen.
8. Proceed with extraction or store weighed tissue at –80 °C.
9. If working with freeze dried tissue the need of pre-cooling and keeping tissue frozen is not required.

3.5 Sample Preparation for GC-MS

1. Weigh approximately 30 mg homogenized tissue powder in a 2 mL Eppendorf tube (round bottom-shaped).
2. Add 500 μL 100 % (v/v) methanol, vortex (Enzymatic activity stops here).
3. Add 20 μL stock $^{13}C_6$-Sorbitol and $^{13}C^{15}N$-Valine (1 mg/mL stock solution) as an internal quantitative standard for the polar phase (*see* **Note 8**).
4. Vortex.
5. Shake 15 min at 70 °C in a Thermoshaker.
6. Centrifuge for 15 min at 11, 337 × *g*.
7. Transfer supernatant to a new reaction tube.
8. Add 500 μL H_2O to the pellet.
9. Vortex.
10. Centrifuge for 15 min at 11, 337 × *g*.
11. Combine both supernatants.
12. Vortex.
13. Take aliquots for derivatization into an insert for the instrument autosampler vial (*see* **Notes 9** and **10**).
14. Dry aliquots under *vacuo* without heating by placing the insert with the extract into a reaction tube. Once dry, it can be placed into the autosampler vial prior to analysis. If these need to be stored then store in bag with Silica gel at room temperature (*see* **Note 11**).

3.6 Sample Preparation for LC-MS

1. Weigh approximately 50 mg of ground tissue into 2 mL reaction tubes (e.g., Eppendorf).
2. Add 500 μL of a 2:3:3 (v/v/v) mixture of water:acetonitrile:isopropanol.
3. Sonicate for 15 min at room temperature.
4. Centrifuge at 11, 337 × *g*.
5. Transfer supernatant to a new reaction tube.
6. Re-extract pellet with the addition of another 500 μL of a 2:3:3 v/v/v mixture of water:acetonitrile:isopropanol.

7. Sonicate for 15 min at room temperature.
8. Centrifuge at 11, 337 ×*g*.
9. Combine both supernatants.
10. Transfer to autosampler vial for analysis. If these need to be stored, freeze extract at –20 °C (*see* **Note 12**).

4 Notes

1. To avoid any position or time of harvest effects on the variability of the data set always randomize samples when harvesting.
2. Always select a leaf at the same position (e.g., always the upper or a lower leaf), if not use randomized positions of leaves to be harvest.
3. If working with freeze-dried material avoid any water and moisture. As soon as tissue is in contact with moisture enzymes will become active.
4. For many separation and detection techniques the pigments contained in plant tissues, such as chlorophyll and carotenoids, disturb the analysis and may be removed from the extract.
5. It is important to ensure that tissue is *always frozen* throughout any processes such as weighing or transfer between reaction tubes. At warmer temperatures, enzymatic activity can occur and cause spurious changes in metabolites that are unrelated to the treatment. If larger tissue pieces are to be harvested (e.g., fruits), cut into smaller pieces prior to shock freezing so the freezing process is continuous and quick throughout the tissue and any enzymatic reactions are avoided in the middle of the tissue piece.
6. Some metabolites are sensitive and may alter their structure under light exposure. Any water and moisture, as mentioned in **Note 3**, will activate enzymes.
7. In order to relate metabolite levels between samples, the weight of each sample has to be exactly recorded. These weights (either frozen or freeze-dried samples) are then used for subsequent normalization of metabolite levels per gram of fresh or dry weight.
8. Internal standards are ideally compounds not present in the biological extract and are included prior to or during metabolite extraction. They encounter for any analytical error during extraction, derivatization, separation, and detection processes. For instance, stable isotope-labeled internal standards having identical chemical properties as the metabolites under analysis are ideal.

9. The volume for each aliquot of extract depends on the tissue type, the amount of tissue extracted and volume of extraction solvents used. This needs to be optimized for every tissue type under analysis to have best separation and detection within the linear range of the instrument of choice.
10. It is suggested to always prepare more than one aliquot of each sample for backup. Once the samples have been derivatized for GC-MS, they cannot be stored. Therefore, if there is equipment failure during a run; the experiment can be saved if identical samples are available.
11. If long-term storage of dried extracts is required, they need to be kept under argon to avoid oxidation and degradation of metabolites. Dried extracts need to be kept in the dark and under dry conditions for the same reasons as mentioned in **Note 6**.
12. Liquid extracts for LC-MS analysis are ideally analyzed straight away; however if necessary can be stored at −20 °C or lower. However this should be avoided since there may be metabolite alterations during the thawing process.

References

1. Villas-Boas SG, Roessner U, Hansen M, Smedsgaard J, Nielsen J (eds) (2007) Metabolome analysis. Wiley, New Jersey, NJ
2. Beckles DM, Roessner U (2011) Plant metabolomics - applications and opportunities for agricultural biotechnology. In: Altmann A, Hasagawa PM (eds) Plant biotechnology and agriculture: prospects for the 21st century. Elsevier, Waltham, MA
3. Roessner U, Beckles DM (2012) Metabolomics in salinity research. Sergey Shabala and Tracey Ann Cuin (eds.), *Plant Salt Tolerance: Methods and Protocols*, Methods in Molecular Biology, vol. 913, Humana Press, USA
4. Jacobs A, Lunde C, Bacic A, Tester M, Roessner U (2007) The impact of constitutive expression of a moss Na^+ transporter on the metabolomes of rice and barley. Metabolomics 3:307–317
5. Callahan DL, De Souza D, Bacic A, Roessner U (2009) Profiling of polar metabolites in biological extracts using diamond hydride-based aqueous normal phase chromatography. J Sep Sci 32:2273–2280

Chapter 3

Detection of Polar Metabolites Through the Use of Gas Chromatography–Mass Spectrometry

David P. De Souza

Abstract

Gas chromatography–mass spectrometry (GC–MS) is a highly reproducible and sensitive analytical technique that has had significant use in the area of metabolite profiling. GC–MS is able to detect a wide variety of metabolites, with highly differing chemistries. In general, extracted biological samples are volatilized prior to separation on a capillary column with a stationary phase suited to the analysis of the compounds of interest. Separated compounds are eluted into a mass spectrometer equipped with an electron impact ionization source, thereby generating a quantifiable mass spectral fingerprint. This chapter describes a method for the trimethylsilyl derivatization of polar metabolites, followed by detection and relative quantification using a gas chromatograph coupled to a single quadrupole mass spectrometer. Using this method will enable the profiling of the greatest range of polar metabolites.

Key words Gas chromatography, Mass spectrometry, Methoximation, Trimethylsilylation, Quadrupole, Polar metabolites

1 Introduction

Gas chromatography–mass spectrometry (GC–MS) is one of the principal analytical platforms used in the field of metabolomics. With the advent of capillary columns and improved computing allowing faster data acquisition rates, GC–MS has emerged as a robust technology for the efficient separation and quantification of a few hundred metabolites from a single sample extract [1].

A typical metabolomics workflow is composed of three major steps: (1) sample harvest and metabolic quenching, (2) metabolite extraction, and (3) instrument analysis. Individual differences of varied sample types will determine the techniques used at each of these steps. For sample harvest and metabolic quenching, many methods have been published in the literature for various biological systems [2–5]. Many of these publications also contain sample extraction methods, which can be tailored to specific metabolites, i.e., polar vs. lipidic metabolites. Highly reproducible sample

Ute Roessner and Daniel Anthony Dias (eds.), *Metabolomics Tools for Natural Product Discovery: Methods and Protocols*, Methods in Molecular Biology, vol. 1055, DOI 10.1007/978-1-62703-577-4_3, © Springer Science+Business Media, LLC 2013

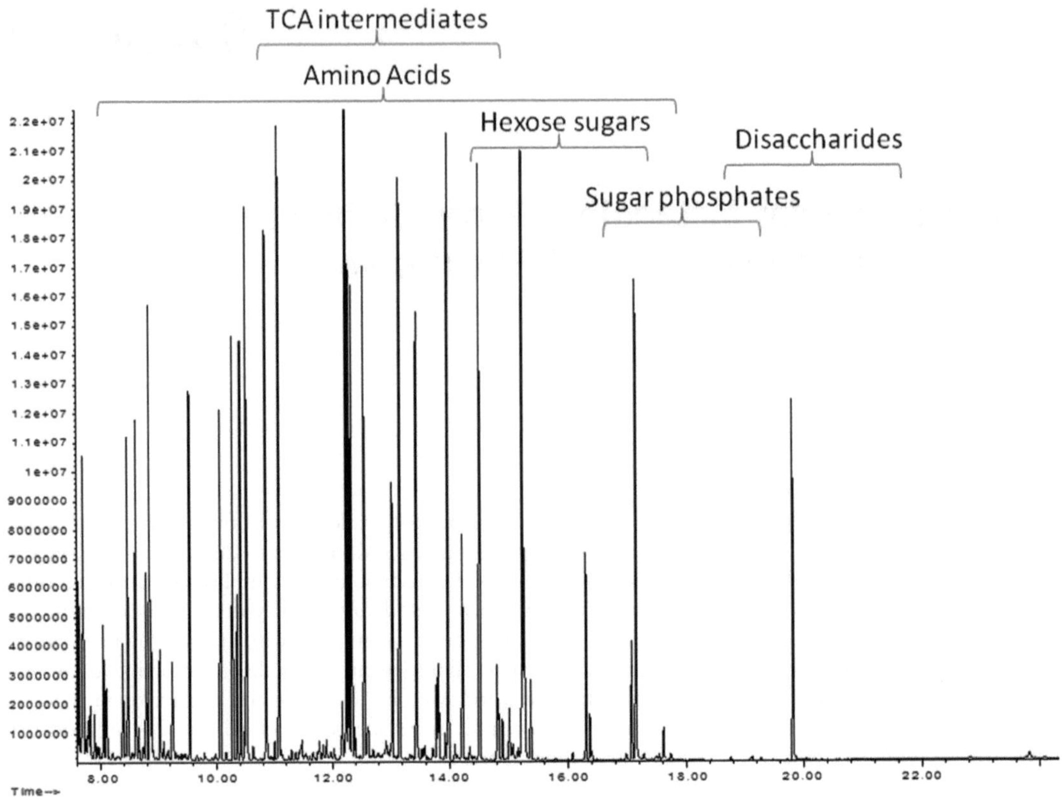

Fig. 1 Chromatographic profile of a metabolite mix containing 45 metabolites spanning the major polar metabolite classes. The profile was captured using the GC–MS described in this chapter. Regions in which certain metabolite classes elute are annotated

preparation is critical for metabolite profiling, but well-optimized instrument analysis parameters are equally important to extract the most information out of these samples.

The GC–MS method described here aims to detect as many polar metabolites as possible. While there is no one method suited to all sample types and metabolite analyses, the method described is a general purpose method that balances productivity (i.e., relatively short run time) with good separation of compounds (Fig. 1). Once this method has been used and understood, analysts would be able to optimize conditions to suit the requirements of their particular sample. The chromatography here uses a fairly nonpolar stationary phase that is suitable to the wide range of chemical types that can be seen in polar metabolite extracts from biological samples. For more lipidic molecules, such as fatty acids, a more polar stationary phase would be much more suitable.

Similar GC–MS methodologies have been applied to the study of small molecule natural products in plants and fungi. These studies have used GC–MS to identify compounds that could be utilized as bioactive compounds for the treatment of bacterial infections [6, 7].

Alternatively, these studies have also identified mycotoxins that can effect crop quality [8] and also have an impact on health. Together these findings show the utility of GC–MS as an analytical technique for the study of small molecule natural products.

2 Materials

2.1 Trimethylsilyl Derivatization

1. 100 % methanol (GC grade).
2. Methoxyamine reagent: Methoxyamine hydrochloride in pyridine (30 mg/mL) (*see* **Note 1**).
3. Trimethylsilane (TMS) reagent: *Bis*-(trimethylsilyl)trifluoroacetamide (BSTFA) + 1 % trimethylchlorosilane (TMCS).

2.2 GC–MS Analysis

1. 2 mL Autosampler vials.
2. 250 μL pulled point vial inserts (Agilent Part Number: 5183-2085).
3. Methanol (GC or HPLC grade).
4. 1:1 (v/v) ethyl acetate:acetone.
5. Hexane (Analytical grade).
6. Christ Alpha RVC 2-33 Vacuum concentrator.
7. Eppendorf Thermomixer Comfort.
8. GC Chemstation Instrument analysis software (E.02.01.1177), Gerstel Maestro software (Version 1.3.20.3/3.5).
9. General purpose split/splitless liner with glass wool, tapered and deactivated (Agilent Part Number: 5183-4711).
10. Merlin microseal (Agilent: 5182-3442).
11. Agilent gold seal (Agilent: 5188-5367).
12. Agilent Factor Four 5 ms column (30 m × 0.25 mm internal diameter × 0.25 μm film thickness, with 10 m integrated Eziguard—Part no. CP9013).
13. Ultra high purity helium (BOC 220G, Coregas Helium 5.0).

3 Methods

3.1 Metabolic Derivatization

1. Transfer 75 μL of aqueous phase (top layer—containing polar metabolites) (*see* **Note 2**) into a 200 μL pulled point insert.
2. Dry sample in a cleaned speed vacuum (35 °C).
3. Continuously add 50 μL of aqueous phase (approximately every 20 min) and repeat **step 2** until all has been transferred.
4. Add 10 μL of GC–MS grade methanol (*see* **Note 3**).
5. Dry sample in speed vacuum (35 °C).

6. Transfer insert into 2 mL autosampler vial.
7. Add 20 μL of methoxyamine reagent, then cap the vial (*see* **Note 4**).
8. Incubate for 2 h on shaker (400 rpm) at 37 °C.
9. Add 20 μL of TMS reagent.
10. Incubate for 1 h at 37 °C in shaker, then at room temp for 1 h.
11. Inject 1 μL of derivatized sample following steps in Subheading 3.2 below.

3.2 GC–MS Analysis

The following conditions are suitable for polar metabolite profiling of biological samples. While this is a good general method, it is always advisable to perform comprehensive method development to ensure the best analytical conditions possible. Use these conditions as a starting point, and pay close attention to *see* **Notes 5–11** as they contain essential tips for suitable alterations to the method. Take note that the GC conditions are strongly linked to the MS conditions, and if you change one, then there should be relative changes in the other.

3.2.1 Autosampler Conditions

The parameters described here (*see* **Note 5**) are solely for the injection of sample and pre/postinjection washing steps. If using an online derivatization robot (Gerstel MPS2, CTC Combipal), then washing steps before and after addition of reagent will also be required.

3.3 Injection Parameters

1. One syringe wash with *n*-hexane.
2. Four plunges of sample.
3. 0 s pre-inject delay.
4. 1 μL (*see* **Note 6**) injected Splitless (*see* **Note 7**).
5. 0 s post-inject delay.
6. Three syringe washes with *n*-hexane.
7. Five syringe washes with 1:1 ethyl acetate/acetone.

3.3.1 Inlet and Carrier Gas Conditions (Agilent 7890A GC)

1. Ensure the following parameters (*see* **Note 8**) and components are installed, enter the settings within Chemstation under "GC Edit Parameters." For installation of components refer to Agilent 7890A GC manual.

3.3.2 Inlet

1. General purpose split/splitless liner with glass wool, tapered and deactivated (Agilent Part Number: 5183-4711).
2. Split/splitless inlet fitted with Merlin microseal (Agilent: 5182-3442).
3. Agilent gold seal (Agilent: 5188-5367).
4. Purged at 50 mL/min after 60 s with UHP Helium.
5. Gas saver at 20 mL/min after 120 s with UHP Helium.

3.3.3 Carrier Gas

1. Ultra high purity helium.
2. Constant column flow rate of 1 mL/min.
3. Inlet purged at 20 mL/min for 60 s.
4. Gas saver 15 mL/min after 60 s.

3.4 Gas Chromatographic Conditions (Agilent 7890A GC)

Apply the following oven program settings within Chemstation method. As mentioned earlier, these settings are a good starting point, but can be changed to suit the respective sample. Faster oven ramps will result in sharper, narrower peaks, but may result in a loss of baseline resolution. In comparison, a slower oven ramp may improve baseline resolution, but at the cost of signal intensity, where broadening of peaks results in lower signal to noise ratios.

3.4.1 Column

1. Length = 10 m Eziguard + 30 m VF-5 ms column (*see* **Note 9**).
2. Internal diameter = 250 μM.
3. Film thickness = 0.25 μM.

3.4.2 Oven Program

1. 35 °C—2 min hold.
2. 15 °C/min ramp to 325 °C.
3. Hold for 3 min (*see* **Note 10**).

3.5 Mass Spectrometer (MS) Conditions (Agilent 5975C Inert EI/CI MSD)

Apply the following parameters within Chemstation's "MS SIM/ Scan Parameters." These parameters are extremely dependent on the speed of chromatography, with particular reference to eluted peak width. These parameters are suited to the oven ramp described previously. *See* **Note 11** for further details regarding changing mass spectrometer parameters.

3.5.1 Mass Spectrometer Conditions

1. Transfer line = 280 °C.
2. Ionization Chamber = 250 °C.
3. Quadrupole temperature = 150 °C.
4. EM gain factor = 1 V.
5. Detector threshold = 150.
6. Detector *m*/*z* range = 50–600 *m*/*z*.
7. Detector sampling rate = 2^1 (5.02 scans/s under above conditions) (*see* **Note 11**).

3.6 Data Analysis

All mass spectrometry vendors will have software that is capable of dealing with metabolite profiling data. These software packages are very comprehensive, but can also be quite costly. The alternative is to investigate freely available software to peak find and deconvolute your data. These include PyMS [9], which is primarily used for untargeted data analysis. For a more targeted approach, Metab [10] is appropriate, while for Agilent-specific data, CompExtractor can be used [11]. Data extraction is only part of the process of metabolomics data analysis, as techniques need to be applied to make sense of the data. Typically, multivariate statistics is used to make sense of the many variables measured, for which a good overview on techniques can be read in Want et al. [12].

4 Notes

1. Methoxyamine hydrochloride is a very hygroscopic material, which is best stored in a desiccator. Prepare the methoxyamine solution freshly, rather than making up a large batch and storing aliquots. Make sure the pyridine you use is fresh and has not oxidized (should be colorless, not a yellow color). If possible purchase pyridine with molecular sieves in the container, which will remove any dissolved water from the pyridine.
2. Quite often extraction methods will include a biphasic partition to isolate polar metabolites. This is not an absolutely necessary step, but is preferable. References in the introduction contain excellent metabolite extraction methods for a wide variety of biological samples.
3. Methoxyamine hydrochloride is added as the first derivatization step to enable linearization of sugar molecules. This results in a simplified chromatogram. Without methoximation a hexose sugar, such as glucose, will give four chromatographic peaks after TMS derivatization, whereas after methoximation and TMS derivatization only two peaks would be observed. The methoximation step is not absolutely mandatory, but does significantly simplify the chromatogram. If methoximation is not performed, then the equivalent volume of TMS reagent should be substituted instead.
4. The final evaporation of samples with methanol is absolutely mandatory. *Do not skip this step.* The derivatizing agents used, especially the TMS reagent, are very hygroscopic. This means that the reagents will absorb any water that is present, and subsequently will drop out of solution. This will result in a wasted sample, which cannot be recovered.
5. The type of autosampler on your GC–MS will determine how you perform the derivatization of your samples. If only a sample injector is available, then all samples will have to be "batch" derivatized. This means that all the samples will be derivatized at once, but will be incubating for different periods of time due to run order when analyzed by GC–MS. This can result in sample variation, as some metabolites derivatize and/or degrade at different rates. It is preferable to have a GC–MS set up with an autosampler robot of some description, i.e., Gerstel

MPS2, CTC Combipal. These robots allow online derivatization of samples and "in time" injection of the sample. This means that samples are derivatized for exactly the same period of time prior to analysis, therefore reducing the amount of unwanted variation. Good practice is to randomize the sample analysis sequence so that no one group of samples is located at the start or the finish of the analysis sequence. This practice is observed irrespective of "batch" or "in time" derivatization.

6. 1 μL of sample is the maximum that can be injected into the GC inlet for this method. Conversion of 1 μL sample in pyridine solvent into the gas phase will equate to a gaseous volume of approximately 880 μL. The capacity of the inlet liner specified in Subheading 3.3.2, **step 1** is a maximum of 900 μL. Loading more than 1 μL of liquid sample will result in overloading of the inlet liner, and subsequent contamination of the inlet itself.
7. This method recommends that samples be injected "splitless," which involves the entire amount of sample to be transferred onto the column. Samples can be reinjected in "split" mode if some peaks are chromatographically overloaded. Split ratios are determined by the analyst, and can range from 10:1 to 100:1 split, which would equate to loading 1/10th–1/100th of the sample.
8. All conditions described in the following sections are based on the use of an Agilent 7890A GC coupled to a 5975C MSD. All these conditions are directly applicable to other GC–MSDs.
9. The column specified here is an Agilent column, with a 5 ms stationary phase. This phase is relatively nonpolar and well suited to the broad range of chemical classes observed when analyzing polar metabolites. There are many other equivalent columns manufactured by other vendors, that are equally suitable—just make sure to use the 5 ms phase. The integrated guard column is also important as it allows cutting of the GC–MS column with very little change to retention time. This is important as regular maintenance requires the cutting of the GC column to remove any contaminants that have coated the beginning and end of the column.
10. The oven ramp described is a relatively mid-range oven ramp, neither slow nor fast. The chromatographic peaks obtained will tend to be 3–3.5 s wide providing all other conditions are set as described. The type of chromatography used is very sample dependent, where those in the area of plant metabolomics generally prefer a slower oven ramp in order to give better separation of sugar metabolites. Where this is not an issue, faster oven ramps can be used to give sharper peaks, better sensitivity

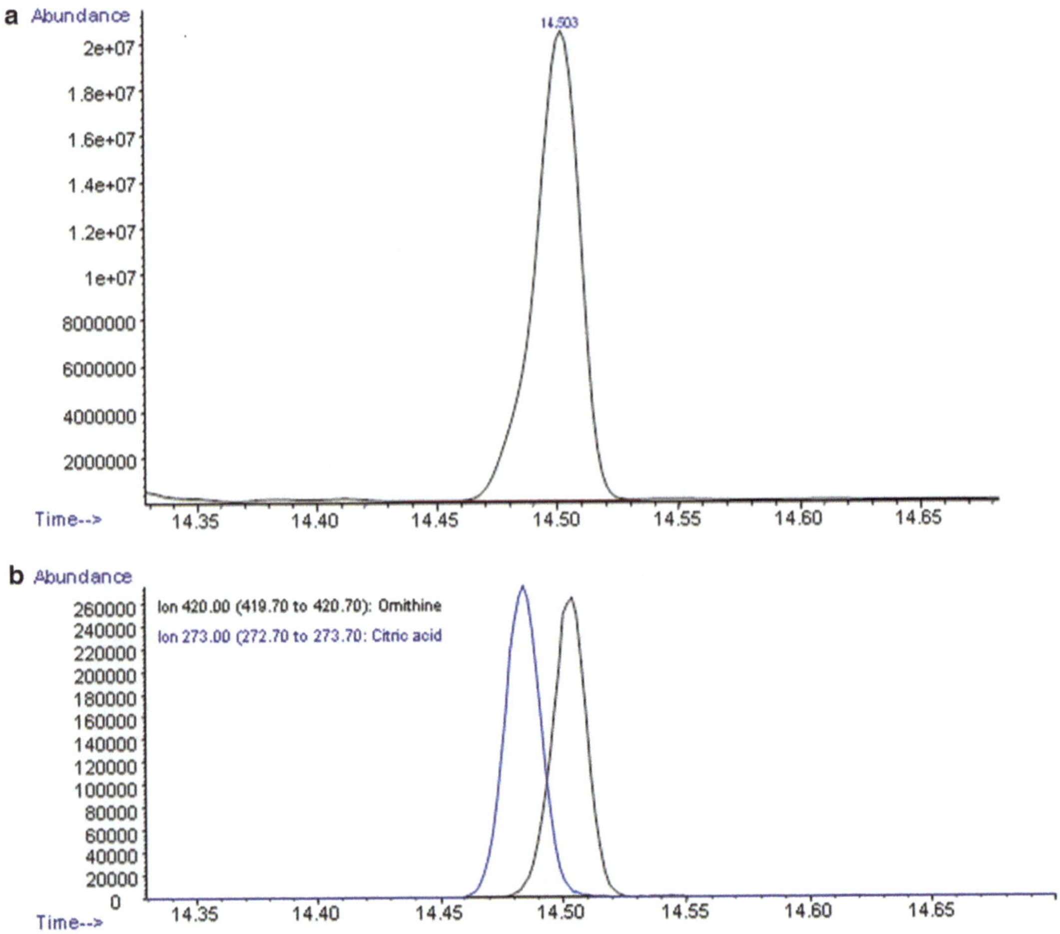

Fig. 2 A peak that shows a classical co-elution of metabolites. Panel (**a**) illustrates what appears to be a single peak in the total ion chromatogram. Panel (**b**) illustrates clear co-elution of the metabolites ornithine and citrate, when unique ions for each compound are extracted. This is only achievable if enough scans have been collected across the TIC peak to enable deconvolution

and reduce analysis time. The oven ramp described can be easily adjusted as required.

11. As stated in **Note 8**, the settings described are specifically for an Agilent GC–MSD. The scan rate stated should be able to be achieved on any other vendor's instrument. The important factor is to aim to have at least 10–20 scans across each chromatographic peak. This ensures enough sensitivity and the ability to deconvolute where there may be co-eluting peaks (Fig. 2).

References

1. Roessner U, Luedemann A, Brust D, Fiehn O, Linke T, Willmitzer L, Fernie A (2001) Metabolic profiling allows comprehensive phenotyping of genetically or environmentally modified plant systems. Plant Cell 13:11–29
2. Sheedy JR, Ebeling PR, Gooley PR, McConville MJ (2010) A sample preparation protocol for 1H nuclear magnetic resonance studies of water-soluble metabolites in blood and urine. Anal Biochem 398:263–265
3. Meyer H, Liebeke M, Lalk M (2010) A protocol for the investigation of the intracellular Staphylococcus aureus metabolome. Anal Biochem 401:250–259
4. De Souza DP, Saunders EC, McConville MJ, Likic VA (2006) Progressive peak clustering in GC-MS Metabolomic experiments applied to Leishmania parasites. Bioinformatics 22: 1391–1396
5. Villas-Boas SG, Hojer-Pedersen J, Akesson M, Smedsgaard J, Nielsen J (2005) Global metabolite analysis of yeast: evaluation of sample preparation methods. Yeast 22:1155–1169
6. Ambikapathy V (2011) GC-MS determination of bioactive compounds of enicostemma littorale (Blume). Asian J Plant Sci Res 1:56–60
7. Joseph B (2011) Effect of bioactive compounds and its pharmaceutical activities of sida cordifolia (Linn.). Int J Biol Med Res 2: 1038–1042
8. Tan KC, Trengove RD, Maker GL, Oliver RP, Solomon PS (2009) Metabolite profiling identifies the mycotoxin alternariol in the pathogen Stagonospora nodorum. Metabolomics 5: 330–335
9. O'Callaghan S, Desouza DP, Isaac A, Wang Q, Hodkinson L, Olshansky M, Erwin T, Appelbe B, Tull DL, Roessner U, Bacic A, McConville MJ, Likic VA (2012) PyMS: a Python toolkit for processing of gas chromatography–mass spectrometry (GC-MS) data. Application and comparative study of selected tools. BMC Bioinformatics 13:115
10. Aggio R, Villas-Boas SG, Ruggiero K (2011) Metab: an R package for high-throughput analysis of metabolomics data generated by GC-MS. Bioinformatics 27:2316–2318
11. Choe S, Woo SH, Kim DW, Park Y, Choi H, Hwang BY, Lee D, Kim S (2012) Development of a target component extraction method from GC-MS data with an in-house program for metabolite profiling. Anal Biochem 426: 94–102
12. Want E, Masson P (2011) Processing and analysis of GC/LC-MS-based metabolomics data. Methods Mol Biol 708:277–298

Chapter 4

A Robust GC-MS Method for the Quantitation of Fatty Acids in Biological Systems

Nirupama Samanmalie Jayasinghe and Daniel Anthony Dias

Abstract

Fatty acids (FAs) are involved in a wide range of functions in biological systems. It is important to measure the exact amount of fatty acids in biological matrices in order to determine the level of fatty acids and understand the role they play. The ability to quantify fatty acids in various systems, especially plant species and microbes has recently paved the way to the mass production of pharmaceuticals and energy substitutes including biodiesel. This chapter describes an efficient method to quantify the total fatty acids (TFAs) in biological systems using gas chromatography-mass spectrometry (GC-MS) and a commercially available standard mix of fatty acid methyl esters (FAMEs) using a *step-by-step* methodology to setup a quantitation method using the Agilent Chemstation software.

Key words Total fatty acids, Trans-esterification, FAMEs, GC-MS, Quantitation

1 Introduction

Fatty acids can simply be described as straight chain saturated or unsaturated carboxylic acids and are the main constituents of highly abundant lipids (i.e., glycerolipids) [1, 2]. Fatty acids are available as either unbound called free fatty acids (FFAs) or bound fatty acids. Bound fatty acids are mainly found in glycerolipids, glycerophospholipids, and sphingolipids. Please *see* Chapter 6 for a further description into lipidomics. Total fatty acids (TFAs) can be defined as a combination of bound and unbound fatty acids in a given biological system. GC-MS is a highly sensitive, efficient, and reproducible analytical technique which can be used to quantitate fatty acids and has been routinely used for decades [3]. However, with respect to metabolomics or the biosynthesis of primary and secondary metabolites, a combination of several analytical platforms must be utilized to understand the metabolome of a particular system. For example, liquid chromatography coupled with mass spectrometry (LC-MS) can be used to analyze lipid classes.

Ute Roessner and Daniel Anthony Dias (eds.), *Metabolomics Tools for Natural Product Discovery: Methods and Protocols*, Methods in Molecular Biology, vol. 1055, DOI 10.1007/978-1-62703-577-4_4, © Springer Science+Business Media, LLC 2013

Whereas high performance liquid chromatography can be used to fractionate lipid classes and these lipid fractions can subsequently be analyzed using GC-MS to measure TFAs. There are two main factors to consider when quantitating TFAs; the extraction of samples and the correct preparation of calibration curves which will be described in this chapter.

2 Materials

In order to analyze TFAs, free and bound fatty acids (lipids) must be extracted from the biological matrix of interest. There are numerous methodologies which describe the quantitative extraction of lipids and fatty acids in the literature; however, the choice of extraction method depends on the tissue type and fatty acids/lipids of interest [2]. Multiple extraction methods must be tested and fatty acid recoveries must be carried out in order to obtain an optimized extraction protocol. Precautions must be taken to minimize degradation and auto-oxidation of fatty acids. Unsaturated fatty acids are susceptible to rapidly oxidize which could give misleading quantitation results of TFAs (*see* **Note 1**).

Preparation of calibration standards can be tedious and time consuming given the number of fatty acids present in biological systems for example the individual preparation of standards, that is, C_4–C_{28} fatty acids is time consuming. Furthermore evaporation of solvents becomes a major issue during the preparation of fatty acid standards as the majority of fatty acids are only soluble in nonpolar solvents such as chloroform, *n*-hexane, or dichloromethane. In order to overcome these issues and minimize technical errors a commercially available quantitative standard mix of fatty acid methyl esters (FAMEs) could be used.

2.1 Preparation of the Calibration Solution and Extraction Solvents for GC-MS Analysis

1. Supelco 37 Component FAME (1 mL) (Sigma Aldrich, Part No. 47885-U).
2. Internal standard (INSTD) (i.e., Myristic acid-1-^{13}C).
3. Chloroform (HPLC grade).
4. Methanol (HPLC grade).
5. Clear screw top 2 mL glass vials and screw caps.
6. Butylated hydroxytoluene (BHT).

2.2 Extraction of Lipids and Fatty Acids

1. Mortar and pestle or automated grinder.
2. Liquid nitrogen.
3. 2 mL Eppendorf tubes.
4. Analytical balance.

5. Spatula.
6. Calibrated pipettes and pipette tips 5–1,000 µL.
7. Centrifuge.

2.3 Derivatization

1. 0.2 N Methanolic solution of *m*-trifluoromethylphenyl trimethylammonium hydroxide (Meth Prep II reagent) (GRACE, Part No.18007).
2. Thermomixer.
3. 350 µL glass vial inserts.

2.4 GC-MS Requirements and Specifications

Please *see* Chapters 2 and 3 for a general GC-MS instrument setup.

3 Methods

3.1 Preparation of the Internal Standard (INSTD)

1. Prepare a 2:1 chloroform:methanol solution (10 mL) containing 60 µM INSTD for samples.
2. Prepare a 2:1 chloroform:methanol solution (10 mL) containing 120 µM INSTD for calibration standards.

3.2 Preparation of Extraction Solvent

1. Prepare methanol containing 0.1 % (v/v) BHT.

3.3 Extraction of Lipids and Fatty Acids

It is extremely important to quench metabolism immediately after harvesting tissue to ensure that lipolytic and other enzymatic reactions are terminated to obtain a "snap-shot" of steady-state levels of fatty acids and lipids. Keep the tissues frozen until enzymatic activity has quenched by heating with methanol. Below is an example for a TFA extraction using a modified and optimized "Folch" method [4].

1. Homogenize tissue with liquid nitrogen using prechilled mortar and pestle. Do not defrost tissue.
2. In availability of automated grinders follow manufactures instructions.
3. Weigh approximately 10 mg of homogenized tissue into a 2 mL Eppendorf tube. Make sure sample tissues do not defrost during the weighing process.
4. Add 200 µL of 100 % (v/v) methanol and vortex for about 30 s.
5. Incubate at 70 °C for 15 min at 750 rpm using a thermomixer.

6. Cool down the sample to room temperature and spin down shortly in a centrifuge.
7. Add 400 μL of chloroform containing 0.1 % v/v BHT in the fume cupboard, vortex.
8. Centrifuge for 15 min at 15,900 × *g*.
9. Transfer the supernatant into a new 2 mL Eppendorf tube.
10. Aliquot 200 μL of the supernatant into a glass vial insert.
11. Evaporate to dryness under a stream of N_2 or under vacuum without heating (*see* **Note 2**).
12. Seal and store dried samples in a −20 °C freezer until further analysis.
13. Reconstitute the dried pellet in 25 μL of 2:1 (v/v) chloroform:methanol containing 60 μM INSTD (Myristic acid-1-^{13}C) prior to GC-MS analysis.

3.4 Preparation of Calibration Standards for GC-MS Analysis

It is a requirement of the software that the concentration of INSTD must be kept constant across all the samples and calibration standards as relative responses are used to create calibration curves. The responses of the FAME standards are normalized to the response of INSTD (relative response).

1. As presented in Table 1 below prepare calibration standards of FAMEs containing the INSTD ranging from 5 to 200 μM based on the Supelco FAME mix (*see* **Notes 3–5**).
2. Store these stock calibrations for future use at −20 °C (*see* **Note 6**).

Table 1
Reagent volumes needed for preparation of calibration standards stock

FAME mix volume (μL)	2:1 Chloroform: methanol (μL)	2:1 Chloroform: methanol containing INSTD (μL)	Final volume (μL)
10	490	500	1,000
20	480	500	1,000
30	470	500	1,000
50	450	500	1,000
75	425	500	1,000
100	400	500	1,000
125	375	500	1,000
150	350	500	1,000

3.5 Derivatization and GC-MS Analysis

GC-MS is only capable of analyzing volatile compounds or that are readily derivatizable. Therefore lipids must be hydrolyzed to obtain fatty acids which are bound to a glycerol backbone (e.g., Triglycerides) followed by derivatization to their corresponding methyl esters. The commercially available Meth Prep II reagent is capable of a one step *trans*-esterification of glycerolipids, glycerophospholipids, and sphingolipids and methylation of FFAs in which the reaction takes place in the inlet of the GC-MS [5]. It is important to mention that precautions must be taken to minimize impurities due to the high sensitivity of mass spectrophotometer (*see* **Note** 7).

1. Add a 5 μL aliquot of Meth Prep II reagent into a glass vial containing a glass insert and then add 25 μL of a calibration stock. Seal immediately.
2. Add a 5 μL aliquot of Meth Prep II into sample inserts from of Lipids and FFA extraction. Seal immediately.
3. The concentration of INSTD in both calibration mix and sample is 50 μM after the addition of Meth Prep II reagent.
4. Mix the samples and calibration standards containing Meth Prep II reagent at 37 °C for 30 min at 750 rpm using a thermomixer (*see* **Note 8**).
5. Inject 1 μL into GC-MS using a 30 m VF-5 ms column, 0.2 μm film thickness with a 10 m Integra guard column using the following temperature program as shown Table 2.
6. A drawback associated with using this column is co-eluting *cis* and *trans* FAMEs which are unable to be resolved. This can be overcome by using specific columns such as the Omegawax 250, PAG, SP-2380, and/or SP2560 which either improve or completely resolve these isomers.
7. Eluting FAMEs are transferred to the mass spectrometer through the transfer liner which is set at 280 °C.
8. Mass spectrometer temperature is set to 230 °C.
9. Eluting compounds are ionized and fragmented using electron impact (EI) ionization of 70 eV.

Table 2
Temperature program used to separate FAMES

	Rate (°C/min)	Value (°C)	Hold time (min)	Run time (min)
Initial		50	1	1
Ramp 1	15	230	3	16
Ramp 2	10	325	3	28.5

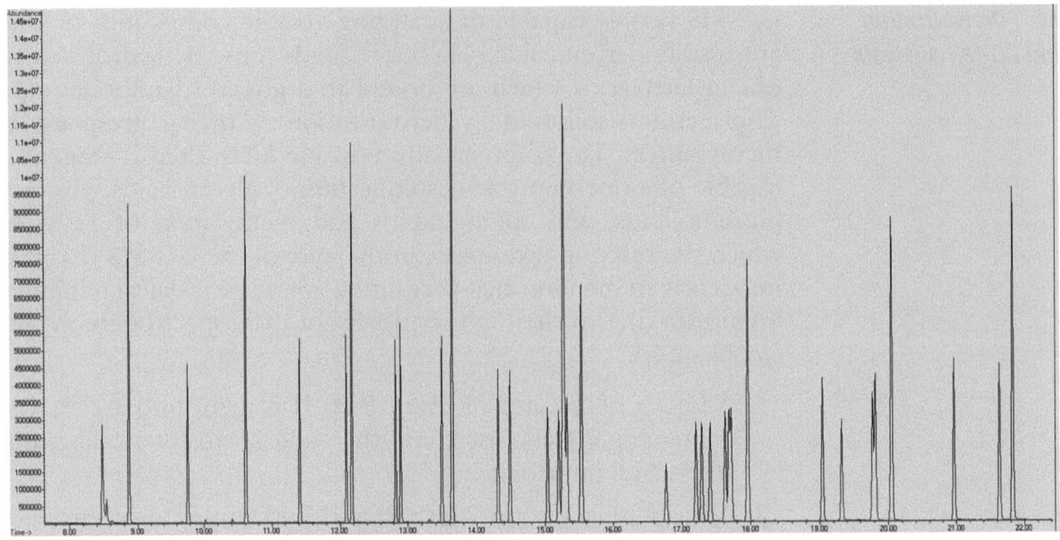

Fig. 1 Total ion chromatogram of a selected calibration Supelco 37 FAMEs mix

10. Resulting ions are filtered through the quadrupole and mass spectra are recorded at 2 scan/s[1] with an *m/z* 50–600 scanning range (Fig. 1).
11. Inject three hexane blanks between every sample and calibration standards (*see* **Note 9**).

4 Post Data Analysis

4.1 Using Agilent Chemstation to Quantify FAMEs

MSD Chemstation Data Analysis software (Version E.02.01.1177) is used to set up calibration curves and also quantitate fatty acids. There are several approaches which should be considered when setting up a quantitation method and quantitating fatty acids (i.e., Automated or Manual setup). For the purposes of this chapter, manual setup will be summarized. When setting up a quantitation method manually, using Chemstation the following workflow should be considered.

1. Optimize integration parameters (integration parameter file).
2. Create quantitation database.
3. Specify calibration level(s).
4. Update retention times in quantitation data base using *Easy ID tool.*
5. Update responses.

6. Review and edit quantitation results using *QEdit tool.*
7. Edit compounds.
8. Report results.

4.1.1 Optimize Integration Parameters

1. Select and load the actual data acquisition method from the *Method* menu bar in Chemstation.
2. Go to the *Chromatogram* menu bar and click on *select integrator*. This will prompt you to a *Temporarily Select Integrator* dialog box.
3. Choose *RTE integrator*.
4. Then select *MS Signal Integration Parameters* from *Chromatogram* menu bar to edit and optimize the integration parameters (Fig. 2) (*see* **Note 10**).
5. Click on *Load* to load the existing ".P" file which has integration parameters particular to the data acquisition method or if you want to custom setup integration parameters, change the parameters manually by following the guidelines displayed as illustrated below in (Fig. 2).
6. Make sure the settings are saved before going to the next step.

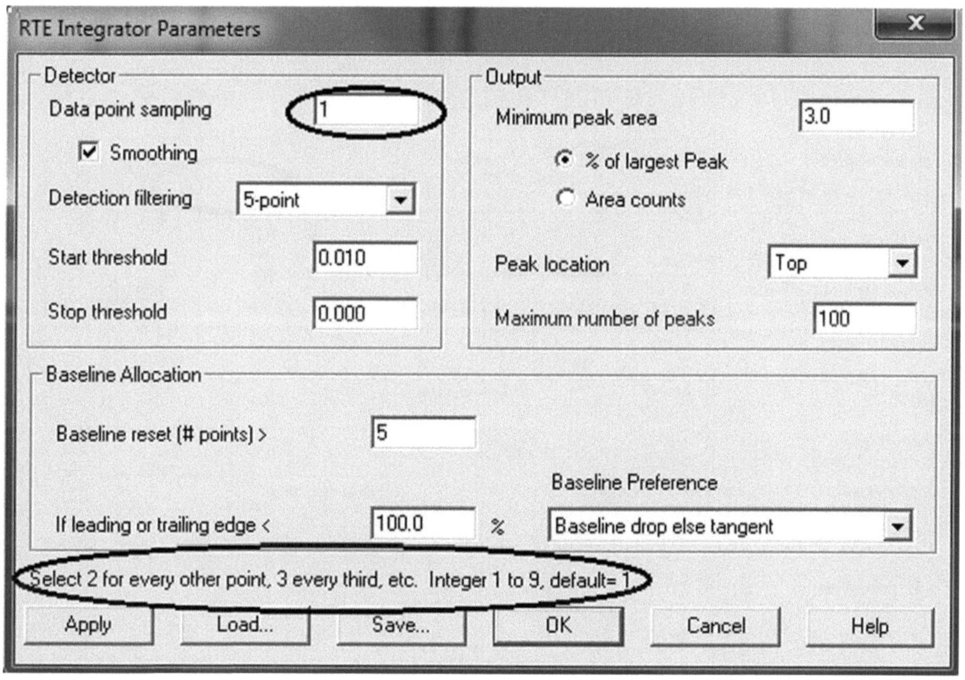

Fig. 2 RTE integrator parameter dialog box and *circled* are an example for guidelines (i.e., guidelines for data point sampling parameter)

4.1.2 Create Quantitation Data Base (QDB)

The compound database contains the details about the compounds of interest. Mainly these include, compound name, retention time, target ion, qualifier ion(s), data points, and identification selection criteria.

1. To create a QDB go to the *Calibrate* menu bar in Chemstation and select *Set Up Quantitation*. This will open the *Quantitation Database Globals* dialog box (Fig. 3). This is the fundamental point in supplying basic information to create the QDB.
2. Enter *Calibration Title*.
3. Keep the default values for *Locating Peaks* (*see* **Note 11**).
4. Specify *New Compound Info* (Fig. 3).
5. Browse and upload the optimized integration parameter file (Subheading 4.1.1). This will be located in the data acquisition method folder. Make sure use *RTEINT* is ticked (Fig. 3).

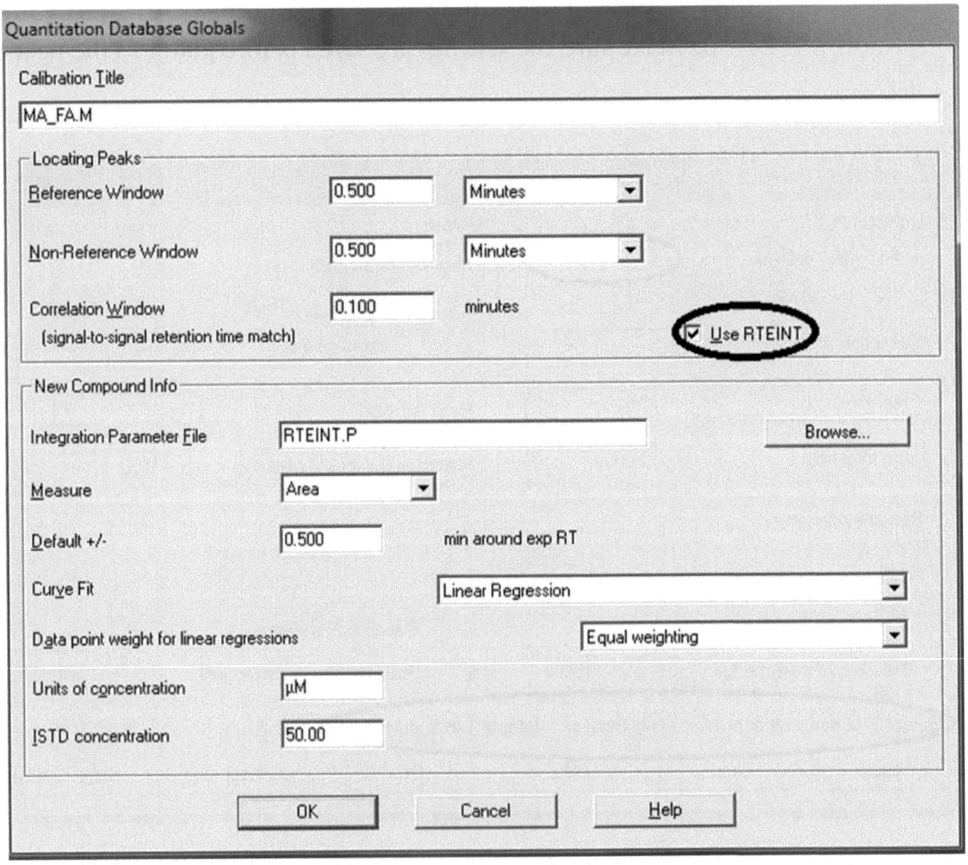

Fig. 3 Quantitation database globals dialog box

6. Click *OK*.
7. *Edit compounds* dialog box will be automatically displayed (Fig. 4).
8. Click *insert above* on Edit Compounds dialog box to add a new target compound (Fig. 4) (*see* **Note 12**).
9. This will open up *Quant setup* dialog box (Fig. 5).
10. Navigate through the TIC and locate the target FAME. Then *double right click* on compound peak to add to the list. Make sure that INSTD is added first.
11. *Double right click* on the peak, this will record the retention time in the *Quant Setup* dialog box.
12. Type a name for the target FAME.
13. Select a specific ion from the spectrum and *click both mouse buttons* together on the ion to insert target ion. Area under target ion will represent the amount of the particular FAME (*see* **Note 13**).
14. Then select at least one qualifier ion from the spectrum and repeat **step 12** to insert the ion (Fig. 5) (*see* **Note 14**).
15. Presence of qualifier ions in the correct amounts relative to the target ion gives evidence of the correct target compound identification. Target ion: qualifier ion response ratio will be automatically calculated. For example, C18:0, Retention Time (RT): 12.44, Target 298.3, Q1 255.2, Target ion: qualifier ion response ratio 108.7, etc. Please refer to Table 3 which lists selected specific target ions and qualifiers for Supelco FAMEs mix.
16. Press *Save* after completing an entry and repeat above-mentioned steps to continue adding more compounds (Fig. 4).

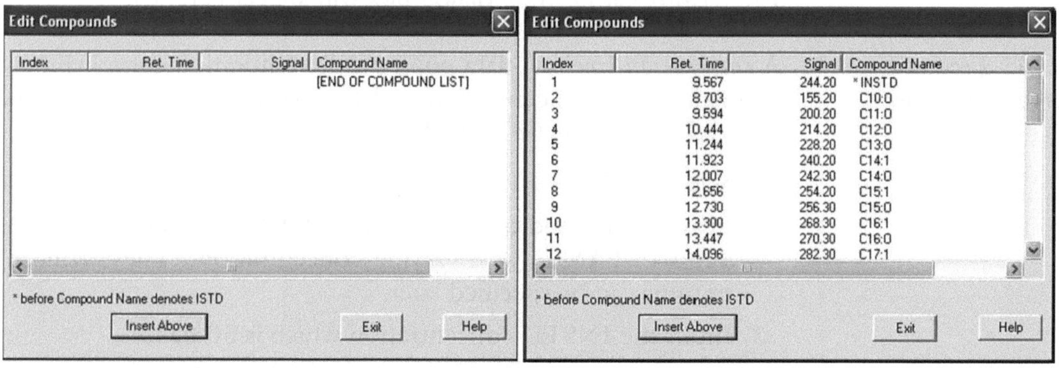

Fig. 4 Edit compounds dialog box (before and after the addition of compounds)

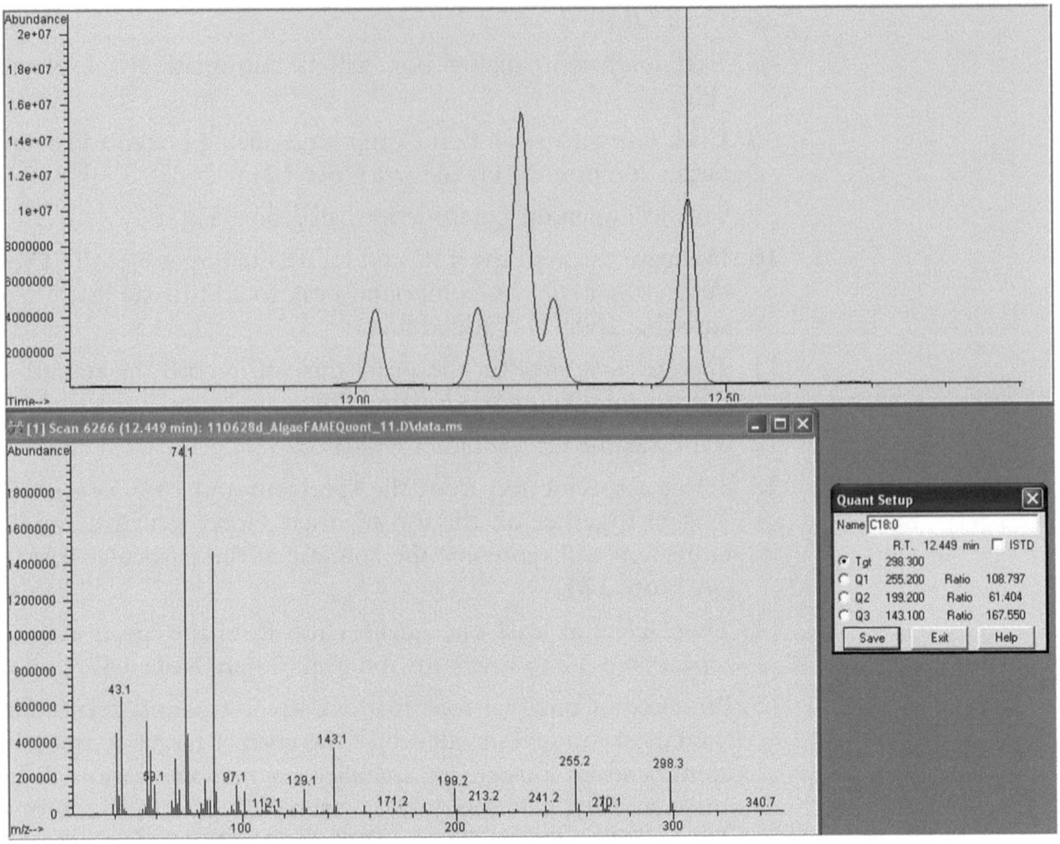

Fig. 5 Quant setup dialog box and adding C18:0 to compound list

17. Make sure all the compounds are added after the INSTD as all compounds followed by the INSTD are analyzed relative to INSTD.
18. Press *exit* on *Edit compounds* dialog box to go to next step.
19. *Update Calibration* dialog box will appear (Fig. 6).

4.1.3 Specification of Calibration Levels

A *calibration level* is a data point on the calibration curve which is defined by the concentration and the corresponding area of the selected target ion (*see* **Note 15**).

1. Select *Add Level* in the *Update Calibration* dialog box (Fig. 6).
2. Leave compound concentration at 0.000 as each level contains a mix of FAMEs with varying concentrations. These concentrations can be specified later.
3. Input the INSTD concentration which is 50 μM.
4. A *Level ID* can be a name or a number which represents each calibration level. For example FAME mix volumes of Table 1 which are 10, 20…75…, 150 μL can be used as level IDs where *10* represents lowest, *75* mid, and *150* represents the highest level of the calibration.

Table 3
List of target ions and qualifier ions

Compound	Target	Qualifier 1	Qualifier 2	Qualifier 3
C10:0	186.2	155.2	143.1	87.1
C11:0	200.2	169.2	157.1	143.1
C12:0	214.2	183.2	171.2	143.1
C13:0	228.2	197.2	185.2	143.1
C14:1	240.2	208.2	166.2	124.1
C14:0	242.3	211.2	199.2	143.1
C15:1	254.2	222.2	180.2	110.1
C15:0	256.3	225.2	213.2	143.1
C16:1	268.3	236.3	194.2	152.2
C16:0	270.3	239.3	227.2	143.1
C17:1	282.3	250.3	208.2	166.2
C17:0	284.3	253.3	241.2	199.2
C18:3n3	292.1	194.1	150.1	107.1
C18:2n6c	294.3	263.1	150.1	123.1
C18:1n9c	296.3	264.0	–	–
C18:2n6t	294.3	266.2	–	–
C18:3n6	292.3	222.4	–	–
C18:1n9t	235.2	123.1	166.2	111.1
C18:0	298.3	255.3	199.2	143.1
C20:4n6	203.2	150.2	133.1	119.1
C20:5n3	201.2	180.1	133.1	119.1
C20:3n3	320.3	289.3	222.2	177.2
C20:2	322.2	391.3	150.1	123.1
C20:1	324.3	292.3	250.3	208.2
C20:3n6	320.3	289.3	264.1	149.1
C20:0	326.4	283.3	199.2	143.1
C21:0	340.4	297.3	199.2	143.1
C22:6n3	159.1	145.1	131.1	119.1
C22:2	350.4	319.3	109.1	95.1
C22:1n9	320.4	352.4	278.3	236.3
C22:0	354.4	311.3	255.3	143.1
C23:0	368.4	325.4	269.3	143.1
C24:1	348.4	380.4	306.3	264.3
C24:0	382.5	339.4	283.3	143.1

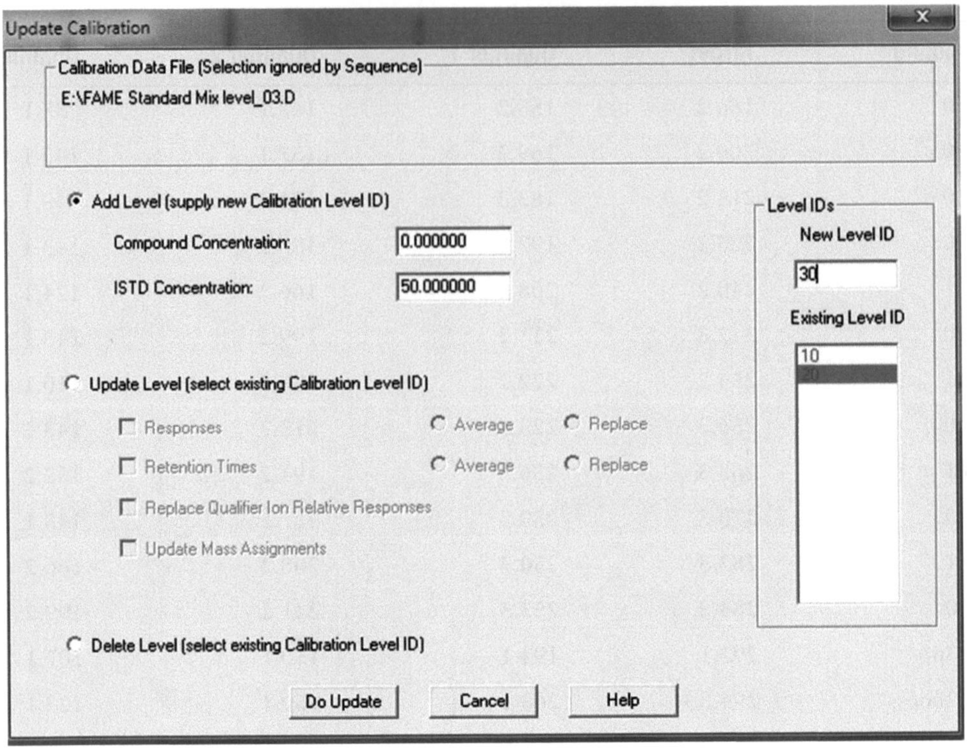

Fig. 6 Specification of levels dialog box

5. In order to add a *Level ID*, specify a *level ID* in *New Level ID* box. Note that you will have a FAMEs calibration data file opened in the background and make sure that you have entered the corresponding level ID.
6. Click *Do Update*. You will be prompted whether to *quantitate the data file*.
7. Press *No*.
8. The new level is now added to the existing level ID.
9. *Edit compounds* dialog box will be prompted. Click *OK* to save the new level ID.
10. Open the next calibration level data file.
11. Select *Update* on the *Calibration* menu bar.
12. Select *update one level* and repeat the above steps to add next level ID.
13. Repeat these steps until all the levels are added.

4.1.4 Update Retention Times in the Quantitation Data Base Using the Easy ID Tool

The *Easy ID* tool is located under the *View* menu bar which is used to update retention times of the individual FAMEs on a compound by compound basis. The “start and end” retention time of each peak can be specified using this tool. Therefore signals are only extracted from the user specified window.

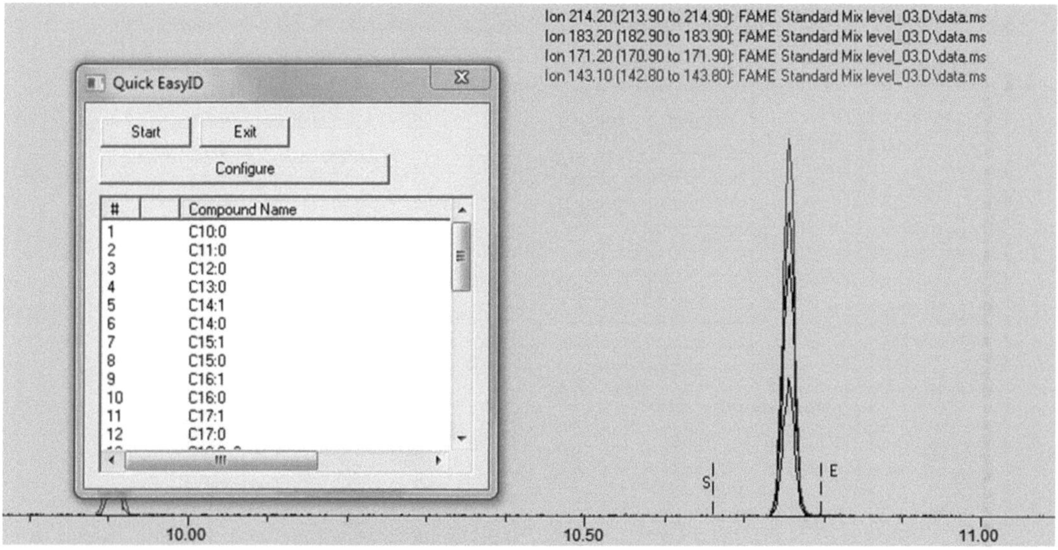

Fig. 7 Quick EasyID window and start and end of C14:0 peak

1. Open a FAME mix calibration data file. Ideally a data file of mid-range concentration. For example *Level 75*.
2. Select *Easy ID* under the *View* menu bar in Chemstation. This will direct the user to Easy ID view from data analysis view (Fig. 7). Each compound will be represented by specified Target ions and Qualifier ions (selected ion chromatogram).
3. *Right double click* on the first compound in the *Quick EasyID* dialog box.
4. Hold the *shift* key and *left double click* at the start of the peak followed by a *right double click* at the end of the peak to update retention times (Fig. 7).
5. *Right double click* on the selected ion chromatogram to navigate to the next peak. Repeat the **step 4**. Apply this same approach for rest of the peaks. Press *exit*.
6. Be aware that this step will only change the retention time window for each compound. Therefore integration has to be performed for every level to update responses of all the FAMEs.

4.1.5 Update Responses

1. Open the lowest level calibration data file (i.e., Level 10).
2. To update each calibration level with corresponding responses, qualifier ion response ratios and mass assignments select *Update* on the *Calibration* menu bar.
3. Select *update one level.*
4. You will be asked whether to quantitate data file. Press *Yes.*
5. *Update Calibration* dialog box will be prompted.
6. Select *Update Level* on the dialog box (Fig. 6).

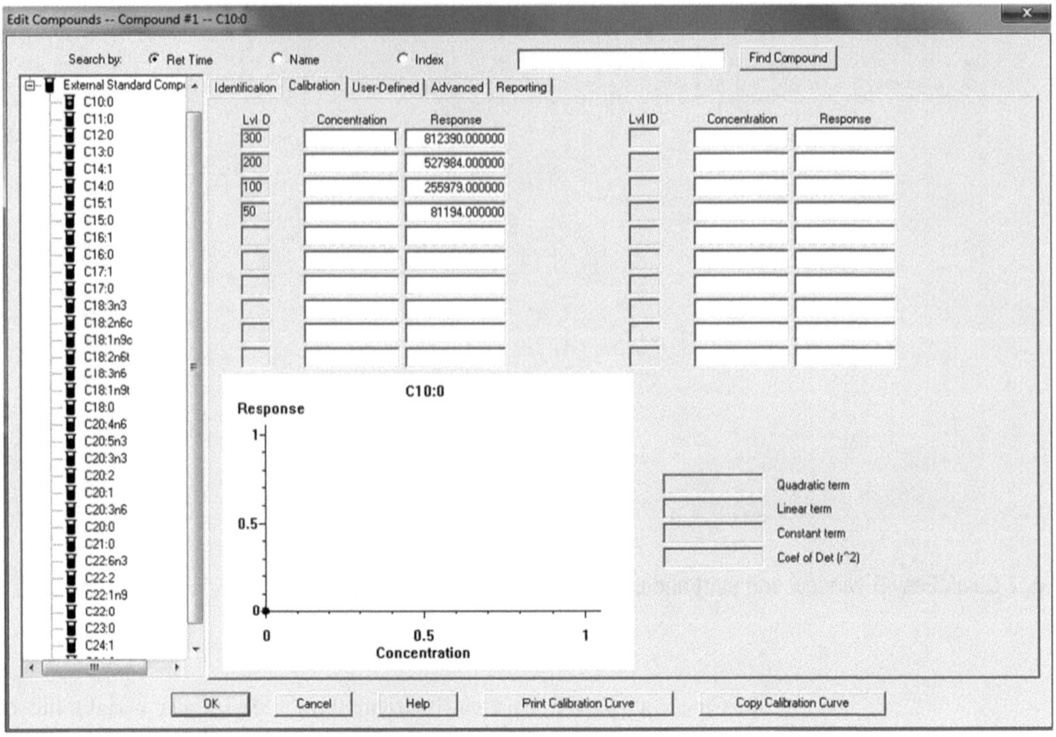

Fig. 8 Edit compounds dialog box with responses

7. Select the corresponding *Existing Level ID* (i.e., 10) for the data file.
8. Select *replace Responses* and *Average Retention Times.* Click *Do Update.*
9. The *Edit compounds* dialog box will appear every time a level is updated (Fig. 8). This also means that new corresponding responses are added for each FAME. Press *OK.*
10. Go to *Method* menu bar and select *save* to save changes.
11. Repeat these steps for all levels.

4.1.6 Review and Edit Quantitation Results Using the QEdit Tool

The *QEdit tool* which is located under the *View* menu bar of Chemstation is used to review and modify integration. Low intensity signals and overlapping compounds result in poor integration. This can be reviewed and edited using the QEdit tool (Fig. 9).

1. Select *QEdit Quant Results* under the *View* menu bar. This will direct the user to the QEdit display.
2. *Left double click* on the first compound in the QEdit dialog box. Review the integration. If poor integration was performed click and drag the right mouse button to draw an integration line under the peak.

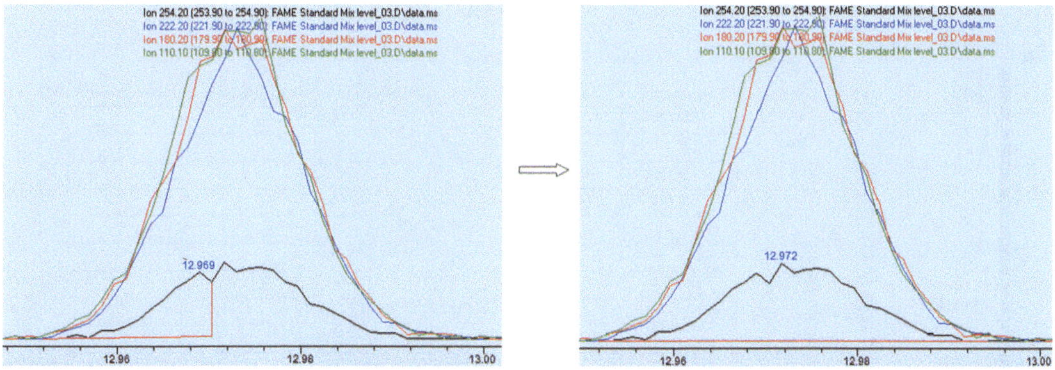

Fig. 9 Poor integration of the C17:0 FAME (*left*) and corrected integration after editing (*right*)

3. Follow this same approach for the rest of the FAMEs. A cross in front of each compound in the QEdit dialog box indicates that the peak is located and integrated.
4. Save QEdit results. *Do not automatically re-quantitate* if signals are manually integrated using QEdit the responses will be restored back to the original by re-quantitation.
5. Repeat this for all the levels.
6. After reviewing and editing the quantified results, update the responses with the new quantified results. To do this follow steps under *Update Responses* (Subheading 4.1.5). Click "NO" when asked whether to re-quantitate data file.
7. After updating the last level, do not close the *Edit compounds* dialog box which contains updated responses for all FAMEs (Fig. 8). Go to the next step below, *Edit compounds.*

4.1.7 Edit Compounds

The *Edit Compounds* dialog box can be defined as *Quantitation Database* or as the summary of all the steps followed so far (Fig. 8). Here you can edit parameters of individual FAMEs if required.

1. To add the concentration of individual FAMEs at each calibration level, go to *Calibration* in *Edit compounds* dialog box (Fig. 10). Then type in the calculated concentrations for each relevant calibration level. These concentrations can be calculated based on % w/v of each fatty acid specified by the manufacturer and reagent volumes used in Table 1. Addition of concentrations will create calibration points.
2. Additional parameters can be changed under *Identification* in *Edit Compounds* dialog box, e.g., Quant signal (i.e., target ion and TIC), Curve fit, and weight.

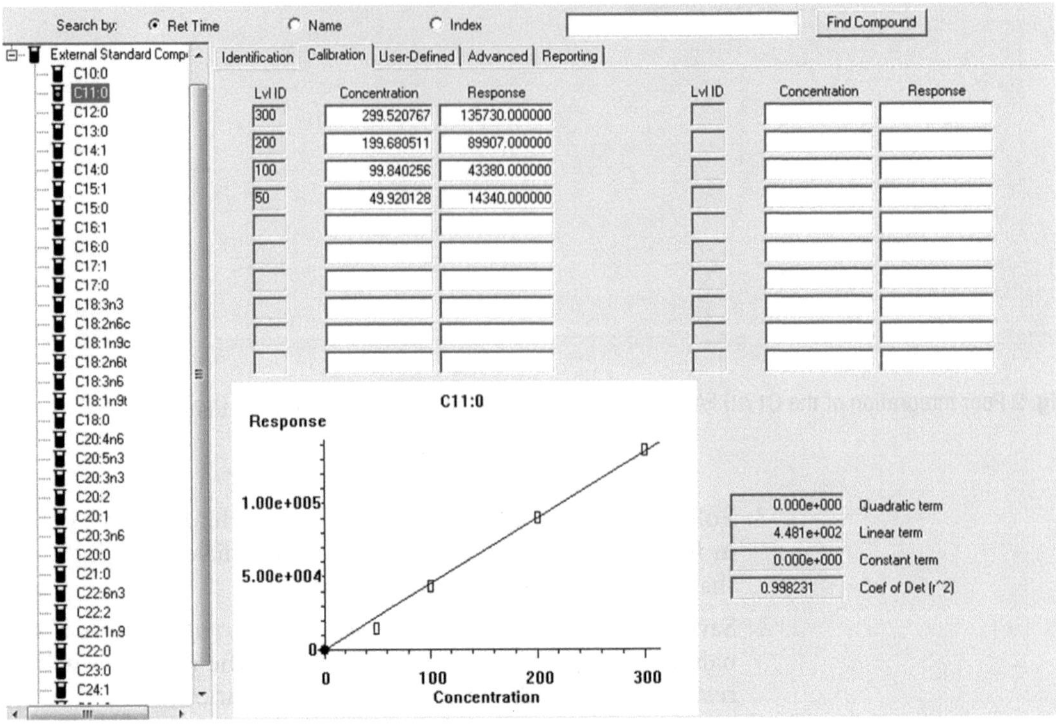

Fig. 10 Quantitation database with updated concentrations and responses

3. Setting up of the quantitation method is now complete. Make sure the method is saved. This quantitation method will be automatically saved under the actual data acquisition method.
4. This will be the template for all sample batches to be run under the same data acquisition method (*see* **Note 16**).
5. Make sure that you run the calibration standard mix with every sample batch.
6. Follow **steps 1–8** under *Update responses* every time you are analyzing a new batch.
7. Use QEdit to review quantified results and update the responses again to apply any changes made to the integration.

4.1.8 Report Results

1. Open the actual data acquisition method from *Method* menu bar of chemstation. This method will contain the quantitation method you created.
2. Open a sample data file.
3. Select *Calculate* under *Quantitate* menu bar in Chemstation.
4. This will produce a quantitation report. Consider this report as a raw report and review the integrations using *QEdit tool.* Reintegrate as necessary by following steps of *Review and edit quantitation results using the QEdit tool* (Subheading 4.1.6).

5. Then click on *Generate report* under *Quantitate* menu bar on Chemstation. These results can be exported as different formats under *Export reports* menu bar (i.e., file.csv).
6. These concentrations can be directly back calculated to obtain mass of individual FAs and expressed as ratio to the initial mass of materials used (*see* **Note 17**).

5 Notes

1. Antioxidants can be used to minimize oxidation of unsaturated fatty acids. For example, 1 % (v/v) of Butylatedhydroxytoluene (BHT).
2. Heating can cause modification of fatty acids and oxidation of unsaturated fatty acids.
3. Chloroform is a toxic chemical, always work in the fume cupboard.
4. Chloroform is extremely volatile so care must be taken when preparing stock solutions as evaporation can lead to increased concentration of compounds which will give misleading quantitation and nonlinear calibration curves.
5. The response factors of each FAME is different and therefore the linear range is not constant. However, a linear range of 5–200 μM range covers the lowest and highest detection limits for all FAMEs (C_{10}–C_{24}).
6. The FAME standard calibration stock must be stored in a sealed glass vial in the dark at −20 °C for future usage. Long-term storage durations can alter the composition of the stock. Perform an initial quality check by acquiring a GC-MS of the stock standard prior to preparation of the calibration assay.
7. Use only high purity MS grade chemicals and carry out regular maintenance of your GC-MS.
8. This can be either carried out "offline" using a thermomixer or "online" using an auto-sampler depending on your instrumental setup.
9. FAMES tend to stick onto the column and may result in carry over. It is suggested to acquire several *n*-hexane washes between sample injections to minimize over quantitation.
10. Use mid-range loadings to set up parameters and fine tune parameters using lower loading.
11. Global settings are not selective enough for compound identification. Most of the parameters need to be adjusted for individual compounds (i.e., FAME peak widths vary from one to another). Therefore most of the parameters need to be optimized at a later stage.

12. When you are adding compounds make sure that you know the elution order of the FAMEs. The most convenient way of identifying a FAME is by looking for the molecular ion in the mass spectrum. Also note that the elution is also column specific. For example the column specified in this chapter; unsaturated FAMEs are eluted before the saturated FAMEs with the same chain-length.
13. Use the molecular ion as the target ion for co-eluting peaks in most cases. For example co-eluting FAMEs which are C18:1n9c, C18:2n6t, and C18:3n6 can be resolved by using the molecular ions specified in Table 3.
14. For most of the FAMEs, molecular ions can be assigned as target ions. However it is likely that the abundance of the molecular ion is low and therefore a fragment ion would have to be selected.
15. User can specify the quant signal (i.e,. Target ion, and TIC). These options are available under *Identification* of *Edit compounds* dialog box (Fig. 8). For example, if the target ion is chosen as the quant signal the response will be the area response of the target ion.
16. Make sure Retention Time is Locked (RTL) and prepare calibration standards and samples in the same manner. Keep the INSTD concentration constant between samples and calibration standards and make sure to prepare all the relevant changes to quantitation method if any additional parameters have been altered, e.g., the INSTD concentration.
17. Reagent volumes, extraction solvent volumes, and dilutions must be taken into account when calculating the actual concentration of each FAME in the original sample.

References

1. Christie WW, Han X (2010) Lipid analysis – isolation, separation, identification and lipidomic analysis, 4th edn. Oily Press, Bridgewater
2. Kumari P, Reddy CRK, Jha B (2011) Comparative evaluation and selection of a method for lipid and fatty acid. Anal Biochem 415:134–144
3. Roessner U, Wagner C, Kopka J, Trethewey RN, Willmitzer L (2000) Simultaneous analysis of metabolites in potato tuber by gas chromatography-mass spectrometry. Plant J 23: 131–142
4. Folch J, Lees M, Stanley GHS (1957) A simple method for the isolation and purification of total lipides from animal tissues. J Biol Chem 226:497–509
5. Goetz N, Burgaud H, Berrebi C, Bore P (1984) Analysis of the lipid content of single hair bulbs. Comparison with the content of the sebaceous gland and with surface lipids. J Soc Cosmet Chem 35:411–422

Chapter 5

A Workflow from Untargeted LC-MS Profiling to Targeted Natural Product Isolation

Damien L. Callahan and Candace E. Elliott

Abstract

Liquid chromatography-mass spectrometry (LC-MS or HPLC-MS) is an extremely sensitive analytical technique that enables the detection of metabolites with a vast range of chemistries and molecular masses. Extracts from any biological starting material are first fractionated chromatographically on a stationary phase suitable for the retention of the target molecules. The eluent is then transferred directly to the ionization source for MS detection. There is a vast range of chromatographic separation methods and MS configurations. This chapter describes a method for the detection of a broad range of metabolites using reversed phase (C18) LC-MS as well as a method for the isolation of targeted metabolites of interest.

Key words Liquid chromatography, Ultra-high pressure-liquid chromatography, Preparative chromatography, Reversed phase, Mass spectrometry, Metabolomics

1 Introduction

The advancements in chromatographic stationary phases, ultra-high pressure LC systems, and high mass resolution MS instrumentation [1, 2] have resulted in an analytical technique that can routinely detect thousands of metabolites from complex biological extracts (Fig. 1). Due to the complexity of these extracts, identification of metabolites is a colossal challenge. However, through the use of chromatography, metabolites of interest can be isolated and further characterized by scaling up the separation from an analytical to a preparative scale [3]. The interfaces between LC systems and the MS instruments are also under continual improvement. Electrospray ionization sources (ESI) are now more compatible with the higher flow rates utilized in ultra-high pressure chromatography (UPLC).

There are three crucial phases in this technique: sample extraction [4], chromatographic separation, and MS detection [5]. Each step has specific considerations, for example, the extraction protocol

Ute Roessner and Daniel Anthony Dias (eds.), *Metabolomics Tools for Natural Product Discovery: Methods and Protocols*, Methods in Molecular Biology, vol. 1055, DOI 10.1007/978-1-62703-577-4_5, © Springer Science+Business Media, LLC 2013

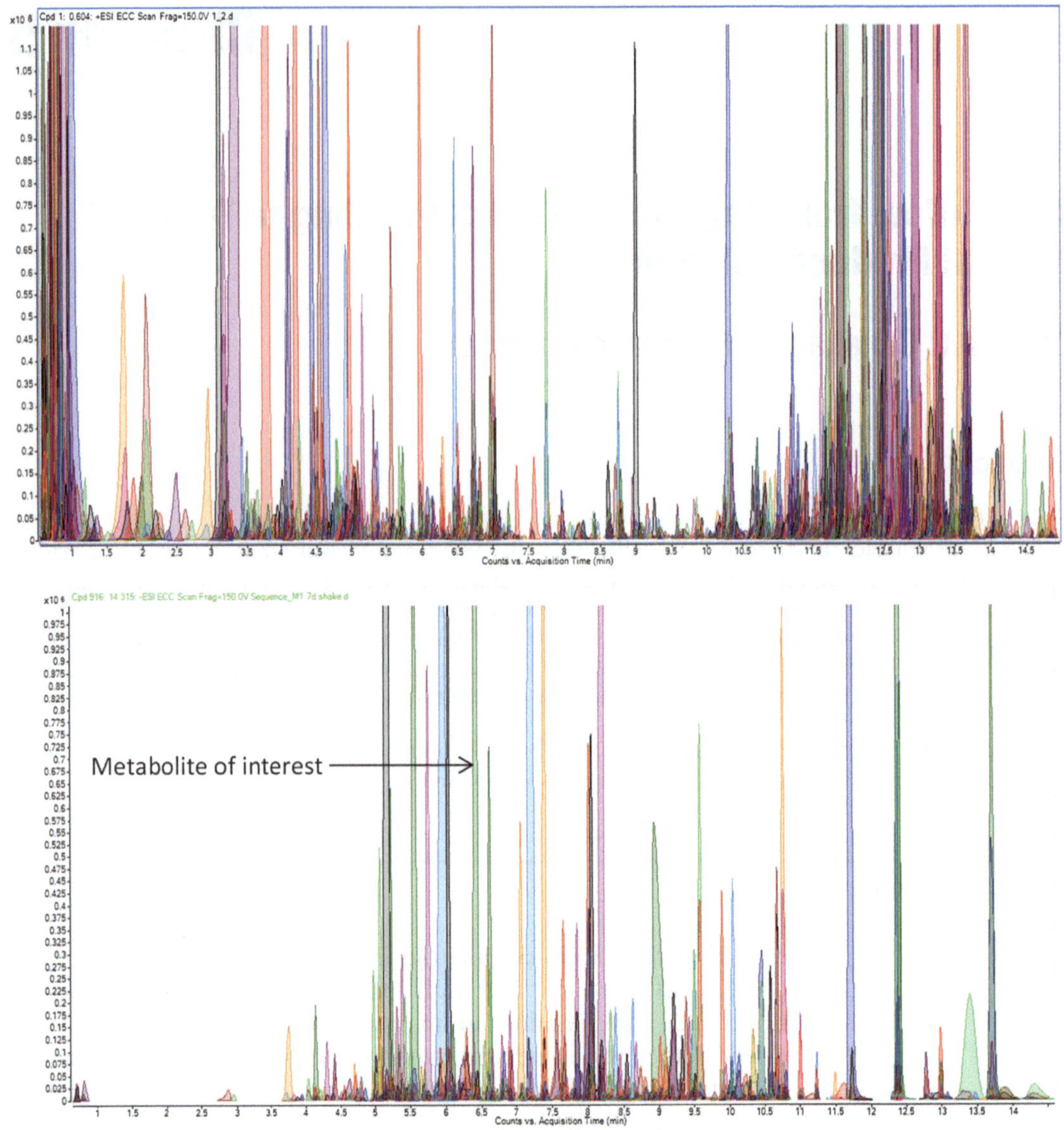

Fig. 1 Each chromatographic peak represents a metabolite, also known as "mass feature," which has a defined retention time, mass and peak area and height. The mass feature does not have to be an identified metabolite. (*Top*) Positive ion LC-MS mass feature extracted chromatogram from a leaf extract of *Thlaspi caerulescens* with 2,396 metabolites detected; (*bottom*) negative ion LC-MS mass feature extracted chromatogram targeting secondary metabolites from an ethyl acetate extract of the fungus *Leptosphaeria maculans*. Note that both chromatograms have been zoomed in to show low abundant mass features

should be optimized to match the chemistry of the metabolites of interest, the chromatographic stationary phase must be able to retain the metabolites, and the ionization mode (polarity) and source type must be appropriate for the molecules being detected by the mass spectrometer. Direct infusion of samples (fingerprinting) is also an approach taken in metabolomics [6]. However, chromatography is necessary for resolving the many isomers present in

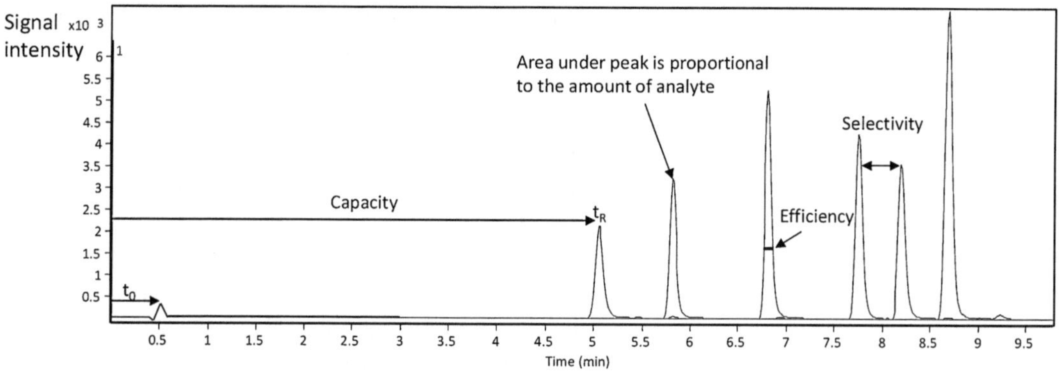

Fig. 2 An annotated chromatogram showing common chromatographic terms: capacity reflects the interaction of compounds with stationary phase, selectivity describes the ability of system to discriminate between compounds, efficiency relates to peak width, i.e., wider peaks require more separation, t_0 = elution time of unretained peak, t_R = retention time—determines sample identity

biological extracts. Without pre-separation these metabolites are analyzed as one, and subsequent purification of target metabolites is not possible.

In the discovery phase of metabolite profiling, the aim is to detect as many metabolites as possible [7]. A compromise must be made between high throughput and time-consuming complex sample preparation and chromatographic analysis time. The details outlined here provide a method with efficient separation and minimal analysis time. The chromatography is based on the most commonly used stationary phase, reversed phase, which separates molecules based on hydrophobic interactions. Due to the great range of LC and MS specifications, the method can be adapted to suit most configurations and in some cases alternatives have been included to suit other systems. For example, there is a trend for LC systems to operate with higher maximum pressures which enables users to utilize the increased chromatographic efficiency (Fig. 2) obtained from using smaller particle sized columns. There are a range of other chromatographic options such as hydrophilic interaction chromatography for the retention and analysis of polar compounds [8]. This chapter describes an LC-MS application and details on how to isolate a targeted natural product.

2 Materials

2.1 Chemicals

1. Due to the sensitivity of mass spectrometry all attempts to minimize the presence of chemical contaminants must be taken (*see* **Notes 1–3**). Ideally, glassware such as measuring cylinders and mobile phase bottles should be exclusively used for the preparation of mobile phases. If the use of plastic is necessary,

for example pipettes and tubes, they must be solvent compatible. Phthalates from plastics and detergents such as polyethylene glycol are typical contaminants that can interfere with ionization and detection. An extensive list of common ESI contaminants can be found in Keller et al. [9].

2. MS grade mobile phases and additives should be used and 18 MΩ deionized water must be used throughout the procedure. Organic solvents and acid modifiers are a health hazard and must be handled in a fume cupboard with appropriate personal protective equipment and medical safety data sheets must be read before handling chemicals.
3. Waste must be handled by following all local waste disposal regulations. Also, the ESI source must be vented out of the laboratory.
4. The following chemicals are required for this method: acetonitrile, isopropanol, formic acid, deionized water, and MS calibrants.

2.2 HPLC Requirements and Specifications

1. Mobile phase A (1 L):0.1 % formic acid in water: ensure that the glass mobile phase bottle has been cleaned and well rinsed. Using a graduated measuring cylinder, transfer 1 L of deionized water to the mobile phase bottle and add 1 mL of MS grade formic acid. Glass ampoules of MS grade formic acid are ideal for this. Mix well before placing on the LC system (*see* **Note 1**).
2. Mobile phase B (1 L): 0.1 % formic acid in acetonitrile: As above, transfer 1 L of MS grade acetonitrile into mobile phase container and add 1 mL of MS grade formic acid (*see* **Notes 2** and **3**).
3. Needle wash: 50 % isopropanol in water.
4. A binary LC system equipped with online degasser, temperature controlled (optional) autosampler with an operating pressure maximum of ≥600 bar (8,700 psi; ideal system).
5. C18 reversed phase fully end-capped silica stationary phase with dimensions 2.1 mm i.d. × 100 mm length and 1.8 μm particle size silica is required (*see* **Note 4**).
6. For an LC system with a maximum pressure ≤400 bar (5,800 psi), the following columns can be used, however, the mobile phase gradient (Tables 1 and 2) will need modification in order to compensate for the lower efficiency, selectivity, and capacity of these columns (e.g., C18 reversed phase fully end-capped silica stationary phase with dimensions 2.1 mm i.d. × 100 mm length and 3 μm silica) or [2.7 μM fused core silica columns, e.g., Poroshell (Agilent) or Core-Shell (Phenomonex)].

Table 1
Standard gradient elution timetable for mobile phases A (water) and B (acetonitrile) both containing 0.1 % formic acid

Time (min)	% A	% B
0	95	5
10	0	100
12	0	100
12.1	95	5
17	95	5

Table 2
Gradient elution targeting more polar metabolites using mobile phases A (water) and B (acetonitrile) both containing 0.1 % formic acid

Time (min)	% A	% B
0	99	1
2	99	1
5	70	30
12	100	100
14	100	100
14.1	1	1
19	1	1

7. Semi-preparative C18 reversed phase LC column 10 cm i.d. × 250 cm length with 5 μm particle size silica.
8. A matching guard column or frit filter should be attached to all LC columns for protection against blockages.
9. A thermostat-controlled column oven is recommended to remove retention time shifts due to fluctuations in ambient room temperature.

2.3 Mass Spectrometric System

1. The most appropriate MS for analysis of complex biological extracts is a fast scanning (>2 spectra/s), high resolution accurate mass (<2 ppm) instrument that is capable of collision-induced dissociation (CID) or MS/MS. Quadrupole Time of Flight (QTOF) MS and some Fourier Transform instruments

have these capabilities. The high mass resolution enables detection of compounds with similar molecular masses. The fast scanning is required when using small particle size columns. These columns should result in peak widths that are typically 6 s or less. Other low resolution MS hardware such as quadrupole or ion-trap instruments can still be applied; however, the loss in resolution and scan speed is a significant compromise.

2. Electrospray ionization is the most common ionization source used. This soft ionization technique produces a low amount of in-source fragmentation of molecules. The ionization source used must be matched to the molecules of interest. In some cases atmospheric pressure chemical ionization (APCI) may be appropriate, for example, when using high flow rates or for dirty sample matrices as it is less prone to ion-suppression. APCI is a harder ionization technique as it produces more fragment ions than ESI. ESI sources are continually being developed and can now deal with flow rates up to 1 mL/min. However, sensitivity is still compromised at the higher flow rates.

3 Methods

3.1 Liquid Chromatography

1. Purge LC system with new mobile phases by either running pumps individually at maximum flow rate for at least 3 min (after opening the purge value) or by using the inbuilt purge function of the LC pump.
2. Attach column and ensure that the system volume and dead volumes are minimized by using the shortest practical tubing lengths and correctly seated tubing connections (Fig. 3). Dead volumes are areas where the remixing of separated compounds can occur, such as tubing connections, causing a reduction in column selectivity and efficiency (Fig. 2) (*see* **Notes 5–9**).
3. Check for leaks on all fittings, the column pressure is a good indicator for poor pump priming (air in pump heads) or leaks (*see* **Note 10**).
4. Check that the LC system pressure fluctuation is within instrument specifications for example <2 %.
5. The LC column should be first flushed with at least 20 column volumes of mobile phase B, making sure flow through is diverted to the waste, then equilibrated with initial mobile phase conditions (*see* **Note 11**).
6. Running conditions for a 1.8 μm particle size column are: flow rate 0.4 mL/min, column temperature 35 °C, initial mobile phase composition 5 % B, gradient details in Table 1. For more polar compounds, it may be necessary to use the following

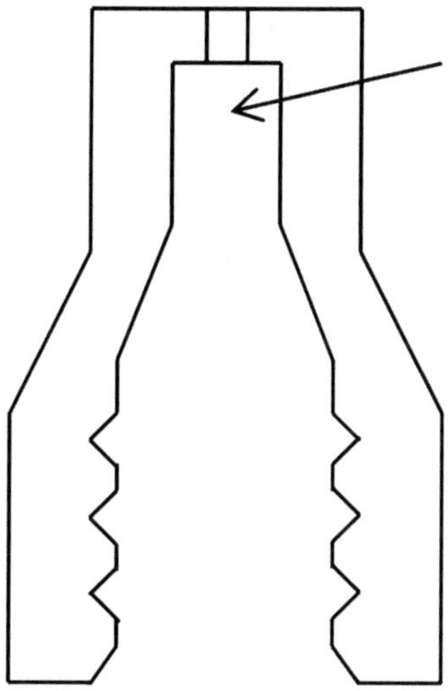

Fig. 3 Typical port for LC fitting. The *arrow* points to the area where dead volume can occur if tubing is incorrectly seated

gradient in order to retain and separate these metabolites: flow rate 0.4 mL/min, column temperature 20 °C, initial mobile phase composition 1 % B, gradient in Table 2.

3.2 Mass Spectrometry

1. Check the ESI source chamber is clean.
2. The gas flow rates, temperatures, and pressure, lens voltages such as the declustering potential or fragmentor voltage and capillary or ESI needle voltage should be optimized for each system. Manufacturers provide guides for source settings under different flow rates. The following conditions are for an Agilent 6520 ESI-QTOF-MS: gas flow 10 L/min, gas temperature 315 °C, nebulizer pressure 45 psi, capillary voltage 4,000 V, fragmentor 140 V (*see* **Notes 12–14**).
3. Set the correct scan rate and mass range for data acquisition. The scan rate should provide at least 13 points across a chromatographic peak (*see* **Notes 15** and **16**).

3.3 System Suitability

1. Calibrate the mass accuracy and check both the mass accuracy and resolution are within the operating specifications for the MS (*see* **Notes 17–20**).
2. Inject blank solvents and note the minimum and maximum column pressures throughout the analysis and background ions. At least three blank injections may be required to observe a stable background trace.
3. Ideally analyze a quality control standard with a known concentration. Mixed QC standards can be purchased such as fatty acids or amino acids, alternatively an *in*-house mixture may be

more appropriate. This will provide information on column performance through peak shape and retention time as well as MS mass accuracy, resolution, and sensitivity. The test standards and blanks also provide the user with information on chemical noise and system contaminants from previous experiments analyzed on the MS system.

4. Quality control during the entire analysis is very important. A standard practice is to use a pooled biological quality control (PBQC) samples [10]. An aliquot from each sample is pooled, mixed, and analyzed throughout the run, for example, every ten samples. This can be coupled with matrix-free sample and blanks. The data from the quality control samples enable the user to validate the data being produced throughout the whole analysis.
5. Sample injection order should be randomized to remove systematic variation throughout the analysis (*see* **Note 21**).

3.4 Data Analysis

Most commercial vendors now have metabolite profiling software that allows chromatographic peak deconvolution (Fig. 1). This provides a list of compounds, also known as mass features, containing the retention time, mass, and intensity and possible library matches. Typically, the next step in the analysis involves retention time alignment and data transformation, for example, log transformation [11]. These data matrices can then be imported into metabolite profiling or statistical software packages where data can be visualized using multivariate statistics or other more rigorous tests. A number of LC-MS data analysis packages such as MZmine [12], XCMS [13], and Metalign [14] are freely available.

3.5 Isolation of Metabolites by Preparative Chromatography

Once the metabolite of interest has been identified, isolation may be necessary for further characterization. The example described below is an isolation of the toxic secondary metabolite sirodesmin, an epipolythiodioxopiperazine, produced by the fungus *Leptophaeria maculans*, which causes "blackleg disease" in canola (*Brassica napus*) plants [15]. The metabolite profile using a standard gradient elution (Table 1) of the ethyl acetate extract of *L. maculans* is illustrated in Fig. 1. This method was scaled up to a semi-preparative LC column to isolate pure sirodesmin (Fig. 4). One key element is the ability to scale up the sample preparation and to access a preparative LC system with a fraction collector. In this example sirodesmin was isolated to use as a standard for the quantitative analysis of other samples and for bioassays to measure toxicity.

1. The culture filtrate is collected from 12-day-old liquid still cultures of *L. maculans* and grown in V8 juice by filtration through Miracloth to remove mycelia.

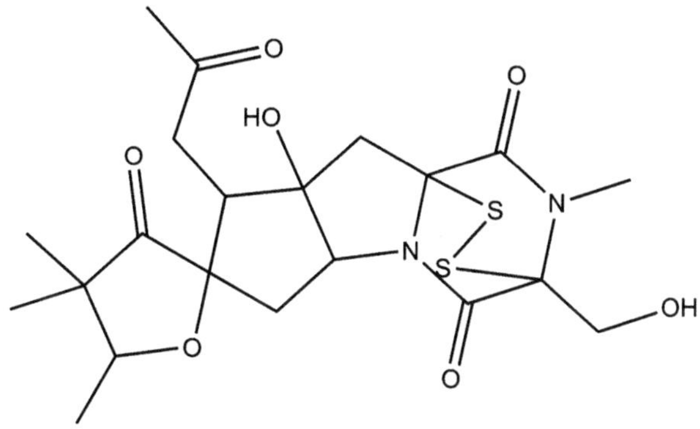

Fig. 4 Structure of the secondary metabolite Sirodesmin isolated from *L. maculans*

2. The culture filtrate (800 mL) is then extracted twice with ethyl acetate (400 mL) and separated by gravity in a 1 L separation funnel.
3. The organic extracts are then combined and dried using a rotary evaporator.
4. The crude extract is then resuspended in acetonitrile (3 mL) with vigorous shaking for 1 min followed by the addition of 3 mL of water and further mixing.
5. The precipitate is then removed by centrifugation at 1,430 × *g* for 5 min then filtration through a 0.2 μm Millex-LG PFTE syringe filter to remove any remaining particulate matter. This crucial step removes compounds that are incompatible with the mobile phase and also matches the injection solvent with the mobile phase gradient.
6. Fractionation is then carried out using a Phenomonex Luna 10 × 250 cm 5 μm (ODS) reversed phase semi-preparative column and an Agilent 1100 HPLC system with a UV/Vis detector and fraction collector.
7. Use the same mobile phases as detailed in Subheading 2.2.
8. For separation, a linear gradient from 10 to 100 % B over 90 min with a 100 % B hold for a further 5 min is used to ensure all metabolites are eluted, followed by a 9 min re-equilibration (Table 3).
9. Flow rate is 4 mL/min and 2 mL of filtered extract is injected via a loop injection.
10. Absorbance is monitored at 220 and 450 nm using an Agilent 1100 series multi-wavelength detector.
11. Time-based fractions (2.4 mL) are then collected every 0.6 min from 20 to 70 min.

Table 3
Gradient elution timetable for the isolation of secondary metabolites using a semi-preparative column

Time (min)	% A	% B
0	90	10
90	0	100
95	0	100
96	90	10
105	90	10

12. The fraction containing sirodesmin (Fig. 4) elutes between 35.6 and 36.2 min as determined by flow injection MS of each fraction.
13. The fractions from the three separate injections are then pooled and freeze-dried.
14. The identity and purity are confirmed by high resolution LC-MS, ^{1}H-NMR, and ^{13}C NMR.
15. No evidence of impurities was observed by either technique, suggesting the fraction was of high (>90 %) purity.
16. If necessary the isolated fraction could be reinjected for additional purification.

4 Notes

1. High-purity (MS grade) solvents and mobile phase additives are crucial for reducing chemical noise. Mobile phase additives must be volatile.
2. Solvent compatible filters should be used to remove particulates from mobile phase solvents after buffer salts have been added. It is not necessary to filter the solvents described here if MS grade solvents and additives are used; in fact filtering can introduce more chemical contaminants.
3. To produce a consistent solvent background, prepare all the mobile phase required for an analysis. This is important when carrying out profile comparisons as the changing chemical noise makes it more difficult for downstream data processing.
4. Ensure the column dimensions, in particular the internal diameter and particle size, are compatible with the LC system: nano (nL/min), capillary (μL/min), UPLC (200–5000 μL/min), and preparative systems (2–10 mL/min).

5. Ensure that the total internal system volume (t_o) (Fig. 2) and dead volumes (Fig. 3) are minimized. System volume is particularly important with low flow rates. To avoid large dead volumes, fittings specifically designed for the manufacturer's port should be used.
6. Back pressure is the key hurdle to overcome when using columns with small particle sizes. Ensure that your LC system can cope with the expected back pressure before purchasing such a column.
7. Stainless steel fittings are required for system pressures over 400 bar.
8. The stationary phase chemistry must be matched to the chemistry of the molecule(s) of interest. For example, highly polar compounds are not retained using reversed phase chemistry and therefore will co-elute with many other polar metabolites (Fig. 5). This can cause ion-suppression and can render preparative isolation of the pure metabolite impossible.
9. The sample injection solvent must be compatible with the mobile phase gradient. This is particularly important for hydrophilic interaction chromatography. The injection of samples

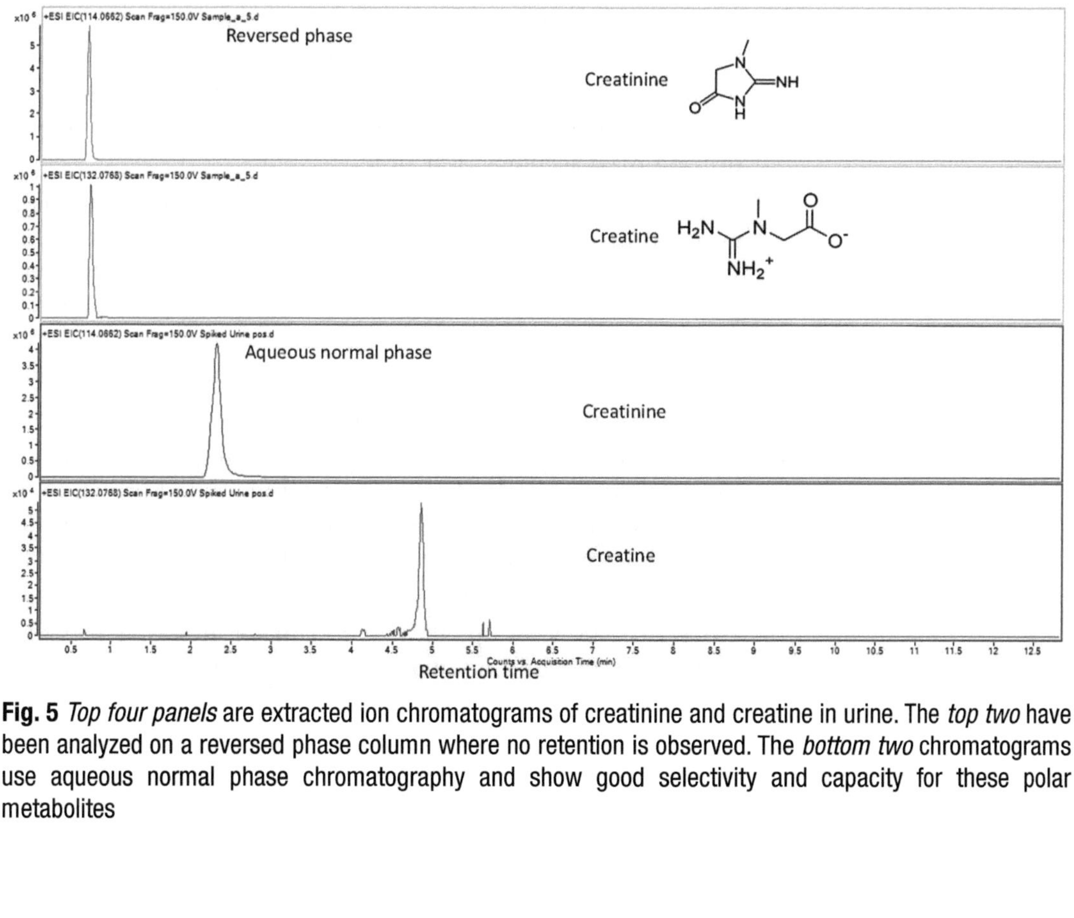

Fig. 5 *Top four panels* are extracted ion chromatograms of creatinine and creatine in urine. The *top two* have been analyzed on a reversed phase column where no retention is observed. The *bottom two* chromatograms use aqueous normal phase chromatography and show good selectivity and capacity for these polar metabolites

dissolved in solvents that do not match the initial mobile phase condition can result in poor retention and poor peak shape of early eluting compounds.

10. Save the column pressure profiles of each sample if possible. This is very useful information for trouble shooting. Shifts in retention time could be due to a back-pressure problem or a slow leak, and pumps may lose prime due to the presence of air in the pump. These issues could be diagnosed by overlaying pressure profiles.
11. Precondition column with the analysis of at least five injections of the pooled biological QC sample. This conditioning ensures that retention times are more reproducible for the early analyzed samples.
12. The electrospray source settings should be optimized in each different MS platform by infusion of a standard using a syringe attached to a T-piece after the column.
13. Ensure that the ionization source is clean and check for any new chemical contaminants that may be present from previous users. The system may need a large number of blank runs in order to produce a constant background.
14. The flow rate must be compatible with the ionization source. New sources with heated nebuliser gases can cope with higher flow rates due to a greater desolvation capacity. For older ESI sources flow rates of 0.2–0.4 mL/min are recommended.
15. Accurate mass instruments must be properly tuned and calibrated. TOF MS instruments in particular, are more prone to drifting mass accuracy over time.
16. Check spectral quality by looking at high and low abundant peaks in profile mode. Peaks should be symmetrical with no splitting and ringing effects. Poorly tuned MS instruments produce noisy data resulting incorrect assignment of mass features, e.g., Fig. 6.
17. For large sample sets consider collecting spectral data in centroid mode only as this reduces file sizes.
18. Determine if the target metabolites ionize in positive or negative ionization mode; both may be required. Formation of adducts such as Na^+, K^+ in positive ion or Cl^-, $HCOO^-$ in negative and neutral losses must be considered when analyzing data and identifying metabolites.
19. Once LC-MS is set up, allow it to run for at least 1 h before calibration, preconditioning, QC checks and initiation of sample work list.

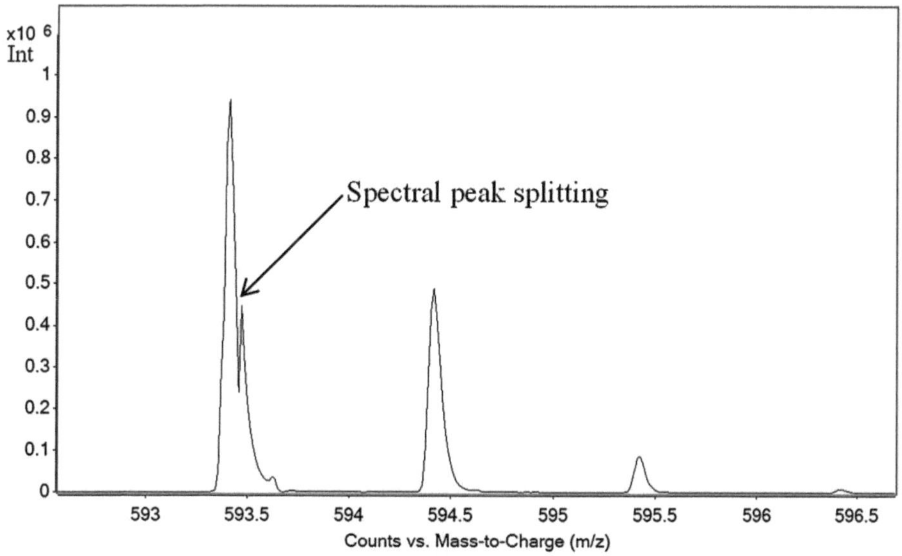

Fig. 6 A spectrum from a poorly tuned QTOF MS instrument. Peaks split above 600,000 counts. This results in nonlinear responses and incorrect chromatographic deconvolution due to the noisy spectral data

20. Sample preparation considerations: how much time is spent before injection of sample preparation? Will there be a simple single phase extraction, liquid/liquid extraction, or solid phase extraction?
21. Be aware of the current accepted practice for reporting on metabolomics data, for example, the Metabolomics Standards Initiative—MSI [16].

References

1. Zhou B, Xiao JF, Tuli L et al (2012) LC-MS-based metabolomics. Mol Biosyst 8:470–481
2. Watson JT, Sparkman OD (2008) Introduction to mass spectrometry. Instrumentation, applications and strategies for data interpretation, 4th edn. Wiley, Chichester, NY
3. Nakabayashi R, Kusano M, Kobayashi M et al (2009) Metabolomics-oriented isolation and structure elucidation of 37 compounds including two anthocyanins from *Arabidopsis thaliana*. Phytochemistry 70:1017–1029
4. Vuckovic D (2012) Current trends and challenges in sample preparation for global metabolomics using liquid chromatography-mass spectrometry. Anal Bioanal Chem 403:1523–1548
5. Lu W, Bennett BD, Rabinowitz JD (2009) Analytical strategies for LC-MS-based targeted metabolomics. J Chromatogr B Analyt Technol Biomed Life Sci 871:236–242
6. Ellis DI, Dunn WB, Griffin JL et al (2007) Metabolic fingerprinting as a diagnostic tool. Pharmacogenomics 8:1243–1266
7. De Vos RCH, Moco S, Lommen A et al (2007) Untargeted large-scale plant metabolomics using liquid chromatography coupled to mass spectrometry. Nat Protoc 2:778–791
8. Rojo D, Barbas C, Ruperez FJ (2012) LC-MS metabolomics of polar compounds. Bioanalysis 4:1235–1243
9. Keller BO, Sui J, Young AB et al (2008) Interferences and contaminants encountered in modern mass spectrometry. Anal Chim Acta 627:71–81
10. Zelena E, Dunn WB, Broadhurst D et al (2009) Development of a robust and repeatable UPLC-MS method for the long-term metabolomic study of human serum. Anal Chem 81:1357–1364

11. van den Berg R, Hoefsloot H, Westerhuis J et al (2006) Centering, scaling, and transformations: improving the biological information content of metabolomics data. BMC Genomics 7:142
12. Pluskal T, Castillo S, Villar-Briones A et al (2010) MZmine 2: modular framework for processing, visualizing, and analyzing mass spectrometry-based molecular profile data. BMC Bioinformatics 11:1471–2105
13. Smith CA, Want EJ, O'Maille G et al (2006) XCMS: processing mass spectrometry data for metabolite profiling using nonlinear peak alignment, matching, and identification. Anal Chem 78:779–787
14. Lommen A, Kools H (2012) MetAlign 3.0: performance enhancement by efficient use of advances in computer hardware. Metabolomics 8:719–726
15. Howlett BJ, Idnurm A, Pedras MS (2001) *Leptosphaeria maculans*, the causal agent of blackleg disease of Brassicas. Fungal Genet Biol 33:1–14
16. Fiehn O, Sumner L, Ward J, Rhee SY, Dickerson J, Lange M, Lane G, Roessner U, Last R, Nikolau B (2007) Minimum reporting standards for plan biology context in metabolomic studies. Metabolomics 3:195–201

Chapter 6

Lipidomics: Extraction Protocols for Biological Matrices

Thusitha Wasntha Thilaka Rupasinghe

Abstract

Lipidomics is defined as a comprehensive analysis of all lipids in a biological system (lipidome). Lipid profiling is a fast-growing area of research due to the involvements of lipids incorporated with human diseases such as obesity, Alzheimer's diseases, and diabetes as well as in drug discovery. Recently improved liquid chromatography and mass spectrometric techniques has provided the rapid expansion of lipidomics research. Lipid extraction and sample preparation is an important step in lipidomics and this chapter describes lipid extractions from different types of samples such as plasma, tissues, yeast, and cells.

Key words Lipids, Tissue extraction, Cell extraction, Monophasic, Biphasic, LC-MS

1 Introduction

Lipids are the fundamental components of biological membranes and have a variety of functions within cells, such as serving as the main building blocks of cell membranes, energy storage, cell signaling, membrane anchoring, and protein trafficking [1]. In addition, the physiological importance of lipids are substantiated by lipid abnormalities in numerous diseases such as diabetes, obesity, atherosclerosis, and Alzheimer's disease. Lipids also play a vital role in pathophysiology, illustrated by a large number of lipid studies involved in infection diseases [2].

Recently, lipidomics has integrated with other "omics," such as genomics, proteomics, and metabolomics, and now contributes to discover the function of lipids in biological systems. In addition, lipidomics is a powerful tool to illustrate the bioactivity mechanism of lipid-based diseases [1]. It elucidates the functions of all lipids in a biological system such as cell signaling, membrane structuring, cell–cell and cell–protein interacting and also responses to environmental changes over time [1].

Ute Roessner and Daniel Anthony Dias (eds.), *Metabolomics Tools for Natural Product Discovery: Methods and Protocols*, Methods in Molecular Biology, vol. 1055, DOI 10.1007/978-1-62703-577-4_6, © Springer Science+Business Media, LLC 2013

1.1 Lipid Classification

Lipids are defined as hydrophobic or amphipathic small molecules which consist of a number of structurally and functionally distinct molecules that span from nonpolar, neutral to polar compounds. Due to this large diversity of polarity, lipid analysis in complicated biological systems is a huge challenge and different analytical platforms have been utilized [3]. According to the lipid classification and nomenclature scheme introduced by LIPIDMAPS, eight categories of lipids are characterized based on their chemical structure and biosynthetic perspectives [4]. The eight lipid classes include fatty acids (FA), glycerolipids (GL), glycerophospholipids (GP), sphingolipids (SP), sterol lipids (ST), prenol lipids (PR), saccharolipids (SR), and polyketides (PK). Representatives of basic structures for each lipid class are illustrated in Fig. 1. Fatty acids are the most common basic structural element of lipids, having unsaturated and saturated straight hydrocarbon chains. Within the FA lipid class, arachidonic acid is an important fatty acid, which is the precursor of eicosanoids functioning as lipid signaling in inflammatory processes. Glycerolipids (GL) consists of glycerol backbones with various numbers of fatty acids such as monoacylglycerides, diacylglycerides, and triacylglycerides (Fig. 1).

Glycerophospholipids (GP) are the most abundant lipid class reported in biological systems, consisting of glycerol with functional head groups as shown in Table. 1. GPs are subclassified according to their head group as glycerophosphocholines (PC), glycerophosphoethanolamine (PE), glycerophosphoserines (PS), glycerophosphoglycerol (PG), glycerophosphoinositol (PI), and glycerophosphatidic acid (PA). GPs are the major building blocks of cell membranes that are involved in cell signaling, membrane anchoring, and substrate transporting. Spingolipids (SP) consist of amide groups linked with long-chained fatty acids with sphingoid backbones as shown in Table. 2. SPs mainly consist of sphingosine (Sph), ceramides (Cer), spingomyelins (SM), glucosylceramide (Glc Cer), and lactosylceramide (Lac Cer).

1.2 Lipid Analysis

Phospholipid (PL) analysis is important in cell biology as they are the structural building blocks in cell membranes, precursors of intercellular signaling molecules, and important participants in the regulation and control of cellular function and disease. Comprehensive analysis of phospholipid species has been a challenge due to their diversity, availability of a large number of species in biological systems, complicity of identification and separation of lipid classes. Even though "shotgun" lipidomics provides fast analysis, resolution of isobaric lipid species reported is limited [1]. As lipidomics is an emerging field in biology, liquid chromatography-mass spectrometry (LC-MS) is widely and routinely used for lipid-based analyses. LC is a powerful analytical technique which provides online chromatographic separation and retention time is adding another dimension of selectivity compared to the direct infusion of lipids to the mass spectrometer.

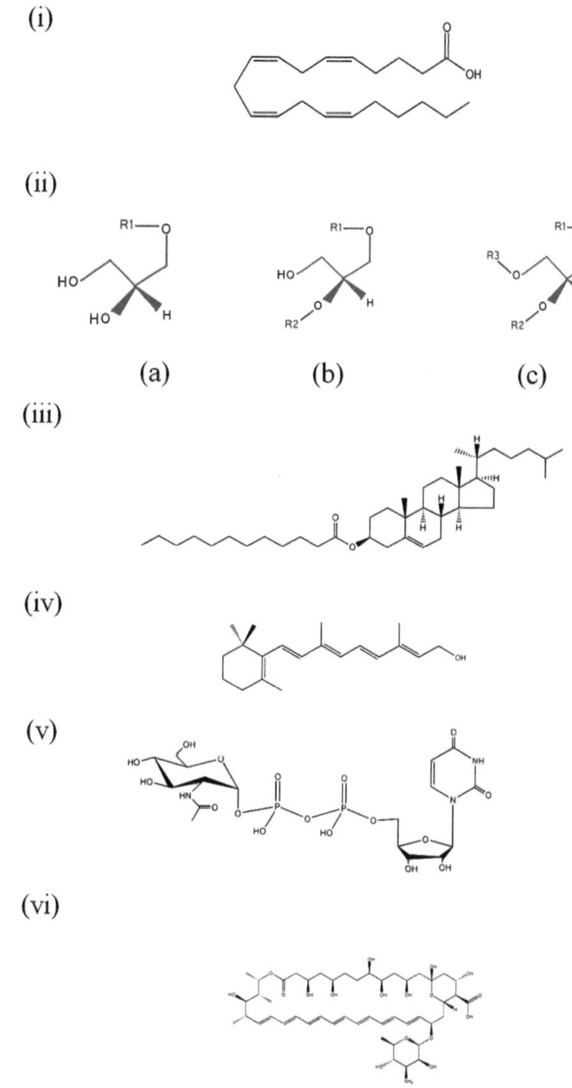

Fig. 1 Representative structures of lipid species for each lipid class. (**i**) Fatty acid (FA): arachidonic acid, (**ii**) Glycerolipid (GL): (**a**) mono acyl glycerol (MAG), (**b**) di acyl glycerol (DAG), (**c**) tri acyl glycerol (TAG), (**iii**) Sterol lipids (SL), (**iv**) Prenol lipids (PR), (**v**) Saccharolipids [SL], and (**vi**) Polyketides

It has been reported, LC-MS has also been used to analyze complex lipid classes like glycerolipids, glycerophospholipids, and glycolipids [5].

1.3 Extraction of Lipids

Lipid extraction methods from biological samples benefit the high solubility of hydrocarbon chains in organic solvents. One of the most common extraction methods for lipids in biological systems is the use of chloroform and methanol in a ratio of 2:1 (v/v) previously reported by Bligh and Dyer [6], and Folch [7]. According to the Bligh-Dyer method, the solvent ratio of chloroform:methanol:water is 8:4:3 by volume which provides the

Table 1
Subclasses of glycerophospholipids (GL): each phospholipid subgroup is decided by the head group (X)

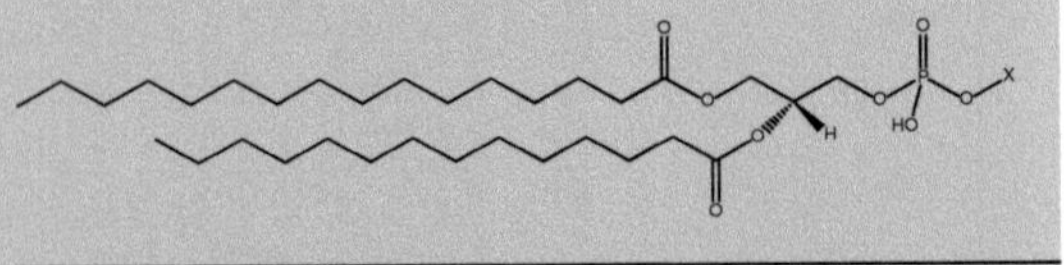

Glycerophospholipids (PL)	X	
Glycerophosphatidic acid/(PA)	Acid	H
Glycerophosphocholine/(PC)	Choline	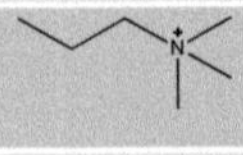
Glycerophosphoethanolamine/(PE)	Ethanolamine	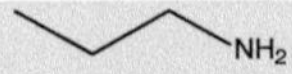
Glycerophosphoinositol/(PI)	Inositol	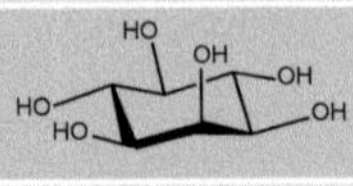
Glycerophosphoglycerol/(PG)	Glycerol	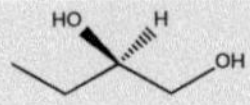
Glycerophosphoserine/(PS)	Serine	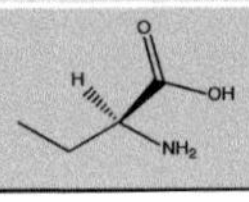

Table 2
Subclasses of sphingolipids (SL): each sphingolipid subgroup is decided by the head group (R)

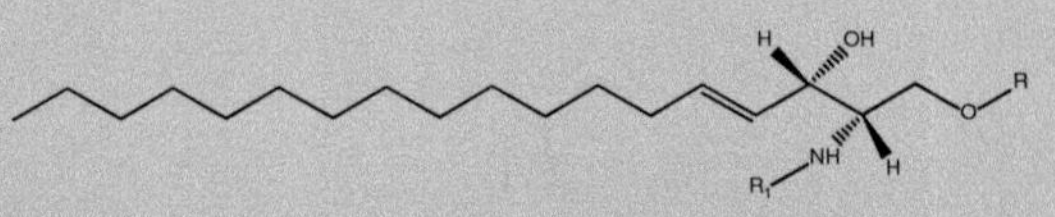

Sphingolipids	R	
Sphingosine	H & (Rl = H)	H
Ceramide (Cer)	H	
Sphingomyelin (SM)	Phosphocholine	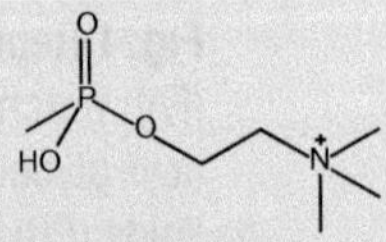
Glucosylceramide (Glc Cer)	Glucose	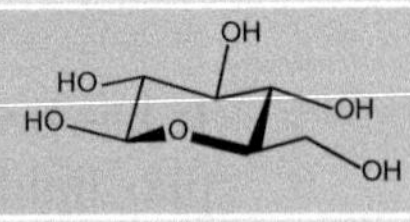
Lactosylceramide (Lac Cer)	Lactose	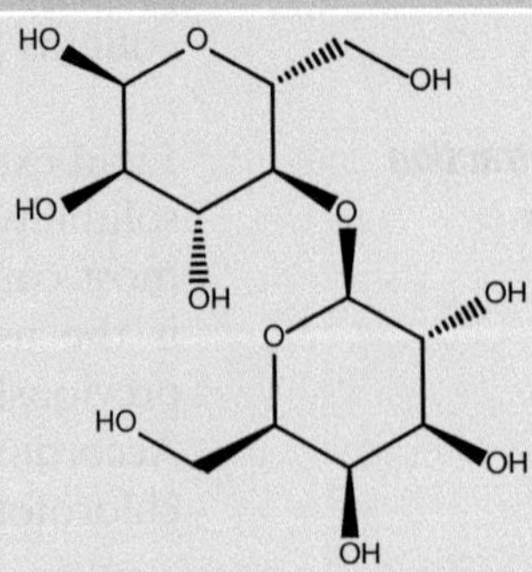

most efficient biphasic partitioning of lipids into the organic phase. Due to the solubility and polarity of the different lipid classes, modification of the Bligh and Dyer method using the chloroform:methanol extraction procedure improves the extraction efficiency. In addition to chloroform:methanol as the extraction solvent, isopropanol–hexane has also been reported for the extraction of prostaglandins [8]. Despite this inexpensive extraction protocol it provides a minimized toxic working environment for sample preparation, however has poor extraction efficiency for gangliosides. In general, lipid extraction must be incorporated with organic solvents in lipidomics, but a single standard protocol for extracting all lipid classes has not been established.

2 Materials

Lipid extraction in different types of sample matrices has been well documented [6, 8]. It is important to validate the extraction protocol for individual lipid classes of interest in selected sample matrices. In a lipidomics approach, validation of extraction protocols involves several steps such as homogenizing, lipid metabolite quenching, extracting, reconstituting, and sample loading for LC-MS analyses. Efficiency of the lipid extraction is validated by spiking known amounts of lipid standards to the sample matrix. In sample preparation for lipid analyses, it is critical to choose the relevant organic solvent for the protein precipitation and remove insoluble particles by centrifuging, rather than filtering. In addition, stability and solubility of lipid species in extraction solvents play a crucial role in method development.

2.1 Sample Harvesting and Quenching of Lipid Metabolism

1. Reaction tubes (e.g., 2 mL safe lock Eppendorf tubes) labeled with identifiable names for each sample to be harvested (*see* **Note 1**).

2.2 Storage of Samples

1. −20 °C freezer if frozen.

2.3 Homogenization of Tissue and Cells Using Mortar and Pestle or Automatic Grinders

1. Mortar and pestle.
2. Automatic grinders such as ball mills, ultra turrax, or cryo-mills.

2.4 Weighing of Tissue Samples

1. New labeled reaction tubes (e.g., 1.5 mL safe lock Eppendorf tubes).
2. Accurate balance (to at least four decimal places).

3. Liquid nitrogen if working with frozen fresh tissue.
4. Spatula.

2.5 Lipid Extraction for LC-MS Profiling

1. Solvents: HPLC grade 2:1 (v/v) chloroform:methanol.
2. Internal standard (e.g., deuterium-labeled lipid standard, Avanti catalog number, 860374P for 16:0 D31–18:1 PE).
3. 1 mM Butylated hydroxytoluene (BHT).
4. Cold phosphate-buffered saline (PBS).
5. 0.1 N HCl.
6. Vortexer.
7. Thermomixer.
8. Bench top centrifuge (Coulter Microfuge 22R centrifuge) for reaction tubes (up to 13,000 rpm).
9. Liquid nitrogen.
10. TurboVap nitrogen evaporator.

3 Methods

3.1 Lipid Extraction

Despite most lipid species are stable at room temperature, it is important to quench lipid metabolism as soon as possible during the extraction process. Lipid species could undergo enzymatic reactions which may result in the oxidation and formation of saturated lipids. During homogenizing, the usage of excess force by the beads in cryomill or fast mortar action may result in the generation of heat and may release acyl fatty acids from lipid species leaving *lyso*-lipid species. It is important to preserve lipid species and avoid any metabolic changes during the lipid extraction process. Samples frozen or quenched with cold organic solvents are the most common methods of metabolic arrest. However, due to the poor solubility of lipids in cold temperature, lipid extractions are carried out at room temperature or higher than room temperature depending on the lipid class of interest. Use of an antioxidant, such as butylatedhydroxytoluene (BHT) is recommended in lipid extractions to avoid lipid oxidation [9]. In a lipidomics approach, monophasic extraction facilitates to extract both apolar and polar lipid species. The following extraction protocol describes the phospholipid extraction in plasma [10], tissue [8], yeast [11], and cultured cells [12] for LC-MS analysis in lipidomics.

3.2 Phospholipid Extraction from Plasma for LC-MS Analysis

3.2.1 Sample Preparation

1. If plasma is frozen, thaw samples and bring to room temperature before aliquoting.
2. For pre-aliquoted plasma with known volumes, centrifuge samples using "short run" function to collect plasma remaining in the lid of Eppendorf tube.

3.2.2 Plasma Extraction for Phospholipids

1. Pipette out known volume (50 μL) of plasma into 1.5 mL Eppendorf tubes.
2. Add 20 times sample volume (1,000 μL) of a 2:1 (v/v) mixture of chilled chloroform:methanol with 1 mM BHT.
3. Add 500 nmol of deuterium-labeled internal lipid standard.
4. Vortex for 1 min at room temperature.
5. Shake for 30 min at room temperature.
6. Centrifuge for 10 min at 13,000 rpm.
7. Transfer supernatant to a new Eppendorf tube.
8. Re-extract protein pellet with the addition of another 500 μL (ten times the plasma volume) of a 2:1 (v/v) mixture of cold chloroform:methanol.
9. Vortex for 1 min at room temperature.
10. Shake for 30 min at room temperature.
11. Centrifuge at 13, 000 rpm.
12. Combine both supernatants.
13. Dry combined supernatant under nitrogen at room temperature.
14. Reconstitute dried plasma lipids in 100 μL of suitable organic solvent for LC-MS analysis.
15. Transfer to autosampler vial for analysis.
16. If long-term storage is required, store dried extracts (before reconstituting) at −20 °C.

3.3 Lipid Extraction from Tissue for LC-MS Analysis

3.3.1 Tissue Preparation

1. Prepare liquid nitrogen in appropriate storage containers (e.g., stainless steel container).
2. Wash tissue using ice-cold phosphate-buffered saline (PBS) to remove contaminants such as blood, then place back into reaction tube.
3. Freeze in liquid nitrogen (*see* **Note 2**).

3.4 Weighing of Tissue

1. Use frozen tissues for weighing (*see* **Note 3**).
2. Prepare new labeled reaction tubes (*see* **Note 4**).
3. Precool reaction tubes (cryomill tubes with ceramic beads) in liquid nitrogen.
4. Place tubes on balance and set balance to zero (TARE).
5. Place tissue pieces in cooled tubes using a precooled spatula.
6. Read, weigh, and record.
7. Place tube with weighed tissue back into liquid nitrogen.
8. Proceed with extraction or store weighed tissue in −80 °C.
9. If working with freeze dried tissue, sample procedure applied without the need of precooling and keeping tissue frozen.

3.5 Homogenization and Extraction of Tissue

1. Homogenize tissue using cryomill (automatic grinder). Operate cryomill according to the manufacturer's instructions.

3.6 Sample Preparation for LC-MS

1. Weigh approximately ~10 mg of tissue into a 2 mL cryomill tube with ceramic beads.
2. Add 1,000 μL of a 2:1 (v/v) mixture of ice-cold chloroform:methanol with 1 mM BHT.
3. Add 500 nmol of deuterium-labeled internal lipid standard.
4. Homogenize frozen tissue using cryomill grinder (*see* **Note 5**).
5. Sonicate cold samples for 30 min at room temperature (*see* **Note 6**).
6. Centrifuge at 13,000 rpm for 10 min at 4 °C.
7. Transfer supernatant to a new Eppendorf tube.
8. Re-extract tissue pellet with the addition of another 500 μL of a 2:1 (v/v) mixture of chilled chloroform:methanol.
9. Sonicate for 30 min at room temperature.
10. Centrifuge at 13,000 rpm.
11. Combine both supernatants.
12. Dry combined supernatant under nitrogen at room temperature.
13. Reconstitute dried lipids in 100 μL of suitable organic solvent for LC-MS analysis.
14. Transfer to autosampler vial for analysis.
15. If need to be stored, store dried extract (before resuspending, **step 12**) in at −20 °C.

3.7 Lipid Extraction from Yeast for LC-MS Analysis

3.7.1 Yeast Harvesting (See Note 6)

1. Prepare liquid nitrogen in appropriate storage container.
2. Add 200 μL of chilled distilled water to 0.2 optical density (OD) units of yeast pellets and vortex (*see* **Note 1**).

3.7.2 Homogenization and Extraction of Yeast

1. Homogenize yeast using cryomill (automatic grinder).

3.7.3 Sample Preparation for LC-MS

1. Transfer approximately 0.2 OD units of yeast pellets in 200 μL of ice-cold water into 2 mL cryomill tubes with ceramic beads.
2. Follow **steps 2–14** in Subheading 3.6 for extraction.

3.8 Lipid Extraction from Cells for LC-MS Analysis

3.8.1 Cell Harvesting (See Note 6)

1. Prepare liquid nitrogen in appropriate storage container.
2. Remove culture from incubator.
3. Transfer the required volume of culture to a tube and centrifuge the samples at maximum speed for the shortest amount of time at 0 °C to remove all supernatant (*see* **Note 7**).

4. Add 200 μL of chilled distilled water to known volume and optical density of yeast pellets and vortex, and transfer into 2.0 mL Eppendorf tube.

3.8.2 Metabolic Arrest and Cell Washing

1. Transfer approximately 10^6 number of cell pellets in 500 μL of PBS into 2 mL Eppendorf tubes.
2. Wash cell pellets twice with 500 μL PBS (chilled).
3. Spin at maximum speed at 0 °C to pellet cells and remove all supernatant.

3.8.3 Sample Preparation for LC-MS

1. Resuspend cell pellets into 500 μL of −40 °C methanol: 0.1 N HCl (1:1).
2. Dipping tubes into liquid nitrogen until frozen, then dip them into a dry-ice/ethanol bath until thawed.
3. Repeat this freeze-thaw cycle in **step 3**, ten times.
4. Add 750 μL chloroform:methanol (1:2) containing 1 mM BHT.
5. Add 500 nmol of deuterium-labeled internal lipid standard.
6. Extract metabolites by vortexing mix, vigorously at room temperature for 5 min.
7. Sonicate for 30 min at room temperature.
8. Centrifuge samples at 13,000 rpm, at 0 °C for 5 min.
9. Transfer the supernatant into new 1.5 mL tube (*see* **Note 8**).
10. Add 500 μL chloroform:methanol 0.1 N HCl (ratio 1:2:0.8).
11. Sonicate for 30 min at room temperature.
12. Centrifuge at 13,000 rpm for 10 min.
13. Combine both supernatants.
14. Dry combined supernatant under nitrogen.
15. Resuspend dried lipids in 100 μL of suitable organic solvent for LC-MS analysis. Transfer to autosampler vial for analysis.
16. If long-term storage is required, store dried extracts (before reconstituting) at −20 °C.

4 Notes

1. Use cold temperatures when quenching the lipid metabolites to avoid metabolite degradation.
2. For frozen tissue pieces, cut into smaller pieces which could be place in the cryomill tubes and/or Eppendorf tubes.
3. If working with frozen tissue, prepare liquid nitrogen in appropriate storage container.

4. Randomize sample when harvesting to reduce the systematic technical errors.
5. Homogenization of tissue samples using cryomill must operate using ceramic beads under cold temperature (−18 °C).
6. After quenching and homogenizing lipid metabolites, extract lipids at room temperature to avoid lipids sticking on the reaction tubes.
7. You need to calculate the amount of culture fluid required to collect the correct number of cells or cell pellet weight. The required amount of culture fluid is then mixed with double the amount of 75 % methanol (i.e., 5 mL culture fluid is added to 10 mL of 75 % methanol), to reach a final concentration of 50 % methanol.
8. If a tube size of less than 2 mL can be used, then do so to avoid transferring fluids between tubes. If an ultracentrifuge is available to centrifuge 15 mL or larger tubes, then do so at increased speeds for minimal time. Make sure that the same process is used for all conditions at all-time points.

References

1. Watson AD (2006) Lipidomics: a global approach to lipid analysis in biological system. J Lipid Res 14:2101–2111
2. Mazumdar J, Striepen B (2007) Make it or take it: fatty acid metabolisum of apicomplexan parasites. Eukaryot Cell 10:1727–1735
3. Hu C, Heijden R, Wang M, Greef J, Hankemeier T, Xu G (2009) Analytical strategies in lipidomics and application in disease biomarker discovery. J Chromatogr B 877:2836–2846
4. Fahy E, Subramaniam S, Brown HA, Glass CK, Merrill AH Jr, Murphy RC, Raetz CR, Russel DW, Seyma Y, Shimizu T, Spener F, Vanmeer MS, Nieuwenhze MS, White SH, Witztum JL, Dennis EA (2005) A comprehensive classification system for lipids. J Lipid Res 46:839
5. Kofeler HC, Fauland A, Rechberger GN, Trotzmuler M (2012) Mass spectrometry based lipidomics: an overview of technological platforms. Metabolites 2:19–38
6. Bligh EG, Dyer WJ (1959) A rapid method of total lipid extraction and purification. Can J Biochem 37:911–917
7. Folch J, Lees M, Sloane-Stanley GH (1957) A simple method for the isolation and purification of total lipids from animal tissues. J Biol Chem 226:497–509
8. Sunders RD, Horrocks LA (1984) Simultaneous extraction and preparation for high performance liquid chromatography of prostaglandins and phospholipids. Anal Biochem 143: 71–75
9. Li W, Laird JM, Roychowdhury S, Nagy LE, Zhou R, Crabb JW, Salomom RG (2009) Isolevuglandins covalently modify phosphatidylethanolamines in vivo: detection and quantitative analysis of hydroxylactam adducts. Free Radic Biol Med 47: 1539–1552
10. Barber MN, Risis S, Yang C, Meikle PJ, Staples M, Febbraio MA, Bruce CH (2012) Plasma lysophosphatidylcholine levels are reduced in obesity and type 2 diabetes. PLoS One 7: e41456
11. Schmidt SA, Jacob SS, Ahn SB, Rupasinghe T, Kromer JO, Khan A, Varela C, Staples M, Febbraio MA, Bruce CH (2012) Two strings to the systems biology bow: co-extracting the metabolome and proteome of yeast. Metabolomics. doi:10.1007/s11306-012-0437-1
12. Milne S, Ivanova P, Forrester J, Brown HAH (2006) Lipidomics: an analysis of cellular lipids by ESI-MS. Methods 39:92–103

Chapter 7

Metabolite Analysis of Biological Fluids and Tissues by Proton Nuclear Magnetic Resonance Spectroscopy

John Robert Sheedy

Abstract

NMR-based biochemical profiling of natural products has become popular due to the development of high-resolution instruments (>400 MHz) and cryogenically cooled probes/preamplifiers, by increasing the sensitivity of NMR instruments several fold and reducing instrument noise levels. NMR provides a rapid, nondestructive, high-throughput method that requires minimal sample preparation, therefore maintaining the biological integrity of the sample. One-dimensional (1D) solution-state ^{1}H NMR is used in untargeted sample screening (metabolomics/metabonomics) to gain insight into spectral pattern changes associated with samples of different origins. Metabolomics and metabonomics contextually explains the systematic and quantitative measurement of metabolites that are produced from the biochemical reactions of living systems. This chapter describes some commonly used ^{1}H NMR experiments for identification and quantification of small molecular weight, water soluble metabolites in biological samples, some considerations for choosing the correct NMR experiment, and sample preparation protocols for isolating metabolites from a number of biological sample types.

Key words Nuclear magnetic resonance spectroscopy, Metabolomics, Sample preparation, Liquid–liquid extraction, Biofluid, Tissue

1 Introduction

1.1 Overview of NMR Spectroscopy

Nuclear Magnetic Resonance Spectroscopy (NMR) is a powerful analytical technique that permits the investigation of structure, dynamics, and kinetics of a wide range of biological systems. NMR utilizes the magnetic properties of atomic nuclei, through exploiting the excited states of nuclei by application of an external magnetic field and measurement of the energy state transitions (by measurement of emitted relaxation energy) using radiofrequency electromagnetic radiation [1].

Nuclei with an odd mass, isotopes with unpaired electrons, an odd number of protons and/or an odd number of neutrons, possess a nonzero magnetic moment. When these types of nuclei are placed in an external magnetic field, the nondegenerate energy

Ute Roessner and Daniel Anthony Dias (eds.), *Metabolomics Tools for Natural Product Discovery: Methods and Protocols*, Methods in Molecular Biology, vol. 1055, DOI 10.1007/978-1-62703-577-4_7, © Springer Science+Business Media, LLC 2013

states of the nuclei can be utilized to provide chemical and structural information of biological molecules [2]. This externally applied magnetic field (B_0) (up to 20 T in "high-field" NMR instruments) causes any odd-spin nuclei present in a biological sample to align with the external magnetic field, and subsequent irradiation of the sample with a radiofrequency (RF) pulse stimulates a transition from the ground to excited state. This technique, known as "pulsed-NMR," causes excitation of nuclear spins through a short, controlled application of RF energy, with the resulting excited states of nuclei producing an oscillating magnetic field that induces a current in the receiver coil, where the NMR signal is also detected. The RF-induced current can be digitized and measured (as a function of time) and is termed the free induction decay (FID). The FID undergoes Fourier transformation (FT) to produce a spectrum expressed in the frequency domain which can be interpreted to reveal the physical chemical properties of the sample being analyzed [1, 3].

The practice of NMR also exploits the relaxation properties of molecules, in particular spin–lattice relaxation/T_1 relaxation (transfer of energy from the excited state to the surroundings, or lattice) and spin–spin relaxation/T_2 relaxation (loss of coherence/dephasing of nuclear magnetic moments) as magnetic moments return to a random arrangement parallel to the applied magnetic field [1, 2]. Therefore, NMR can be used to selectively observe particular chemical classes with characteristic relaxation properties. In addition, NMR is a particularly suitable technique for accurate quantification of metabolites, because the intensity of the NMR signal is directly proportional to the number of nuclei present in the sample.

NMR spectroscopy is an ideal chemical technique for generating metabolomics data as it is highly quantitative, highly reproducible, nonselective, and nondestructive [4]. The three most frequently exploited nuclei in NMR metabolomics are ^{1}H, ^{13}C, and ^{31}P, due to their natural abundance in biological compounds (Table 1). Unlike ^{31}P, which is present in a limited number of metabolically abundant molecules, the proton (^{1}H) is ubiquitous. This ubiquity gives rise at one and the same time to enormous analytical potential (because most metabolites yield a ^{1}H NMR spectrum), as well as to serious problems of analytical discrimination (because of the small chemical shift range of the proton often leads to unmanageable overlap of metabolite spectra) [5]. However, resolution of metabolites can be improved due to the increased signal dispersion obtained using high-field NMR instruments to circumvent problems with overlapping signals. Therefore, high-field 1D ^{1}H NMR metabolomics experiments have been used extensively to extract highly detailed and unbiased chemical information from biofluids. Other nuclei exploited for targeted metabolomics applications including ^{13}C, ^{31}P, ^{15}N, and ^{19}F have been successfully used for

Table 1
Natural abundances of some biologically relevant nuclei

Isotope	% Natural abundance	Spin number (*I*)
^{1}H	99.985	½
^{13}C	1.108	½
^{14}N	99.63	1
^{15}N	0.37	½
^{17}O	0.037	5/2
^{31}P	100	½
^{33}S	0.76	3/2

flux analysis and targeted metabolite monitoring, where isotopic labelling of samples or substrates prior to NMR analysis, and subsequent heteronuclear NMR experiments are used to enhance the visualization of preselected metabolite groups in metabolite and metabolic profile analyses [5–8].

There are several solution-state NMR-based experiments for the chemical analysis of biofluids or metabolite extracts from tissues or cell extracts, in most cases employing a 1D or 2D NMR experimental approach, for the chemical and structural elucidation of metabolites in complex mixtures, with ^{1}H NMR experiments being the first pass choice for global metabolite profiling of biofluids [9].

High-field NMR instruments are generally used in metabolomics, as they provide a level of detection appropriate for resolution of metabolites below micromolar concentration ranges, sufficient for differentiation of correlated and uncorrelated NMR signals. For example, a typical ^{1}H NMR spectrum of urine contains hundreds of sharp lines from predominantly low molecular weight metabolites, whereas blood plasma and serum contain both low and high molecular weight components, giving rise to a wide range of signal line widths. Broad signals from protein, lipids, and macromolecules contribute strongly to the NMR spectrum, with the sharp signals from small molecules superimposed on them. Furthermore, the high proton concentration of water present in biofluids, tissue and cell extracts contributes to a large interfering signal in the ^{1}H NMR spectrum [reducing *signal-to-noise* (S/N) ratio and resolution of metabolites], necessitating the implementation of appropriate NMR solvent suppression techniques [9].

1.2 Routine NMR Experiments

To overcome problems in NMR metabolomics studies of biofluids arising from the presence of proteins, lipids, and macromolecules, relaxation-edited and diffusion-edited pulse sequences can be selectively used to observe molecules based on their diffusion or relaxation properties. Generally, molecules with greater molecular weight diffuse more slowly and have relatively short spin-relaxation times, whereas small molecules diffuse quickly and have long spin-relaxation times. The CPMG pulse sequence (usually combined with presaturation for water suppression, commonly denoted as cpmgpr) is based upon the spin-echo pulse (or Hahn spin-echo), modified to reduce the attenuation effects due to molecular diffusion and to reduce the cumulative error resulting from imperfect 180° pulses [10].

Two-dimensional NMR spectroscopy is also necessary in metabolomics analysis of complex samples, as it can serve to increase signal dispersion (resolution) to elucidate the connectivities in molecular bonding networks, assisting in the identification of metabolites. The most commonly implemented pulse sequences in 2D NMR metabolomics are the 2D *J*-resolved, COSY (COrrelationSpectroscopY), and TOCSY (Total COrrelationSpectroscopY) experiments. 2D NMR retains many of the benefits of 1D NMR, as all observed metabolites can in principle be quantified using a single standard, but additionally spreads the overlapping resonances into a second dimension, reducing congestion to improve metabolite identification.

COSY is a 2D NMR experiment used to identify pairs of spin–spin coupled (or *J*-coupled) protons. In the COSY experiment, the conventional 1D NMR spectrum of chemical shift (in ppm) appears along a diagonal ridge running from the lower left to upper right corner of the contour plot. The COSY spectrum gives cross-peak information arising from protons separated by up to three chemical bonds (e.g., H–C–C–H) and is the most common form of 2D NMR, used for identifying molecules through their spin–spin couplings which manifest as unique cross-peak patterns for different metabolites. The TOCSY NMR experiment is analogous to COSY in that *J*-couplings and cross-peaks are resolved to give information about molecular connectivities of the metabolites present in a sample. In the TOCSY spectra, cross-peaks are composed of lines that are all in positive absorption mode, thus preventing the reduction in S/N via destructive interference. In addition, the chemical shift information of one proton within a metabolite's spin system is relayed to all other protons within the spin system [1]. This renders the TOCSY NMR experiment somewhat more robust for individual resonance identification compared to COSY, although both COSY and TOCSY 2D homonuclear NMR experiments are used for structural elucidation of metabolites and serve as tools for unequivocal identification of metabolites in complex mixtures.

1.3 Water Suppression for NMR Experiments

A requirement of NMR analysis of biofluids employs a water suppression technique to suppress as much of the water resonance as possible, so that nearby metabolite signals may be resolved. To generate accurate and interpretable data from the NMR spectrum, an appropriate NMR pulse sequence including a solvent suppression technique must be chosen, while maintaining a flat spectral baseline. Water suppression is best achieved when the sample is well shimmed to achieve a narrow water line width, where the RF transmitter is placed directly on the water resonance. However, the water frequency is slightly pH and temperature dependent, so these factors need to be controlled prior to conducting the NMR experiment [10]. This is a particularly important consideration for complex samples such as urine, where spectral distortion around the water and urea signals should be minimized [11]. Three commonly reported water suppression techniques used in NMR pulse sequences in metabolomics include WET, WATERGATE, and presaturation [9–12].

1.4 Sample Preparation Requirements for NMR-Based Metabolite Analysis

Several sample preparation techniques employ the use of protein precipitation mediated by the addition of solvents, followed by sample incubation on ice and centrifugation to pellet the precipitated protein. In the case of metabolite extraction from biological tissues and cells, a homogenization step is implemented prior to the addition of solvents to liberate metabolites from the in-tact tissue.

Several solvent-based metabolite extraction systems (liquid–liquid extraction—LLE) have been evaluated in the literature, commonly using methanol:chloroform:water, perchloric acid, acetone, or acetonitrile. However, loss of volatile small molecular weight metabolites is a problem with any solvent removal process employing drying or lyophilization. For biofluids such as urine and blood, the endogenous protein, lipids, and macromolecular substances should be removed prior to NMR analysis, as the broad signals arising from these large molecules often overshadow the sharp signals of metabolites.

The power of NMR in metabolomics is due to its unbiased identification of diverse chemical species in a single sample, but also for its ability for accurately determining the concentration of metabolites in a complex sample [13]. This is achieved by the addition of reference compounds during sample preparation, such as 3-trimethylsilylpropionic acid (TSP), 3-trimethylsilylpropane-1-sulfonic acid (TMS), 2,2-dimethyl-2-silapentane-5-sulfonic acid (DSS), formate, or imidazole [9, 14]. TSP and DSS are generally modified to contain deuterium for all methylene groups, so that TSP and DSS-d_6 give rise to only one singlet resonance at 0.00 ppm. TSP and DSS can also be used to quantify other metabolites present in the sample when accurately added in a known concentration.

2 Materials

All devices requiring flushing with water should be prepared using 18 mΩ deionized water. Analysis of biological samples should commence where sample collection storage integrity has been maintained (e.g., initial quenching of metabolism using liquid nitrogen, and sample storage at −80 °C in sealed vials). Deuterated sample storage vials should also be flushed with nitrogen prior to capping. The following sample preparation protocols should be undertaken under standard laboratory conditions. All NMR tubes referred to in the following protocols are 5 mm diameter glass NMR tubes (7- or 8-in. tube height).

2.1 Internal Standard Solution for Liquid–Liquid Extraction

1. Preparation of a 500 mL stock of D_2O containing 1 mM imidazole as an internal standard for metabolite recovery for methanol-*d4* (CD_3OD)/chloroform-d ($CDCl_3$)-based LLE, add 34.0385 mg of anhydrous imidazole to 0.5 L of D_2O. Use magnetic stirrer to ensure imidazole is thoroughly dissolved (*see* **Note 1**).

2.2 NMR Sample Buffer

1. Preparation of a 500 mL stock of 200 mM trisodium phosphate buffer solution, add 16.3940 g of anhydrous Na_3PO_4 to 0.5 L of D_2O. Use magnetic stirrer to ensure Na_3PO_4 is thoroughly dissolved (*see* **Note 2**).
2. Add 0.1 g of sodium azide (NaN_3) to 500 mL of 200 mM $Na_3PO_4{\cdot}D_2O$ solution so that NaN_3 concentration is 0.2 % w/v.
3. Adjust pH of solution to 7 by titrating in >2 mL of deuterium chloride (DCl). Readjust solution to pH 7 with necessary volume of sodium deuteroxide (NaOD) (*see* **Note 2**).

2.3 Reference Solution for Spectral Calibration and Chemical Quantification

1. Preparation of a 50 mL stock of reference solution of 5 mM 2,2-dimethyl-2-silapentane-5-sulfonic acid (DSS-d_6), add 56.0625 mg of DSS-d_6 to 0.05 L of D_2O (*see* **Note 3**).

3 Methods

3.1 Preparation of Urine for Metabolite Identification by ^1H NMR

1. Collect 1 mL of mid-stream urine into sterile specimen container (*see* **Note 4**).
2. Transfer a 1 mL aliquot of urine to a 2 mL o-ring screw cap sample vial (*see* **Note 5**).
3. Centrifuge at 4 °C for 15 min at 5,000 × *g*.
4. Transfer a 500 μL aliquot of supernatant to new o-ring vial.
5. Add 300 μL of ice-chilled $CDCl_3$ and 200 μL of ice-chilled CD_3OD (*see* **Note 6**).
6. Vortex sample for 15 s until an emulsion forms.
7. Rest sample on ice for 15 min.

8. Centrifuge at 4 °C for 15 min at 5,000 × *g* to create a bi-phasic mixture.
9. Add 60 μL of 5 mM DSS-D_2O solution into a 5 mm glass NMR tube.
10. Transfer a 270 μL aliquot of upper aqueous phase containing urine/CD_3OD into NMR tube.
11. Add 270 μL of 200 mM trisodium phosphate/sodium azide buffer ($Na_3PO_4 \cdot NaN_3 \cdot D_2O$), adjusted to pH 7, to 330 μL sample volume in NMR tube (*see* **Note 7**).
12. Collect NMR spectra (*see* **Note 8**).

3.2 Preparation of Blood for Metabolite Identification by ^1H-NMR

1. Add 2 mL H_2O to 1.5 or 3 kDa centrifugal filtration devices to remove glycerol from filtration membrane.
2. Centrifuge filtration devices containing H_2O for 90 min at 5,000 × *g* speed.
3. Discard H_2O from filtration membrane well and collection tube.
4. Repeat **steps 1** and **2** (*see* **Note 9**).
5. Add 500 μL of preprepared blood aliquot (*see* **Note 10**) to filtration membrane well and dilute with 500 μL D_2O/imidazole (*see* **Note 11**).
6. Secure cap onto tube containing filtration membrane and centrifuge at 5,000 × *g* for 1 h at 20 °C (ambient temperature required to maintain viscosity of serum sample for efficient elution of protein-free serum).
7. Collect a 500 μL aliquot of eluent (containing serum/D_2O).
8. In a fresh 2 mL screw cap o-ring tube, transfer 500 μL of eluent to tube and add 300 μL $CDCl_3$ and 200 μL of CD_3OD (*see* **Note 6**).
9. Vortex sample for 15 s.
10. Chill samples on ice for 15 min.
11. Centrifuge samples at 4 °C for 15 min at 5,000 × *g*.
12. Add 60 μL of 5 mM DSS-D_2O solution into a 5 mm glass NMR tube.
13. Transfer a 270 μL aliquot of upper aqueous phase containing blood/CD_3OD into NMR tube.
14. Add 270 μL of 200 mM trisodium phosphate/sodium azide buffer ($Na_3PO_4 \cdot NaN_3 \cdot D_2O$), adjusted to pH 7, to 330 μL sample volume in NMR tube (*see* **Note 7**).
15. Collect NMR spectra (*see* **Note 8**).

3.3 Preparation of Biological Tissues for Intracellular Metabolite Identification by ^{1}H NMR

1. Collect tissue and flash freeze in liquid N_2, store tissue at −80 °C (*see* **Note 12**).
2. Weigh out 50–100 mg tissue (proportionately alter recipe depending on tissue weight).
3. Using a mortar and pestle, softly pound frozen tissue until flat under liquid N_2.
4. Transfer tissue to 2.0 mL tube containing 1.4 mm diameter ceramic lysis beads.
5. Add 4 mL/g (0.2 mL) of CD_3OD and 0.85 mL/g (0.0425 mL/g) of D_2O/imidazole (*see* **Note 13**).
6. Homogenize using automated homogenization device (e.g., Precellys 24-well tissue homogenizer).
7. Collect homogenized aliquot, then add 4 mL/g (0.2 mL) $CDCl_3$ and 2 mL/g D_2O (0.1 mL) and vortex for 60 s (*see* **Notes 6** and **13**).
8. Rest samples on ice for 15 min.
9. Centrifuge sample for 15 min at 5,000 × *g* at 4 °C.
10. Add 60 μL of 5 mM DSS-D_2O solution into a 5 mm glass NMR tube.
11. Transfer a 270 μL aliquot of upper aqueous phase containing homogenate/CD_3OD into NMR tube.
12. Add 270 μL of 200 mM trisodium phosphate/sodium azide buffer ($Na_3PO_4 \cdot NaN_3 \cdot D_2O$), adjusted to pH 7, to 330 μL sample volume in NMR tube (*see* **Note 7**).
13. Collect NMR spectra (*see* **Note 8** and Table 2).

3.4 Preparation of Protein-Containing Culture Media for Metabolite Identification by ^{1}H NMR

1. Collect 1 mL of culture media into an o-ring screw cap vial.
2. Centrifuge at 4 °C for 15 min at 5,000 × *g*.
3. Transfer a 500 μL aliquot of supernatant to new o-ring vial.
4. Add 300 μL of ice-chilled chloroform ($CHCl_3$) and 200 μL of ice-chilled methanol (CH_3OH) (*see* **Note 6**).
5. Vortex sample for 15 s until an emulsion forms.
6. Rest sample on ice for 15 min.
7. Centrifuge at 4 °C for 15 min at 5,000 × *g* to create a bi-phasic mixture.
8. Take a 600 μL aliquot of aqueous phase and transfer to new o-ring tube.
9. Puncture 2–3 holes in the tube cap.
10. Lyophilize or speed-vacuum dry sample for 24 h at 45 °C.
11. Reconstitute sample in 300 μL of D_2O. Vortex briefly.
12. Add 60 μL of 5 mM DSS-D_2O solution into a 5 mm glass NMR tube.

Table 2

Metabolite yields (μmol/L) from mouse quadriceps, *C. tepperi* larvae, and rice leaf, seed and root samples, after homogenization using three different homogenization techniques: (1) Hand-held electric homogenizer; (2) Precellys Automated Homogenizer [at 0 °C]; (3) Precellys Automated Homogenizer [at room temperature]. Higher metabolite yields extracted from tissues were achieved in most cases using the automated tissue homogenization system. Non-plant-based tissues gave an overall higher metabolite yield when homogenized under cold conditions (0 °C), while metabolite extraction from plant-based tissues responded better from homogenization at room temperature

	Electric homogenizer					Automated homogenizer (0 °C)					Automated homogenizer (room temperature)				
Metabolite	Quad.	Chir.	Leaf	Seed	Root	Quad.	Chir.	Leaf	Seed	Root	Quad.	Chir.	Leaf	Seed	Root
Acetate	–	–	–	27.75	–	–	–	–	66.45	–	–	–	–	**70.85**	–
Alanine	103.95	167.35	94.05	22.60	29.20	**189.15**	**398.90**	**151.25**	24.20	**29.60**	154.80	342.00	138.75	**32.30**	22.90
Asparagine	–	–	–	–	**373.05**	–	–	–	–	296.30	–	–	–	–	188.30
Betaine	–	–	–	6.15	–	–	–	–	18.35	–	–	–	–	**24.85**	–
Ethanol	–	46.70	–	**26.90**	–	–	**47.60**	–	22.40	–	–	20.45	–	14.95	–
Glucose	98.10	–	–	–	–	**256.65**	–	–	–	–	210.80	–	–	–	–
Glutamine	–	231.10	–	–	946.70	–	**612.10**	–	–	**1053.45**	–	470.70	–	–	1003.30
Glycine	–	–	–	10.70	–	–	–	–	69.25	–	–	–	–	**115.45**	–
Lactate	697.70	364.55	25.85	–	–	**2759.40**	**595.50**	92.45	–	–	2305.70	541.70	**123.90**	–	–
Leucine	7.85	83.30	–	–	–	**27.80**	**190.40**	–	–	–	12.80	186.85	–	–	–
Pyruvate	–	–	22.80	–	–	–	–	37.60	–	–	–	–	**64.05**	–	–
Succinate	4.50	17.45	12.85	8.75	–	**23.85**	**56.35**	**13.80**	10.75	–	12.50	56.15	13.40	**21.45**	–
Taurine	1680.45	–	–	–	–	**2264.25**	–	–	–	–	2260.45	–	–	–	–
Threonine	–	–	14.25	19.30	–	–	–	89.55	**41.35**	–	–	–	**99.00**	36.75	–
Tyrosine	–	82.15	–	–	59.00	–	**381.50**	–	–	**64.50**	–	355.05	–	–	52.45
Valine	–	–	7.15	–	25.90	–	–	22.95	–	**26.75**	–	–	**32.80**	–	23.60

N.B.: Each value represents an average value for two replicates per sample
Each sample contained approximately 50 mg of tissue. Leaf, seed, and root tissues were from *H. vulgare spp.* (Australian barley)
Values highlighted in *bold* represent the highest recovery of metabolites from tissues using the homogenization method above
Quad.: Quadriceps, mouse skeletal muscle; *Chir.*: *C. tepperi* larvae, native Australian freshwater aquatic worm

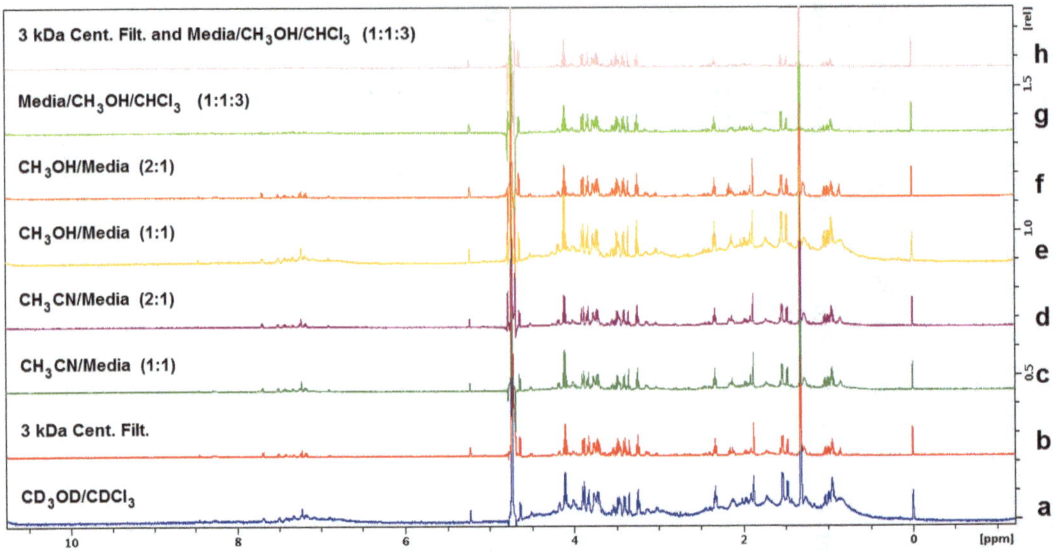

Fig. 1 Comparison of liquid–liquid extraction (LLE)-based techniques for metabolite recovery from embryo culture media, analyzed using ^{1}H-NMR spectroscopy. The following LLE solvent systems were compared—(*a*) methanol-*d*$_4$:chloroform-*d*$_4$:culture media (1:2:3); (*b*) centrifugal filtration of culture media (3 kDa molecular weight cut-off); (*c*) acetonitrile:culture media (1:1); (*d*) acetonitrile:culture media (2:1); (*e*) methanol:culture media (1:1); (*f*) methanol:culture media (2:1); (*g*) chloroform:methanol:culture media (3:1:1); (*h*) centrifugal filtration (3 kDa) followed by chloroform:methanol:culture media (3:1:1). The undulating spectral baseline seen in two of the LLE protocols (*a* and *e*) denotes residual protein in the sample. DSS–protein interactions cause inaccuracies in metabolite quantification using DSS as a reference compound. The absence of the acetate signal at 1.91 ppm in two other LLE protocols (*g* and *h*) shows that some metabolites are unrecoverable from the sample after some LLE protocols

13. Transfer a 270 μL aliquot of D_2O-reconstituted sample into NMR tube.
14. Add 270 μL of 200 mM trisodium phosphate/sodium azide buffer ($Na_3PO_4{\cdot}NaN_3{\cdot}D_2O$), adjusted to pH 7, to 330 μL sample volume in NMR tube (*see* **Note 7**).
15. Collect NMR spectra (*see* **Note 8** and Fig. 1).

3.5 Preparation of Fecal Samples for Metabolite Identification by ^{1}H NMR

1. Weigh out amounts of fecal samples (wet weight) into screw cap o-ring lysis tubes, containing 1.4 mm diameter ceramic lysis beads. Record wet weights of fecal samples.
2. Add 1 mL D_2O/imidazole to each lysis tube, containing fecal samples (*see* **Note 14**).
3. Homogenize samples (e.g., using Precellys 24-well tissue homogenizer) for 5 min (consisting of three repetitions of homogenization of 45 s) at room temperature.
4. Take 300 μL volume of fecal fluid (depending on sample volume availability) from underneath the beads (bottom of lysis tube) and dilute, respectively, with 600 μL D_2O.

5. Centrifuge for 5 min at 5,000 × *g* at ambient temperature to pellet any solid material.
6. Transfer 500 μL supernatant to new screw cap o-ring tube and add 300 μL $CDCl_3$ and 200 μL of CD_3OD.
7. Vortex sample for 15 s.
8. Chill samples on ice for 15 min.
9. Centrifuge samples at 4 °C for 15 min at 5,000 × *g*.
10. Add 60 μL of 5 mM DSS-D_2O solution into NMR tube.
11. Transfer a 270 μL aliquot of upper aqueous phase containing fecal fluid/CD_3OD into NMR tube.
12. Add 270 μL of 200 mM trisodium phosphate/sodium azide buffer ($Na_3PO_4 \cdot NaN_3 \cdot D_2O$), adjusted to pH 7, to 330 μL sample volume in NMR tube (*see* **Note** 7).
13. Collect NMR spectra (*see* **Note 8**).

4 Notes

1. Formate is often used as an internal standard rather than imidazole—however many biological samples (particularly those containing albumin) will have endogenous formate already present in the sample. In these instances imidazole should be used over formate.
2. A non-hydrogen containing phosphate salt is an ideal component for creating an appropriate buffer to minimize nonbiological signals in ^{1}H NMR spectra being collected. Furthermore, titration of deuterated acid/base mitigates addition of unwanted signals for ^{1}H NMR spectral analysis of biological samples. NMR analysis of complex biofluids, tissue or cell extracts can be confounded by pH-dependent chemical shift changes observed amongst samples. This phenomenon is highlighted in NMR analysis of urine compared to other biofluids, as blood and intracellular pH levels are homeostatically regulated, so large differences in metabolite chemical shifts are not observed in serum, plasma, tissue or cell extracts. Conversely, normal human urine pH ranges from 5.5 to 6.5 and the pH under physiological stress may vary anywhere from 4.6 to 8.0. This is correlated to urinary metabolite concentrations, which may vary up to tenfold over the course of a day. Therefore, there is consensus in the published literature that sample buffering using deuterated or non-proton bearing chemicals should be added to all samples for normalizing pH and ionic strength of the sample, minimizing pH-dependent chemical shifts of metabolites and therefore aligning chemical shifts of metabolites over all sample spectra to be analyzed. Phosphate-based

buffers are popularly reported for sample pH standardization (*see* refs. [9, 15]). Furthermore, EDTA has been proposed as a metal chelating agent that can be added to samples, reducing the effects of signal shifts and line broadening associated with the presence of paramagnetic and non-paramagnetic metal ions in a sample, in particular Ca^{2+}, $Fe^{2/3+}$, and Mg^{2+} (*see* ref. [15]).

3. DSS is used as an internal reference compound in NMR experiments for quantifying metabolites in a biological sample. However compounds such as TSP and DSS can bind nonspecifically to serum albumin and other proteins, confounding quantitative analysis. It is essential that all protein is removed prior to sample analysis by NMR using the protocols outlined in this chapter. As TSP and DSS trimethylsilyl functional groups can directly interact with hydrophobic molecules, thus causing line broadening of the trimethylsilyl singlet at 0.00 ppm, all protein needs to be removed prior to metabolite analysis. In this instance, formate can be used if it is not endogenously present in the sample of interest (which generally excludes urine, blood, feces). A newly synthesized chemical, DFTMP (1,1-difluoro-1-trimethylsilanylmethylphosphonic acid), has been proposed as a novel reference compound for NMR metabolomics studies, as it has a sharp singlet resonance at 0.2 ppm on a ^{1}H NMR spectrum, and due to DFTMP containing proton, fluorine and phosphorus nuclei, it can serve as a chemical shift reference point and pH indicator for multinuclear studies (*see* ref. [14]).
4. Collection of an immediate urine purge may result in concomitant collection of high protein levels (associated with microalbuminuria) or urethral bacteria. The presence of these substances may result in a poor representation of the urinary metabolome. Mid-stream urine collection minimizes collection of high concentrations of proteins/bacteria.
5. The use of o-ring screw cap vials ensures that upon sample collection and storage, sample evaporation is mitigated.
6. A cold methanol–chloroform-based LLE ensures precipitation of protein from samples, so that compounds such as DSS can be effectively used as an internal reference compound. Pipettes should be first primed separately with methanol or chloroform to prevent loss of solvent volume leaking from the pipette tip during volume transfer of the solvent to the sample. Each solvent system preferentially retains different classes of metabolites, and particularly in bi-phasic solvent-based metabolite extraction protocols metabolites can be separated according to their hydrophilic/hydrophobic chemical properties. In the case of metabolite extraction from cells or tissues, the protein-free solvent can be dried by lyophilization or speed-vacuum and resuspended in 100 % deuterated solvent to eliminate

unwanted signals in the NMR spectrum from the solvent system used in the metabolite extraction process (*see* refs. [16–19]). This is achieved by either solvent-based extraction or ultracentrifugation, where molecular weight cut-off filters (e.g., 1.5 or 3 kDa filters) remove material above a specified molecular weight (*see* ref. [19]). Biofluids such as urine and blood achieve optimal metabolite recovery using an LLE system employing a fully deuterated LLE solvent system (*see* ref. [20]), while metabolite recovery from cell culture systems respond better to methanol-based protein removal, a water/methanol evaporation step, followed by 100 % reconstitution in buffered deuterium oxide.

7. Addition of the sodium phosphate buffer to the NMR tube as the last step in the sample preparation protocol can ensure that any residual sample residing on the inside of the NMR tube away from the main sample volume is effectively washed into the main sample volume.

8. ^1H NMR spectra should be acquired on a high-field NMR instrument fitted with a cryoprobe (e.g., 800 MHz Bruker-Biospin Avance US^2 spectrometer with a 5 mm TCI cryoprobe). Samples should be locked to D_2O, and tuned, matched, and shimmed prior to data acquisition. 90° pulse width and receiver gain should also be optimized for each sample. The standard and most commonly used 1D NMR pulse sequences used in metabolomics are the 1D Nuclear Overhauser Effect SpectroscopY pulse sequence (NOESY, or noesy) and relaxation-edited pulse sequences such as the Carr-Purcell-Meiboom-Gill (CPMG) pulse sequence. Both of these pulse sequences require incorporation of a solvent suppression capability, with presaturation being commonly implemented in metabolomics literature (expressed as *noesypr* or *cpmgpr* in Bruker nomenclature) (*see* refs. [9, 21]). The "noesy with presaturation" (*noesypr*) 1D pulse sequence acquires the first free induction decay (FID) of the 2D NOESY experiment (recycle delay—90°–τ–90°–τ_m–90°) (*see* ref. [9]). This is a widely used pulse sequence by many metabolomics NMR research groups, where τ has duration around 3 μs and τ_m (mixing time) about 100 ms. The low power irradiation during the mixing time prevents the water signal recovery by longitudinal relaxation. This irradiation is not required if the mixing time is shorter (e.g., if $\tau_m = 10$ ms), and a short-pulsed gradient can be added at the beginning of the mixing time to eliminate residual transverse relaxation. The 1D *noesypr* pulse sequence yields high-quality spectral baselines, which is essential for quantitative accuracy (*see* ref. [22]). T_2-edited pulse sequences such as the CPMG pulse sequence can separate metabolites based on T_2 relaxation time. The broad NMR signals from proteins, lipids,

or small metabolites bound to macromolecules are attenuated/eliminated in the T_2-edited experiment, leaving a spectrum of sharp resonances corresponding to the small molecular weight metabolites in the sample (*see* ref. [23]). This is because the quantitative accuracy of the CPMG experiment is greatly affected by T_2 relaxation times for different metabolites and total echo time used in the CPMG pulse sequence to filter broad signals. This quantitative inaccuracy can be overcome using a T_2 correction factor and implementation of a short recycle delay to provide quantitative results within 95–96 % accuracy (*see* ref. [24]). The 2D *J*-resolved (JRES) experiment yields information about the multiplicity and coupling patterns of resonances for metabolite identification. JRES NMR separates the effects of chemical shift and *J*-coupling into two independent dimensions, based around variants of pulse sequence with the following schematic: recycle delay—90°–($\tau_1/2$)–180°–($\tau_1/2$) (where τ_1 is an incremented time delay used to create an indirect time axis for the second dimension). The first projection (or first row) contains data similar to an ordinary 1D spectrum, while projection of the second row no longer contains any multiplets. Each proton appears as a single peak only, and as a consequence the JRES NMR experiment facilitates the separation of signals in crowded areas of the spectrum (*see* ref. [25, 26]). 2D-COSY spectra can be acquired in the magnitude or phase-sensitive modes. Magnitude COSY experiments are of lower resolution but have positive peaks and do not require phasing, thus permitting the development of automated metabolite assignments (*see* ref. [27]). Phase-sensitive COSY spectra are of higher resolution but due to the anti-phase nature of the cross-peaks, this can lead to an S/N reduction, since the individual anti-phase peaks within the COSY cross-peak destructively interfere with each other. The WET (Water suppression Enhanced by T_1 effects) solvent suppression method provides in-phase spectra with flat baselines and minimal perturbation to the rest of the NMR spectrum, using a repetitive cycle of on-resonance selective irradiation of the water peak with pulsed field gradients to de-phase transverse magnetization (*see* ref. [10]). The WATERGATE method (water suppression by gradient excitation) resembles a spin-echo pulse sequence (τ–180°–τ), with a refocusing pulse flanked by two symmetrical gradient pulses. Transverse coherences are de-phased by the first gradient, provided they experience a 180° rotation by the selective pulses or "pulse train". This can be a hard (frequency nonselective) 180° pulse sandwiched by two soft 90° (frequency-selective) pulses or a frequency-selective pulse train (3α–τ–9α–τ–19α–τ–19α–τ–9α–τ–3α, where $62\alpha = 180°$). However, the WATERGATE solvent suppression technique introduces a larger region of peak elimination (compared to techniques like presaturation), thus suppressing any resonances close to the water peak (*see* ref. [1, 12]). Presaturation is a simple method of solvent suppression, where the water resonance is irradiated with a long low power RF pulse, causing transitions from the ground state to the excited state over a bandwidth of about 1 Hz, exceeding the rate of spin–lattice relaxation to equalize the two atomic energy levels induced in the excited state, so no water signal is observed (*see* ref.[1]). Although each one of these techniques has its advantages and disadvantages, the pre-saturation method is perhaps the most widely used water suppression method due to its ease of implementation (*see* refs. [9, 11]). However in all cases, the phenomenon of radiation damping must be considered (the field induced in the receiver coil by large bulk magnetization, where magnetization is driven back to equilibrium much faster than the relaxation process), so metabolite signals (e.g., protons attached to anomeric carbons of sugars) near the water resonance are not considered for quantification, as these signals are attenuated due to radiation damping (*see* ref. [28]). Improvements to water suppression techniques such as SWET have aided in overcoming some of the effects of radiation damping such as saturation transfer, so that metabolites closer to the water peak can be accurately quantified (*see* ref. [29]).

9. It is essential that membranes of the centrifugal filtration devices are thoroughly rinsed with water, so samples are not contaminated with residual membrane-bound glycerol. Glycerol has distinctive aliphatic resonances which may mask metabolites of interest in the sample to be analyzed.

10. Preparation of whole blood for metabolite analysis:

    Sample prepreparation for plasma analysis:

    Collect >1 mL volume of heparinized (non-coagulated) blood into sealed blood collection tube. Refrigerate at 4 °C until further sample preparation can be undertaken.

    Sample prepreparation for serum analysis:

    Collect >1 mL volume of non-heparinized blood into sealed blood collection tube. Allow samples to rest at 4 °C for 30–45 min to allow blood to coagulate. Take a 1 mL sample of coagulated blood and centrifuge 4 °C for 15 min at 5,000×*g*. Collect 500 μL aliquot of supernatant for further sample preparation.

11. The D_2O spiked with imidazole ensures that any variation due to metabolite recovery from the filtration membrane can be normalized for during data processing. Therefore, if imidazole concentration varies between samples, actual metabolite recovery can be corrected for based on the recovery of imidazole.

12. It is essential that tissues are immediately snap-frozen (quenching metabolism) upon collection to mitigate any metabolic activity, such as metabolite decay.
13. This method can be altered according to the amount of tissue present, however for amounts lower than 50 mg, use an 8:2.5:4 mL/g (methanol/water/chloroform) ratio at beginning.
14. Fecal samples vary according to water content. It is advised that a 50–100 mg sample of wet weight feces per 1 mL of water be the maximum wet weight used for sample preparation. For dry fecal samples, adjust the volume of 1 mL D_2O/imidazole to 2 mL D_2O/imidazole. Keep account for the dilution during data analysis.

Acknowledgments

Dr John R. Sheedy would like to thank Dr. Berin Boughton from Metabolomics Australia (The University of Melbourne) and Dr Sara Long from CAPIM (The University of Melbourne) for their contributions to the biological tissue homogenization work. I would also like to thank Professor David Gardner and Dr. George Thouas (Zoology, The University of Melbourne) for providing embryo culture media samples to evaluate using NMR for profiling and quantifying media components.

References

1. Rule G, Hitchins TK (2006) Fundamentals of protein NMR spectroscopy. Springer, Dordecht
2. Macomber RS (1998) A complete introduction to modern NMR spectroscopy. Wiley, New York
3. Cavanagh J, Fairbrother WJ, Palmer AG, Rance M, Skelton NJ (2007) Protein NMR spectroscopy, principles and practice, 2nd edn. Elsevier, New York
4. Mercier P, Lewis MJ, Chang D, Baker D, Wishart DS (2011) Towards automatic metabolomic profiling of high-resolution one-dimensional proton NMR spectra. J Biomol NMR 49:307–323
5. Allen PS, Thompson RB, Wilman AH (1997) Metabolite-specific NMR spectroscopy *in vivo*. NMR Biomed 10:435–444
6. Sprague DB, Gadian DG, Williams SR, Proctor E (1997) Intracellular metabolites in rat muscle following trauma: a ^{31}P and ^{1}H nuclear magnetic resonance study. J R Soc Med 80:495–498
7. Scarfe GB, Lindon JC, Nicholson JK, Wright B, Clayton E, Wilson ID (1999) Investigation of the quantitative metabolic fate and urinary excretion of 3-methyl-4-trifluoromethylaniline and 3-methyl-4-trifluoromethylacetanilide in the rat. Drug Metab Dispos 27:1171–1178
8. Sriram G, Fulton DB, Iyer VV, Peterson JM, Zhou R, Westgate ME, Spalding MH, Shanks JV (2004) Quantification of compartmented metabolic fluxes in developing soybean embryos by employing biosynthetically directed fractional 13C labelling, two-dimensional [^{13}C-^{1}H] nuclear magnetic resonance, and comprehensive isotopomer balancing. Plant Physiol 136:3043–3057
9. Beckonert O, Keun HC, Ebbels TM, Bundy J, Holmes E, Lindon JC, Nicholson JK (2007) Metabolic profiling, metabolomic and metabonomic procedures for NMR spectroscopy of urine, plasma, serum and tissue extracts. Nat Protoc 2:2692–2703

10. Van QN, Chmurny GN, Veenstra TD (2003) The depletion of protein signals in metabonomics analysis with the WET-CPMG pulse sequence. Biochem Biophys Res Commun 301:952–959
11. Saude EJ, Slupsky CM, Sykes BD (2006) Optimization of NMR analysis of biological fluids for quantitative accuracy. Metabolomics 2:113–123
12. Liu M, Mao X, Ye C, Huang H, Nicholson JK, Lindon JC (1998) Improved WATERGATE pulse sequences for solvent suppression in NMR spectroscopy. J Magn Reson 132: 125–129
13. Dreier L, Wider G (2006) Concentration measurements by PULCON using x-filtered or 2D NMR spectra. Magn Reson Chem 44:S206–S212
14. Reily MD, Robosky LC, Manning ML, Butler A, Baker JD, Winters RT (2006) DFTMP, an NMR reagent for assessing the near-neutral pH of biological samples. J Am Chem Soc 128:12360–12361
15. Asiago VM, Gowda GAN, Zhang S (2008) Use of EDTA to minimize ionic strength dependent frequency shifts in the ^{1}H NMR spectra of urine. Metabolomics 4:328–336
16. Le Belle JE, Harris NG, Williams SR, Bhakoo KK (2002) A comparison of cell and tissue extraction techniques using high-resolution ^{1}H-NMR spectroscopy. NMR Biomed 15: 37–44
17. Wu H, Southam AD, Hines A, Viant MR (2008) High-throughput tissue extraction protocol for NMR and MS based metabolomics. Anal Biochem 372:204–212
18. Kruger NJ, Troncoso-Ponce MA, Ratcliffe RG (2008) ^{1}H-NMR metabolite fingerprinting and metabolomic analysis of perchoric acid extracts from plant tissues. Nat Protoc 3:1001–1012
19. Tiziani S, Emwas AH, Lodi A, Ludwig C, Bunce CM, Viant MR, Günther UL (2008) Optimised metabolite extraction from blood serum for ^{1}H nuclear magnetic resonance spectroscopy. Anal Biochem 377:16–23
20. Sheedy JR, Ebeling PR, Gooley PR, McConville MJ (2010) A sample preparation protocol for ^{1}H nuclear magnetic resonance studies of water-soluble metabolites in blood and urine. Anal Biochem 398:263–265
21. Harker M, Coulson H, Fairweather I, Taylor D, Daykin CA (2006) Study of metabolite composition of eccrine sweat from healthy male and female human subjects by ^{1}H-NMR spectroscopy. Metabolomics 2:105–112
22. Lindon JC, Nicholson JK, Holmes E (2007) The handbook of metabonomics and metabolomics. Elsevier, Amsterdam
23. Tang H, Wang Y, Nicholson JK, Lindon JC (2004) Use of relaxation-edited one-dimensional and two-dimensional nuclear magnetic resonance spectroscopy to improve detection of small metabolites in blood plasma. Anal Biochem 325:260–272
24. Bharti SK, Sinha N, Joshi BS, Mandal SK, Roy R, Khetrapal CL (2008) Improved quantification from ^{1}H-NMR spectra using reduced repetition times. Metabolomics 4:367–376
25. Ludwig C, Viant MR (2009) Two-dimensional J-resolved NMR spectroscopy: Review of a key methodology in the metabolomics toolbox. Phytochem Anal 21:22–32
26. Lin CY, Wu H, Tjeerdema RS, Viant MR (2007) Evaluation of metabolite extraction strategies from tissue samples using NMR metabolomics. Metabolomics 3:55–67
27. Xi Y, deRopp JS, Viant MR, Woodruff DL, Yu P (2006) Automated screening for metabolites in complex mixtures using 2D COSY NMR spectroscopy. Metabolomics 2:221–233
28. Krishnan VV (2006) Radiation damping in microcoil NMR probes. J Magn Reson 179:294–298
29. Wu PSC, Otting G (2005) SWET for secure water suppression on probes with high-quality factor. J Biomol NMR 32:243–250

Chapter 8

NMR Spectroscopy: Structure Elucidation of Cycloelatanene A: A Natural Product Case Study

Sylvia Urban and Daniel Anthony Dias

Abstract

The structure elucidation of new secondary metabolites derived from marine and terrestrial sources is frequently a challenging task. The hurdles include the ability to isolate stable secondary metabolites of sufficient purity that are often present in <0.5 % of the dry weight of the sample. This usually involves a minimum of several chromatographic purification steps. The second issue is the stability of the compound isolated. It must always be assumed when dealing with the isolation of natural products that the compound may rapidly degrade during and/or after the isolation, due to sensitivity to light, air oxidation, and/or temperature. In this way, precautions need to be taken, as much as possible to avoid any such chemical inter-conversions and/or degradations. Immediately after purification, the next step is to rapidly acquire all analytical spectroscopic data in order to complete the characterization of the isolated secondary metabolite(s), prior to any possible decomposition. The final hurdle in this multiple step process, especially in the acquisition of the NMR spectroscopic and other analytical data (mass spectra, infrared and ultraviolet spectra, optical rotation, etc.), is to assemble the structural moieties/units in an effort to complete the structure elucidation. Often ambiguity with the elucidation of the final structure remains when structural fragments identified are difficult to piece together on the basis of the HMBC NMR correlations or when the relative configuration cannot be unequivocally identified on the basis of NOE NMR enhancements observed. Herein, we describe the methodology used to carry out the structure elucidation of a new C_{16} chamigrene, cycloelatanene A (**5**) which was isolated from the southern Australian marine alga *Laurencia elata* (Rhodomelaceae). The general approach and principles used in the structure determination of this compound can be applied to the structure elucidation of other small molecular weight compounds derived from either natural or synthetic sources.

Key words NMR spectroscopy, Red alga, *Laurencia elata*, Relative configuration, Structure elucidation

1 Introduction

Red algae belonging to the genus *Laurencia* (Ceramiales, Rhodomelaceae) are a prolific source of secondary metabolites, predominantly producing sesquiterpenes, diterpenes, triterpenes, C_{15} acetogenins, acetylenes, and chamigrenes [1]. *Laurencia* is a common genus occurring along southern Australian coasts, with numerous species being found in this region [2]. The search for bioactive natural products from this genus of red algae has been an

Ute Roessner and Daniel Anthony Dias (eds.), *Metabolomics Tools for Natural Product Discovery: Methods and Protocols*, Methods in Molecular Biology, vol. 1055, DOI 10.1007/978-1-62703-577-4_8,

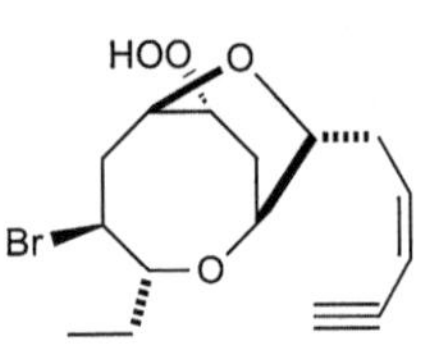

(1)

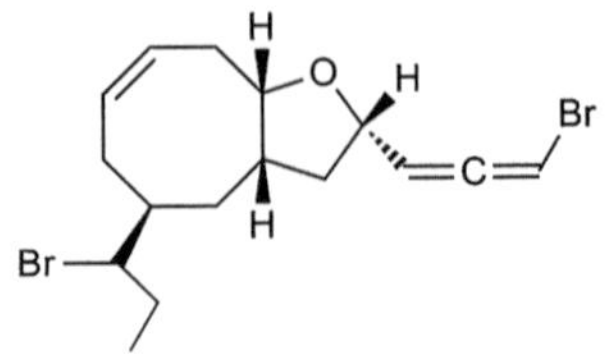

(2)

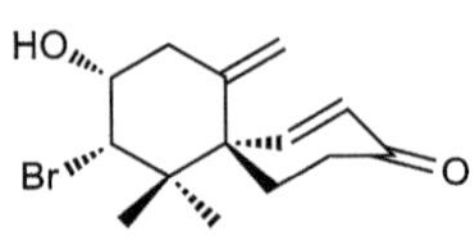

(3)

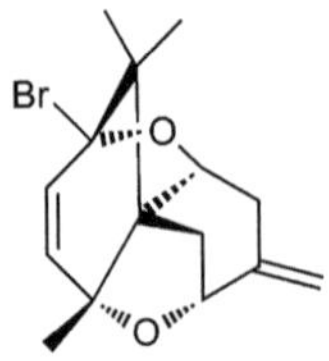

(4)

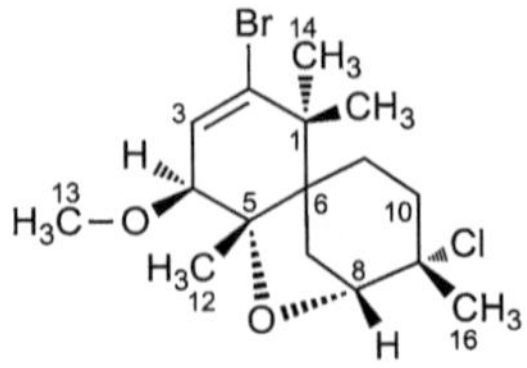

(5)

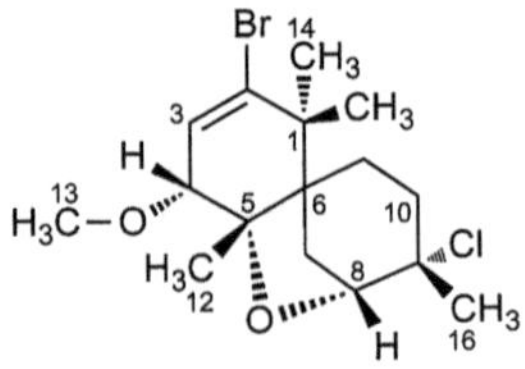

(6)

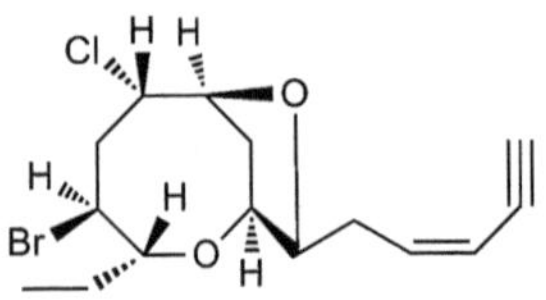

(7)

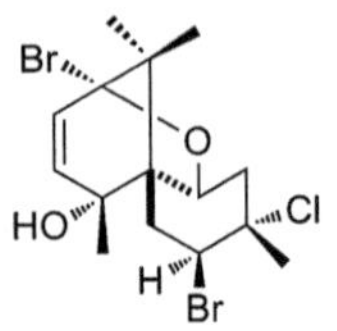

(8)

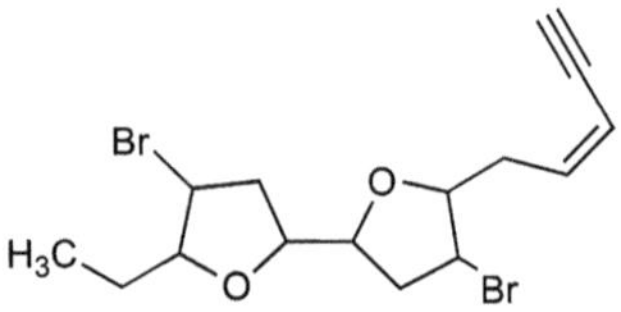

(9)

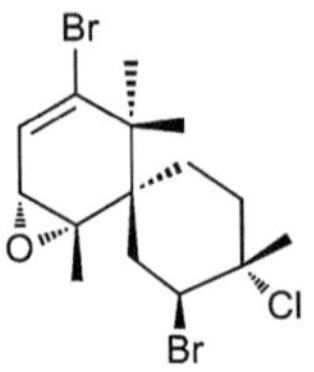

(10)

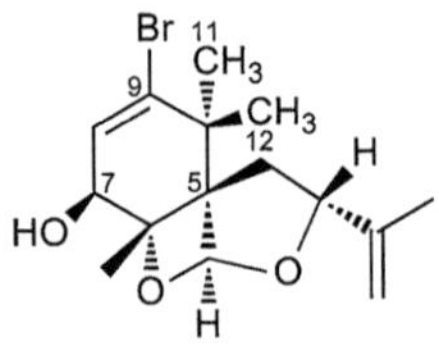

(11)

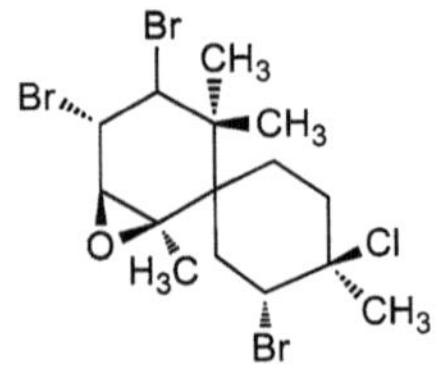

(12)

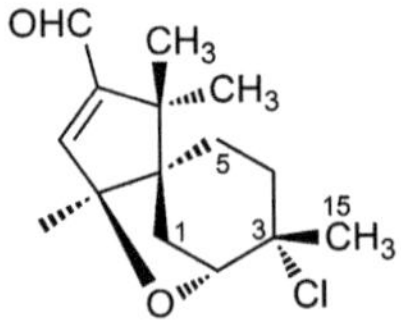

(13)

active area of research since the early 1970s [3] with the discovery of many interesting structural classes including C_{15} acetogenins, such as laurendecumenyne A (**1**) [4] and pannosallene (**2**) [5]. Many of the chamigrenes reported from the genus are halogenated and contain a *spiro* center, along with cyclohexane moieties such as ma'ilione (**3**) [6] and the sesquiterpene bromo diether (**4**) [7].

Recently, the Marine And Terrestrial Natural Product (MATNAP) research group at RMIT University, which studies the chemistry and biological activity of southern Australian marine organisms, examined a specimen of the red alga, *Laurencia elata* collected from St. Paul's Beach, Sorrento, Victoria, Australia. The crude extract of the alga displayed slight antitumor activity as well as antifungal and antiviral activities. This resulted in the isolation and structure determination of two new C_{16} chamigrenes, cycloelatanene A (**5**) and cycloelatanene B (**6**) as well as the isolation of several previously described compounds including (3*Z*)-chlorofucin (**7**), pacifenol (**8**), and elatenyne (**9**). The structures of these secondary metabolites were secured by detailed spectroscopic techniques including NMR spectroscopy. The protocol for the structure elucidation of the new C_{16} chamigrene, cycloelatanene A (**5**) using NMR spectroscopy will be detailed.

2 Materials

1. Glass vials containing freeze-dried, isolated cycloelatanene A.
2. 5 mm NMR tubes and caps.
3. Deuterated chloroform ($CDCl_3$) containing 0.1 % d_5-pyridine.
4. Liquid nitrogen.
5. Glass Pasteur pipettes, natural rubber teats, parafilm, and nitrile gloves.
6. For off-line NMR processing use of Mestre-C, Mestre-C Nova software (www.mestrelab.com). For direct NMR processing use of either VNMR-J (Agilent NMR systems) or TopSpin (Bruker NMR systems) NMR processing software.

3 Methods

3.1 Preparation of Cycloelatanene A for NMR Analysis

1. Following purification via semi-preparative reversed phased HPLC [8] cycloelatanene A should be immediately evaporated under reduced pressure using a rotary evaporator to remove the acetonitrile. The residual water from the mobile phase (*see* **Note 1**) is then freeze-dried (overnight) (*see* **Note 2**).
2. Using a glass Pasteur pipette with a natural rubber teat attached (*see* **Note 3**) add ~600 μL of $CDCl_3$ to completely dissolve cycloelatanene A.
3. Pipette the sample of cycloelatanene A into a 5 mm NMR tube.
4. Cap the NMR tube and wrap a 1 cm×2 cm parafilm strip around the cap and the top of the NMR tube. This will limit evaporation and possible oxidation of cycloelatanene A.

3.2 ^1H NMR Spectroscopy (1D NMR Experiment)

1. Acquire a ^1H NMR spectrum for cycloelatanene A (Fig. 1) using the following settings: number of transients = 64; spectral width 0–12 ppm, then adjust the spectral width accordingly

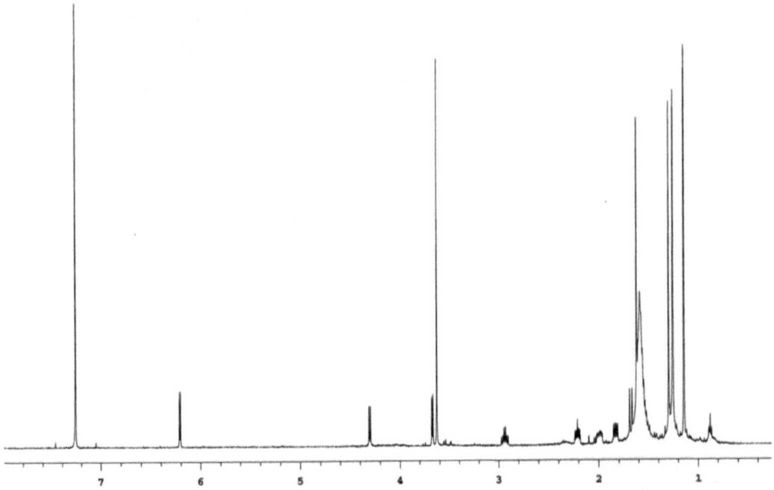

Fig. 1 ^{1}H NMR spectrum (500 MHz, $CDCl_3$) of cycloelatanene A (**5**)

and reference to the solvent signal $CHCl_3$ in deuterated chloroform at δ 7.26 ppm (*see* **Note 4**).

2. In addition also integrate the ^{1}H NMR spectrum to obtain a total hydrogen count.
3. After acquiring the ^{1}H NMR spectrum of cycloelatanene A list all ^{1}H NMR chemical shifts to two decimal places in a single column from highest resonance to the lowest resonance.
4. List the coupling constants (J = Hz) for all ^{1}H NMR chemical shifts.
5. The multiplicity or splitting of the signals needs to be listed. This may be obvious (e.g., a singlet for a methoxy moiety, *s*) or ambiguous (e.g., second order splitting that is often observed for aromatic systems, *m*).

3.3 ^{13}C NMR Spectrum (1D NMR Experiment)

1. Acquire a ^{13}C NMR spectrum for cycloelatanene A using the following parameters: number of transients = 2,000 (longer acquisition results in greater signal to noise and is dependent on the amount of compound isolated); spectral width 0–200 ppm, then adjust the spectral width accordingly (*see* **Note 5**) and reference the central solvent peak triplet to δ 77.0 ppm.
2. After acquiring the ^{13}C NMR spectrum and referencing the spectrum, list all ^{13}C NMR chemical shifts to one decimal place.
3. At this stage the ^{1}H NMR chemical shifts which are directly correlated to certain ^{13}C chemical shifts are unknown. However, it is apparent that cycloelatanene A contains 16 distinct carbons; four of them *tentatively* identified as being qua-

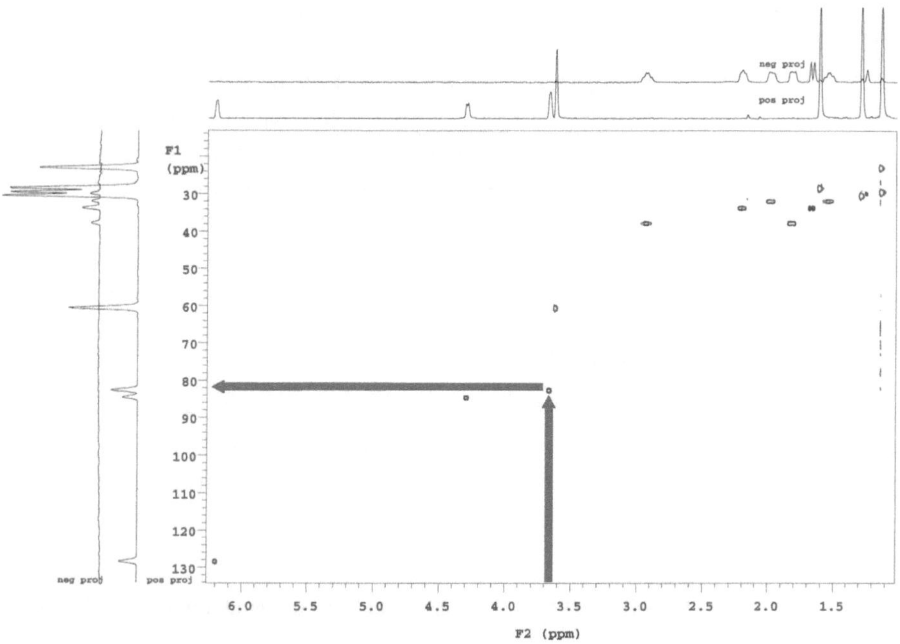

Fig. 2 gHSQCAD NMR spectrum (500 MHz, $CDCl_3$) of cycloelatanene A (**5**)

ternary carbons (*see* **Notes 6** and 7). This immediately supports the fact that there is no element of symmetry in the structure of cycloelatanene A.

3.4 gHSQCAD NMR Spectrum (2D Through Bond NMR Experiment)

1. Acquire a gHSQCAD NMR spectrum for cycloelatanene A using the following parameters: number of transients = (longer acquisition results in great signal to noise and is dependent on the amount of material isolated); spectral width in the ^{13}C dimension is 0–160 ppm (beyond this range only quaternary carbons (C) appear which are not observed in this experiment) then adjust the spectral width accordingly; the spectral width in the ^{1}H dimension is 0–12 ppm, which should then be adjusted accordingly with referencing to the solvent signals at δ 7.26 and 77.0 ppm (*see* **Note 8**).
2. Contours observed in the gHSQCAD NMR spectrum depict which ^{1}H NMR chemical shifts are directly correlated to which ^{13}C NMR chemical shifts. For example, Fig. 2 illustrates a $^{1}J_{CH}$ correlation indicating that the ^{1}H NMR chemical shift at δ 3.67 is directly connected to the carbon at δ 82.4.
3. Positive, contours for a particular $^{1}J_{CH}$ correlation indicate that the multiplicity of the carbons are either methyls (CH_3) or methines (CH), while negative contours for a particular $^{1}J_{CH}$ correlation indicate that the multiplicity of the carbons are

methylenes (CH_2) (*see* **Note 9**). In this way the direct $^1J_{CH}$ correlation observed in the gHSQCAD NMR experiment (Fig. 2) between the proton at δ 3.67 to the carbon at δ 82.4 can be concluded to be a deshielded methine (CH). Quaternary carbons (C) are not observed in this NMR experiment as they do not have an associated proton resonance (Fig. 2).

4. List under "δC[a], mult" which ^{1}H NMR chemical shifts are directly correlated to which ^{13}C NMR chemical shifts. Use the following to denote multiplicity of the carbons: *s*=quaternary, *d*=methine carbon, *t*=methylene carbon, and *q*=methyl carbon (Table 1).
5. List all quaternary carbons (C) in the table.
6. "a"=Carbon assignments based on HSQCAD NMR experiments. In cases where sufficient quantities (>several mgs) of a

Table 1
^{1}H (500 MHz) and ^{13}C (125 MHz) NMR of cycloelatanene A (5) in $CDCl_3$

δH	Integration	(*J* in Hz)	δC[a], mult
–	–	–	45.4, *s*
–	–	–	139.1, *s*
–	–	–	85.7, *s*
–	–	–	48.8, *s*
–	–	–	74.7, *s*
6.21, *d*	1H	(6.0)	128.2, *d*
3.67, *dd*	1H	(6.0, 0.5)	82.4, *d*
2.21, *ddd*	1H	(12.0, 7.5, 4.0)	33.4, *t*
1.68, *d*	1H	(12.5)	
4.30, *dd*	1H	(7.5, 1.0)	84.3, *d*
2.94, *ddd*	1H	(13.5, 13.0, 8.0)	37.5, *t*
1.83, *dd*	1H	(13.5, 7.0)	
1.98, *m*	1H		31.6, *t*
1.54, *ddd*	1H	(13.5, 12.5, 6.5)	
1.29, *s*	3H		30.1, *q*
3.63, *s*	3H		60.3, *q*
1.13, *s*	3H		29.2, *q*
1.14, *s*	3H		22.7, *q*
1.62, *s*	3H		28.1, *q*

[a]Carbon assignments based on HSQCAD NMR experiments

compound are obtained, the alternative DEPT experiments can be acquired. Both a DEPT 90 and DEPT 135 NMR experiment need to be acquired to deduce the multiplicities of the carbon chemical shifts. The DEPT 90 NMR experiment shows only methine (CH) carbons in a positive orientation, while the DEPT 135 NMR experiment displays both methine (CH) and methyl (CH_3) carbons in a positive orientation and methylenes (CH_2) in a negative orientation.

3.5 COSY NMR Spectrum (2D Through Bond NMR Experiment)

1. Acquire a gCOSY NMR spectrum for cycloelatanene A using the following parameters: number of transients = (longer acquisition results in great signal to noise and is dependent on the amount of compound isolated); spectral width is initially set to 0–12.0 ppm and then adjusted accordingly in both ^{1}H NMR dimensions with referencing to the solvent $CHCl_3$ in deuterated chloroform at δ 7.26 ppm.
2. Contours or "cross peaks" in the gCOSY NMR experiment indicate which ^{1}H NMR chemical shifts are adjacent and therefore coupled to neighboring ^{1}H NMR chemical shifts (*see* **Note 10**). Figure 3 shows a COSY correlation from the ^{1}H NMR chemical shift at δ 2.21 to the proton at δ 4.30 and vice versa. It is also possible to observe long range proton to proton correlations in the COSY NMR experiment.

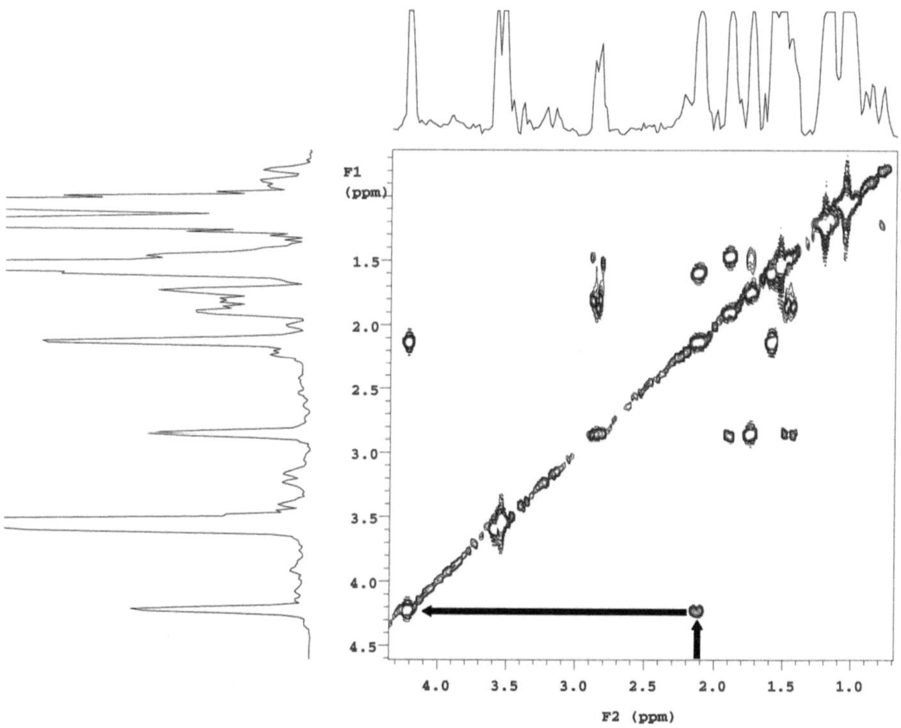

Fig. 3 gCOSY NMR spectrum (expansion) of cycloelatanene A (**5**)

Table 2
^{1}H (500 MHz), ^{13}C (125 MHz), and gCOSY NMR of cycloelatanene A (5) in $CDCl_3$

δH	Integration	(*J* in Hz)	δC[a], mult	gCOSY
–	–	–	45.4, *s*	–
–	–	–	139.1, *s*	–
–	–	–	85.7, *s*	–
–	–	–	48.8, *s*	–
–	–	–	74.7, *s*	–
6.21, *d*	1H	(6.0)	128.2, *d*	3.67
3.67, *dd*	1H	(6.0, 0.5)	82.4, *d*	6.21
2.21, *ddd*	1H	(12.0, 7.5, 4.0)	33.4, *t*	1.68, 4.30, 1.98(w)
1.68, *d*	1H	(12.5)		2.21
4.30, *dd*	1H	(7.5, 1.0)	84.3, *d*	2.21
2.94, *ddd*	1H	(13.5, 13.0, 8.0)	37.5, *t*	2.94, 1.83, 1.98, 1.54
1.83, *dd*	1H	(13.5, 7.0)		2.94, 1.54
1.98, *m*	1H		31.6, *t*	2.21, 2.94, 1.54
1.54, *ddd*	1H	(13.5, 12.5, 6.5)		2.94, 1.83, 1.98, 1.62(w)
1.29, *s*	3H		30.1, *q*	–
3.63, *s*	3H		60.3, *q*	–
1.13, *s*	3H		29.2, *q*	–
1.14, *s*	3H		22.7, *q*	–
1.62, *s*	3H		28.1, *q*	1.54(w)

[a]Carbon assignments based on HSQCAD NMR experiments; *w* weak correlation

3. List under "gCOSY" which ^{1}H NMR chemical shifts correlate to other ^{1}H NMR chemical shifts (Table 2).
4. Place a "-" for all quaternary carbons (C) in the table.

3.6 HMBC NMR Spectrum (2D Through Bond NMR Experiment)

1. Acquire a gHMBC NMR spectrum for cycloelatanene A using the following parameters: number of transients = (longer acquisition results in great signal to noise and is dependent on the amount of compound isolated); spectral width in the ^{13}C dimension is 0–200 ppm, which is then adjusted accordingly; spectral width in the ^{1}H dimension is 0–12.0 ppm, also readjusted accordingly with referencing to the solvent peaks at δ 7.26 ppm, 77.0 ppm.
2. Contours in the gHMBC NMR experiment are referred to as correlations and indicate which ^{1}H NMR chemical shifts are

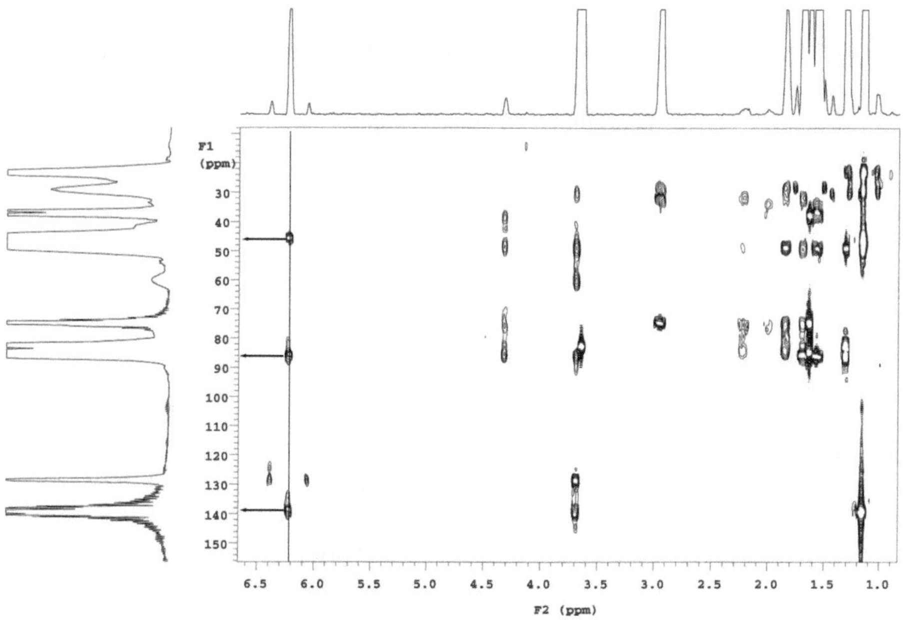

Fig. 4 gHMBC NMR spectrum of cycloelatanene A (**5**)

coupled to which ^{13}C NMR chemical shifts generally up to two or three bonds away. In this respect, this is the most important of the 2D through bond NMR experiments as it allows for structure fragments to be linked to provide a final structure. In the case of cycloelatanene A the HMBC NMR spectrum (Fig. 4) illustrates that the proton with chemical shift at δ 6.21 displays correlations to the carbons at 45.4 (three bond correlation), 85.7 (three bond correlation), and 139.1 (two bond correlation) ppm. While it is typical for HMBC NMR correlations to be usually located either two or three bonds away, in aromatic systems, these correlations can be up to four bonds away (*see* **Note 11**).

3. List under "gHMBC" which ^{1}H NMR chemical shifts correlate to which ^{13}C NMR chemical shifts (Table 3).
4. Place a dash "-" for all quaternary carbons (C) in the table.

3.7 Single Irradiation 1D NOE NMR (1D Through Space NMR Experiment)

1. Acquire single irradiation 1D NOE NMR spectra for cycloelatanene A using a spectral width in the ^{1}H dimension of 0–12.0 ppm, which can then be adjusted accordingly.
2. Each ^{1}H NMR chemical shift will need to be individually irradiated and will appear as a negative peak in the selective 1D NOE NMR experiment (Fig. 5). If NOE NMR enhancements are observed for a particular ^{1}H NMR chemical shift these will appear as positive peaks in the NMR spectrum (*see* **Note 12**). In the case of cycloelatanene A the proton with chemical shift

Table 3
^{1}H (500 MHz), ^{13}C (125 MHz), gCOSY, and gHMBC NMR of cycloelatanene A (5) in $CDCl_3$

δH	(*J* in Hz)	Integration	δC[a], mult	gCOSY	gHMBC
–	–	–	45.4, *s*	–	–
–	–	–	139.1, *s*	–	–
–	–	–	85.7, *s*	–	–
–	–	–	48.8, *s*	–	–
–	–	–	74.7, *s*	–	–
6.21, *d*	(6.0)	1H	128.2, *d*	3.67	45.4, 139.1, 85.7
3.67, *dd*	(6.0, 0.5)	1H	82.4, *d*	6.21	139.1, 128.2, 85.7, 48.8, 30.1, 60.3
2.21, *ddd*	(12.0, 7.5, 4.0)	1H	33.4, *t*	1.68, 4.30, 1.98(w)	48.8, 84.3, 74.7, 31.6
1.68, *d*	(12.5)	1H		2.21	84.3, 74.7, 31.6
4.30, *dd*	(7.5, 1.0)	1H	84.3, *d*	2.21	85.7, 48.8, 74.7, 37.5
2.94, *ddd*	(13.5, 13.0, 8.0)	1H	37.5, *t*	2.94, 1.83, 1.98, 1.54	74.7, 31.6, 1.62
1.83, *dd*	(13.5, 7.0)	1H		2.94, 1.54	48.8, 84.3, 74.7, 31.6, 28.1
1.98, *m*		1H	31.6, *t*	2.21, 2.94, 1.54	48.8, 33.4, 74.7, 37.5
1.54, *ddd*	(13.5, 12.5, 6.5)	1H		2.94, 1.83, 1.98, 1.62(w)	85.7, 48.8, 33.4, 37.5
1.29, *s*		3H	30.1, *q*	–	82.4, 85.7, 48.8
3.63, *s*		3H	60.3, *q*	–	82.4
1.13, *s*		3H	29.2, *q*	–	139.1, 22.7
1.14, *s*		3H	22.7, *q*	–	139.1, 29.2
1.62, *s*		3H	28.1, *q*	1.54(w)	84.3, 74.7, 37.5

[a]Carbon assignments based on HSQCAD NMR experiments; *w* weak correlation

at δ 2.21 was irradiated and NOE enhancements were observed to the signals at δ 1.14, 1.68, 1.29, and 4.30 (Fig. 5).

3. List under "1D NOE" for each ^{1}H NMR chemical shift all observed NOE NMR enhancements (Table 4).

4. Alternatives to the 1D NOE NMR experiment include the 2D NOESY or 2D ROESY NMR experiments. These provide the same information but in a two dimensional format and the experiments generally need to be acquired for a much longer period of time. The 2D ROESY NMR experiment is a phase sensitive experiment so in addition to NOE enhancements it also provides through bond proton to proton correlations.

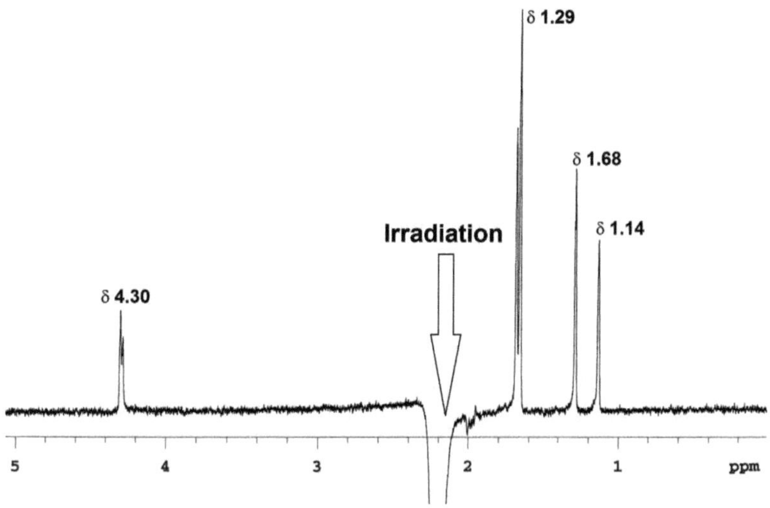

Fig. 5 Selective irradiation of the signal at ^{1}H at δ 2.21 (500 MHz, $CDCl_3$) showing observed NOE NMR enhancements of cycloelatanene A (**5**)

5. The 1D or 2D NOE NMR experiments, in combination with the interpretation of proton coupling constants, are utilized to deduce the relative configuration of a compound. The relative configuration is defined as the configuration at any stereogenic element (center, axis, or plane) with respect to that of any other stereogenic element in the same molecule. In this way these experiments help to confirm whether a particular substituent at a chiral center is positioned up (denoted by dark wedged bonds) versus those that are positioned down (denoted by a hashed bond). NOE NMR experiments are a through space NMR experiment and provide information as to which protons are relatively close in space. These experiments are confined to a particular distance (typically <5 Å), and, as the name suggests, can only provide a relative assignment of the substituents around a chiral center.
6. The absolute configuration of a compound is defined as the spatial arrangement of the atoms in a chiral molecule which distinguishes it from its mirror image and its stereochemical description is denoted as *R* or *S*. The methods available to deduce the absolute configuration of a compound include total synthesis, single crystal X-ray crystallography, chemical degradations to known chiral units or by circular dichroism (CD).
7. The position numbers adopted in the final structural assignment are based upon closely related compounds reported in the literature. The position numbers for the assignment of each chemical shift is made on the basis of the 1D and 2D NMR experiments. For known compounds the position numbering must be adopted in accordance to that previously published.

In cases where a new carbon skeleton is being reported numbering will still need to conform according to closely related literature compounds. Position numbering can be reviewed at any point.

3.8 Structure Elucidation of Cycloelatanene A

In journal publications reporting new natural products, the structure elucidation should conform to a general format. Firstly, a physical description of the appearance of the compound, including its stability must be provided. This is followed by providing high-resolution mass spectrometric data confirming the molecular formula, characteristic UV maxima and associated molar absorptivities, key infrared absorptions and the specific rotation (for chiral compounds). Most of this characterization data is given in the experimental section of the journal. The structure elucidation of cycloelatanene A is detailed below to provide an example of the format that should be adopted. For new structural entities such as cycloelatanene A, the NMR data (Table 4) is used systematically to deduce key structural features of the compound.

Cycloelatanene A (**5**) was isolated as a stable colorless oil for which high-resolution ESIMS established the molecular formula as $C_{16}H_{24}BrClO_2$ (*m/z* 385.0541 $[M+Na]^+$, calc for $C_{16}H_{24}{}^{79}Br^{35}ClNaO_2$, 385.0546 and *m/z* 363.0721 $[M+H]^+$; calculated for $C_{16}H_{25}{}^{79}Br^{35}ClO_2$, 363.0648) indicating four degrees of unsaturation. Analysis of the ^1H, ^{13}C and HSQCAD NMR spectroscopic data indicated the presence of one methoxy moiety (δ_H 3.63, δ_C 60.3); four singlet methyls (δ_H 1.29, δ_C 30.1), (δ_H 1.13, δ_C 29.2), (δ_H 1.14, δ_C 22.7), and (δ_H 1.62, δ_C 28.1); three methylenes [(δ_H 2.21, *ddd*, J=4.0, 7.5, 12.0 Hz), (δ_H 1.68, *d*, J=12.5 Hz, δ_C 33.4); (δ_H 1.98, *m*), (δ_H 1.54, *ddd*, J=7.0, 12.5, 13.5 Hz, δ_C 31.6); (δ_H 1.83, *dd*, J=7.0, 13.5 Hz), (δ_H 2.94, *ddd*, J=8.0, 13.0, 13.5 Hz, δ_C 37.5)]; one olefinic methine [(δ_H 6.21, *d*, J=6.0 Hz, δ_C 128.2)]; two deshielded methines (δ_H 3.67, *dd*, J=6.0, 0.5 Hz, δ_C 82.4), (δ_H 4.30, *dd*, J=7.5, 1.0 Hz, δ_C 84.3); and five quaternary carbons (δ_C 45.4, 139.1, 85.7, 48.8, 74.7).

COSY NMR correlations were observed from the olefinic methine at position 3 (δ_H 6.21, *d*, J=6.0 Hz) to a deshielded oxymethine (δ_H 3.67, *dd*, J=6.0, 0.5 Hz, H4). In turn this oxymethine at position 4 showed key HMBC NMR correlations to the methyl group at (δ_C 30.1, C12), to the methoxy moiety at (δ_C 60.3, C13), to an olefinic carbon (δ_C 139.1, C2), and to a carbon at 48.8 (C6). This allowed placement of the methoxy substituent to position 4. The olefinic methine at position 3 (δ_H 6.21) displayed HMBC NMR correlations to 45.4 (C1) and to the deshielded quaternary carbon 85.7 (C5). Two geminal methyls (δ_H 1.14) and (δ_H 1.13) showing HMBC NMR correlations to each other and to an olefinic moiety (δ_C 139.1, C2) were also identified. Further interpretation of the carbon resonances for this first structural subunit of cycloelatanene A (**5**) indicated that the carbon

Table 4
Complete ^{1}H (500 MHz) and ^{13}C (125 MHz) NMR of cycloelatanene A (5) in $CDCl_3$

Pos.	δH	Int.	(*J* in Hz)	δC[a], mult	gCOSY	gHMBC	1D NOE
1	–	–	–	45.4, *s*	–	–	–
2	–	–	–	139.1, *s*	–	–	–
3	6.21, *d*	1H	(6.0)	128.2, *d*	3.67	45.4, 139.1, 85.7	3.67, 1.13
4	3.67, *dd*	1H	(6.0, 0.5)	82.4, *d*	6.21	139.1, 128.2, 85.7, 48.8, 30.1, 60.3	–
5	–	–	–	85.7, *s*	–	–	–
6	–	–	–	48.8, *s*	–	–	–
7a	2.21, *ddd*	–	(12.0, 7.5, 4.0)	33.4, *t*	1.68, 4.30, 1.98(w)	48.8, 84.3, 74.7, 31.6	1.68, 4.30, 1.29, 1.14
7b	1.68, *d*	1H	(12.5)		2.21	84.3, 74.7, 31.6	2.21, 4.30, 1.13
8	4.30, *dd*	1H	(7.5, 1.0)	84.3, *d*	2.21	85.7, 48.8, 74.7, 37.5	2.21, 1.68, 1.29, 3.63, 1.14, 1.62
9	–	1H	–	74.7, *s*	–	–	–
10a	2.94, *ddd*	1H	(13.5, 13.0, 8.0)	37.5, *t*	2.94, 1.83, 1.98, 1.54	74.7, 31.6, 1.62	1.83, 1.98, 3.63
10b	1.83, *dd*	1H	(13.5, 7.0)		2.94, 1.54	48.8, 84.3, 74.7, 31.6, 28.1	2.94, 1.54
11a	1.98, *m*	1H		31.6, *t*	2.21, 2.94, 1.54	48.8, 33.4, 74.7, 37.5	1.54. 1.14
11b	1.54, *ddd*	1H	(13.5, 12.5, 6.5)		2.94, 1.83, 1.98, 1.62(w)	85.7, 48.8, 33.4, 37.5	1.98, 1.13
12	1.29, *s*	3H		30.1, *q*	–	82.4, 85.7, 48.8	–
13	3.63, *s*	3H		60.3, *q*	–	82.4	6.21, 3.67, 2.94, 1.29
14	1.13, *s*	3H		29.2, *q*	–	139.1, 22.7	–
15	1.14, *s*	3H		22.7, *q*	–	139.1, 29.2	1..29, 3.63(w)
16	1.62, *s*	3H		28.1, *q*	1.54(w)	84.3, 74.7, 37.5	4.30, 1.83

[a]Carbon assignments based on HSQCAD NMR experiments; *w* weak correlation or enhancement

at position 5 was deshielded by its direct attachment to an oxygen and that the olefinic carbon resonance at position 2 was substituted with a halogen. The halogen was concluded to be bromine rather than chlorine on the basis of the chemical shift of the position 2 carbon and was further confirmed by comparison to related literature compounds such as deoxyprepacifenol (**10**) [9, 10]. On the basis of these combined COSY and HMBC NMR correlations structure fragment A could be assembled (Fig. 6).

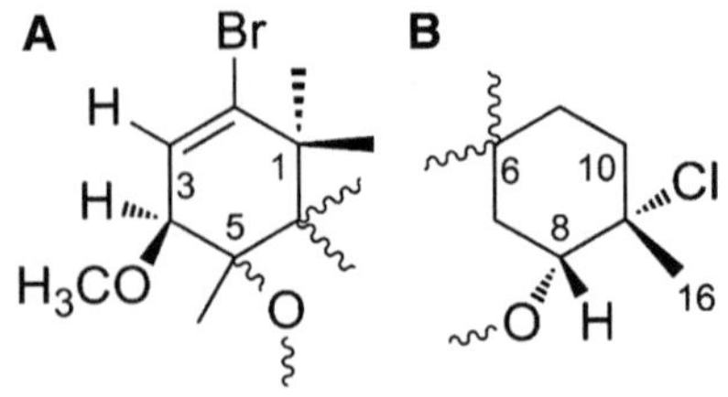

Fig. 6 Structural fragments (**A**) and (**B**) deduced for the structure of cycloelatanene A (**5**)

Upon further inspection of the literature, it was evident that this structural fragment resembled part of the structure of laureacetal-B (**11**) [11]. Chemical shifts comparisons supported this structure fragment. In particular the coupling constant for the deshielded methine at position 4 (δ_H 3.67, *dd*, *J*=6.0, 0.5 Hz) was consistent with that reported for **11** (δ_H 3.98, *d*, *J*=7.0 Hz) [11]. In addition the coupling constant for the olefinic methine at position 3 (δ_H 6.21, *d*, *J*=6.0 Hz) was also consistent with that reported for **11** (δ_H 6.60, *d*, *J*=7.0 Hz), which again supported the relative configuration assignment at position 4 [11]. The absolute configuration of **11** had been previously established by a chemical degradation approach [11]. Both the ^{1}H and ^{13}C NMR assignments for structure fragment A were found to be similar to those for the bromotrimethylcyclohexene moiety present in **11** as well as to those reported for other related chamigrenes [12–14]. A second proposed structure fragment B (Fig. 6) was formulated on the basis of the following COSY and HMBC NMR correlations.

The deshielded oxymethine at position 8 (δ_H 4.30) showed key HMBC NMR correlations to (δ_C 48.8, C6), (δ_C 74.7, C9) and to the methylene (δ_C 37.5, C10). The carbon chemical shift of the position 8 methine, once again, suggested that it was directly attached to oxygen. The quaternary carbon at position 6 (δ_C 48.8) was common to both fragments A and B and concluded to be a *spiro* center on the basis of comparisons to other related *spiro* terpenoids isolated from the genus [12, 15, 16]. The methyl group at (δ_H 1.62) also showed an HMBC NMR correlation to the quaternary carbon at (δ_C 74.7, C9). The carbon chemical shift at position 9 is consistent with the presence of a methyl and a chlorine substituent and is a common structural feature in previously isolated chamigrenes such as the halogenated chamigrene epoxide (**12**) first isolated from the red alga *Laurencia okamurai* [17].

Furthermore, the coupling constant for the oxymethine proton at position 8 (δ_H 4.30, *dd*, *J*=7.5, 1.0 Hz) supported the presence of the chloromethylcyclohexane moiety in fragment B as in the case of laurencial (**13**) (δ_H 4.11, *d*, *J*=6.0 Hz) [18]. Additional HMBC NMR correlations from the methylene at position 11 (δ_H 1.98, *m*) and (δ_H 1.54, *ddd*, *J*=6.5, 12.5, 13.5 Hz) to the position 5 carbon (δ_C 85.7) and to the *spiro* center at position 6

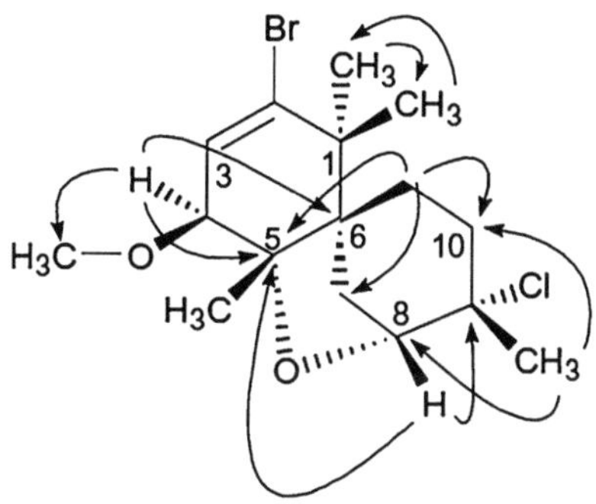

Fig. 7 Key HMBC NMR correlations of cycloelatanene A (**5**)

(δ_C 48.8) and also from the adjacent methylenes at positions C10 (δ_C 37.5) and C7 (δ_C 33.4), respectively (Fig. 6) also supported this second structure fragment.

The key HMBC NMR correlations (Fig. 7) that allowed for the connection of structure fragments A and B were those observed from the oxymethine at position 8 (δ_H 4.30) and the position 11 methylene (δ_H 1.54) to the carbon (δ_C 85.7) at position 5. These correlations concluded that a third five membered ring was also present in the structure of **5** and accounted for the final double bond equivalent.

The relative configuration of cycloelatanene A (**5**) was established on the basis of selective 1D NOE NMR experiments and coupling constants. Key ID NOE NMR enhancements were observed from one of the germinal methyls at position 15 (δ_H 1.14) to the 12-CH_3 (δ_H 1.29) and a weak correlation to the methoxy protons at position 13 (δ_H 3.63). The deshielded methine at position 8 (δ_H 4.30) showed key NOE enhancements to the methyl at 1.14, the deshielded methyls at 1.29 and to the methoxy protons 3.63. In turn, the position 13 methoxy moiety protons displayed an NOE to one of the position 10 methylene protons (δ_H 2.94, H10a). Also observed were key NOE enhancements between one of the position 7 methylene protons (δ_H 2.21, H7a) to both 12-CH_3 (δ_H 1.29) and 15-CH_3 (δ_H 1.14). Another important NOE enhancement was observed from the other position 7 proton (δ_H 1.68, H7b) to the position 14 methyl (δ_H 1.13). These NOE enhancements allowed the complete relative configuration assignment of (**5**) (Fig. 8), with H7a, H8, H10a, H11a, 12-CH_3, 13-CH_3, and 15-CH_3 on the same face of the molecule.

4 Notes

1. Ensure that the water bath temperature in the rotary evaporator is set to ~40 °C and a high vacuum is used (preferably using a reticulated chiller system) and that the round bottom flask is securely clamped.

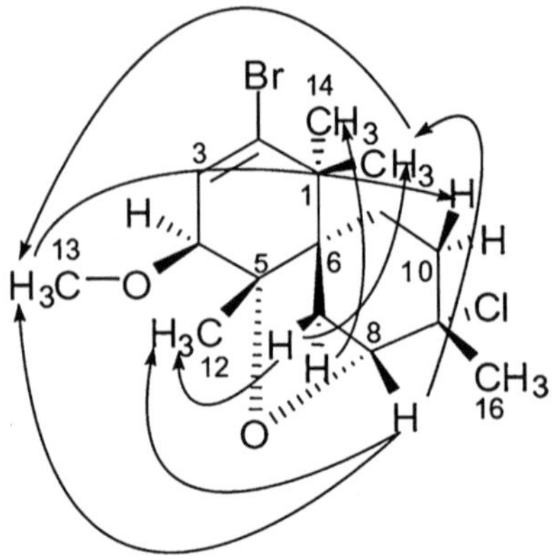

Fig. 8 Key 1D NOE NMR enhancements of cycloelatanene A (**5**)

2. Freeze residual water contained in the round bottom flask by placing it in dish-shaped Dewar flask filled partially with liquid nitrogen. Rotate the round bottom flask so it freezes a thin layer of residual water on the inside surface of the flask. This will assist in an efficient removal of the water during the freeze-drying process.
3. Care should be taken to avoid the use of plastic/rubber teats as solvents may extract phthalates and will therefore contaminate the purified compound. Resonances due to phthalates would be observed in the ^{1}H NMR spectrum at approximately δ 7–7.6 ppm (either side of the $CHCl_3$ 7.26 ppm solvent peak) for the aromatic resonances and at approximately 4 ppm for the deshielded methylene.
4. It is assumed that a fundamental understanding of NMR including shimming, acquisition and data processing is held. When acquiring ^{1}H NMR spectra for the first time a wider spectral 0–12 ppm width should be used and then this should be rescaled accordingly.
5. When acquiring ^{13}C NMR spectra for the first time a wider spectral width should be used (e.g., 0–220 ppm) and then rescaled or readjusted accordingly. This ensures that ^{13}C NMR chemical shifts that resonate further downfield (e.g., ketones) will be observed.
6. Quaternary carbons (C) are much lower in signal intensity as they have a slower relaxation compared to methine carbons (CH) > methylenes carbons (CH_2) > methyls carbons (CH_3).
7. If there is insufficient compound available, a carbon NMR experiment is not necessary as this information can be obtained from the HSQCAD and HMBC NMR experiments.
8. The spectral widths in both the ^{1}H and ^{13}C NMR dimension are based on the previously acquired one dimensional ^{1}H and ^{13}C NMR spectra.

9. If a $^1J_{CH}$ correlation is not observed then it is evident that the remaining ^{13}C NMR chemical shifts are quaternary carbons.
10. It is common to observe ^1H–^1H correlations on methylenes and long range ^1H–^1H COSY correlations with neighboring protons, the latter is usually denoted as "w" for weak correlation and is placed in a footnote under the table.
11. It is common to also observe weak four bond ^1H–^{13}C correlations especially in aromatic systems and these should also be noted as true correlations.
12. If a selective ^1H NMR chemical shift is irradiated and no NOE enhancements are observed, this needs to be considered in the structure elucidation. However, it must be noted that absence of an NOE cannot be used to support a relative configuration assignment.

Acknowledgments

We gratefully acknowledge Mr Robert Brkljaca for his assistance in producing and processing the NMR figures that appear in this chapter.

References

1. Erickson KL, Scheuer PJ (1983) *Marine Natural Products: Chemical and Biological Perspectives*. New York: Academic Press
2. König GM, Wright AD (1997) *Laurencia rigida*: chemical investigations of its antifouling dichloromethane extract. J Nat Prod 60: 967–970
3. Sims JJ, Fenical W (1972) Marine natural products III. Johnstonol, an unusual halogenated epoxide from the red alga *Laurencia Johnstonii*. Tett Lett 3:195–198
4. Nai-Yun J, Xiao-Ming L, Ke L, Bin-Gui W (2007) Laurendecumallenes A-B and laurendecumenynes A-B, halogenated nonterpenoid C_{15}-acetogenins from the marine red alga *Laurencia decumbens*. J Nat Prod 70: 1499–1502
5. Suzuki M, Takahashi Y, Matsuo Y, Masuda M (1996) Pannosallene, a brominated C_{15} nonterpenoid from *Laurencia pannosa*. Phytochemistry 41:1101–1103
6. Francisco MEY, Erickson KL (2001) Ma'iliohydrin, a cytotoxic chamigrene dibromhydrin from a Philippine *Laurencia* species. J Nat Prod 64:790–791
7. Kikuchi H, Suzuki T, Suzuki M, Kurosawa E (1985) A new chamigrene-type bromo diether from the red alga *Laurencia nipponica* Yamada. Bull Chem Soc Jpn 58:2437–2438
8. Dias DA, Urban S (2011) Phytochemical studies of the southern Australian marine alga, *Laurencia elata*. Phytochemistry 72:2081–2089
9. Fronczek F (1989) Redetermination of the absolute configuration of deoxyprepacifenol, from the Mediterranean red alga, *Laurencia majuscula*. Acta Crystallogr Sect C Cryst Struct Commun 45:1102–1104
10. Watanabe K, Umeda K, Miyakado M (1989) Isolation and identification of three insecticidal principles from the red alga *Laurencia nipponica* Yamada. Agric Biol Chem 53:2513–2515
11. Suzuki T, Kurosawa E (1979) New bromo acetal from the marine alga, *Laurencia nipponica* Yamada. Chem Lett (3):301–304
12. Kimura J, Kamada N, Tsujimoto Y (1999) Fourteen chamigrene derivatives from a red alga, *Laurencia nidifica*. Bull Chem Soc Jpn 72:289–292
13. Kurata K, Suzuki T, Suzuki M, Kurosawa E, Furusaki A, Matsumoto T (1983) Constituents of marine plants 53. Laureacetal-D and -E, two new secochamigrene derivatives from the red alga *Laurencia nipponica* Yamada. Chem Lett (4):557–560
14. Suzuki M, Segawa M, Suzuki T, Kurosawa E (1985) Structures of two new halochamigrene derivatives from the red alga *Laurencia nip-*

ponica Yamada. Bull Chem Soc Jpn 58: 2435–2436

15. Jongaramruong J, Blackman AJ, Skelton BW, White AH (2002) Chemical relationships between the sea hare *Aplysia parvula* and the red seaweed *Laurencia filiformis* from Tasmania. Aust J Chem 55:275–280
16. San-Martín A, Darias J, Soto H, Contreras C, Herrera JS, Rovirosa J (1997) A new C15 acetogenin from the marine alga *Laurencia claviformis*. J Nat Prod 10:303–311
17. Ojika M, Shizuri Y, Yamada K (1982) A halogenated chamigrane epoxide and six related halogen-containing sesquiterpenes from the red alga *Laurencia okamurai*. Phytochemistry 21:2410–2411
18. Kurata K, Suzuki T, Suzuki M, Kurosawa E, Furusaki A, Matsumoto T (1983) Constituents of marine plants 52. Laurenical, a novel sesquiterpene α,β-unsaturated aldehyde from the red alga *Laurencia Nipponica* Yamada. Chem Lett (3):299–300

Chapter 9

NMR-Based Metabolomics: A Probe to Utilize Biodiversity

Lúcia P. Santos Pimenta, Hye Kyong Kim, Robert Verpoorte, and Young Hae Choi

Abstract

Metabolomics is a comprehensive profiling tool used to identify qualitatively and quantitatively all the metabolites present in a biological system. As the number of metabolites in a living being is assumed to be around 30,000, it is necessary to use an adequate extraction procedure and a good analytical technique to perform metabolomic analyses. Each analytical platform used in metabolomics has both advantages and disadvantages in terms of the sensitivity and resolution for metabolites to be detected. Of the methods, nuclear magnetic resonance (NMR) spectroscopy has been proved to have several advantages over other MS-based methods. NMR provides an efficient, robust, and nondestructive metabolomics analysis of crude extracts or samples as well as easy quantitation without calibration curves for each metabolite, although it shows relatively low sensitivity than MS-based methods. Thus, NMR-based metabolomics have been often used for the first step to capture the insight of the metabolome of organisms. This chapter presents general steps of NMR-based metabolomic analysis involved in the study of biodiversity.

Key words ^{1}H NMR-based metabolomics, Sample preparation, Multivariate data analysis, Two-dimensional nuclear magnetic resonance (2D-NMR) spectroscopy

1 Introduction

Metabolomics is the qualitative and quantitative measurement of all metabolites of an organism. It is a powerful tool used to understand more comprehensively the highly regulated and complex biochemical networks in the most fundamental cell by measuring the metabolome. Considered the most recent "omics" technologies, the metabolome is postulated to be more closely related to the phenotype of an organism, being more sensitive than the proteome and the transcriptome to external factors that contribute to phenotypic differences between specimens [1]. The metabolome provides end products of gene expression and can define the biochemical phenotype of a cell or tissue. Metabolomics is the characterization of the metabolome and can be used to monitor and assess gene

Ute Roessner and Daniel Anthony Dias (eds.), *Metabolomics Tools for Natural Product Discovery: Methods and Protocols*, Methods in Molecular Biology, vol. 1055, DOI 10.1007/978-1-62703-577-4_9, © Springer Science+Business Media, LLC 2013

function, yielding more information and being more versatile than the other "omics" about functional genomics.

Although metabolomics can give a holistic approach to an organism the metabolic changes involved in response to a biotic or abiotic perturbation, is very difficult to completely achieve. Considering that the total number of metabolites within the plant kingdom, including primary and secondary metabolisms, is estimated to be over 200,000 in total and 30,000 in a single organism, they are present in different ranges of concentration in each organism and there is no single satisfactory metabolomics platform to fulfill the requirements to detect all metabolites. Reproducibility is the most important requirement to achieve in order to develop a metabolomics protocol. The other criterion is the ease of quantification and identification of the metabolites. Among the major problems to fulfill these requirements is how to extract all the metabolites inside a sample, which is practically impossible. Once the sample is treated, for example, by extraction, only the soluble compounds are obtained and insoluble part is not detected [2]. When separation is performed, certain compounds are lost: as such, information will be always lost. Furthermore, all metabolites should be analyzed using an optimized method that can detect all the metabolites extracted at the same time in one single experiment is another difficult task.

Each one of the analytical methods used in metabolomics has its own advantages and limitations, which are related to the differences in polarity, chemical property, stability, and concentration of the high number of metabolites [3]. Gas chromatography (GC) is sensitive but requires volatility of compounds; liquid chromatography (LC) can detect compounds in a certain polarity range. Both MS-based methods have problems of reproducibility, among others on the long-term, as they depend on commercial columns which tend to change through time. Even more importantly, each compound has a different detector response in GC-MS and LC-MS, thus requiring calibration curves for each individual compound if absolute quantification is required. Mass spectrometry (MS) offers high resolution and in terms of sensitivity is the best method, but reproducibility (very much matrix-dependent) and different detector responses are a problem for quantification. Nevertheless, ^{1}H NMR is highly reproducible and easy for quantification as signal intensity is directly related to the molar concentration. However, it has low sensitivity and resolution compared to the MS-based methods [3]. In fact, none of the existing methods are able to achieve the identification and quantification of all metabolites in an organism.

As a significant number of compounds are present in an organism, the data collected will be vastly large. The enormous amount of data collected needs to be processed and treated to obtain biological interpretation. Chemometric tools have been used in data analysis and in most cases, multivariate data analysis, such as Principal Component Analysis (PCA) and Partial Least Squares-Discriminant Analysis (PLS-DA), is employed routinely to extract information from large data sets [4]. A large number of observable variables after multivariate data analysis give the information required. Next step is to identify the metabolites responsible for the biological question and is critical in a metabolomics analysis. However, very few compounds can be identified by the major detection methods used in metabolomics such as NMR and MS [5]. More extensive works for the identification should be elaborated for the future direction of metabolomics. Once the signals are identified as compounds, correlations between metabolites and responses, grouping and all the experimental data can be used in any time, permitting to construct a hypothesis or explain the observations made [4]. The differences between the samples and the identified compounds connected with them will provide a holistic vision about the system. "A single identified metabolite has more meaning than scores of unidentified signals."

In this chapter, we focus mainly on NMR-based metabolomic analysis, including five general practical steps: sample harvesting, sample preparation (grounding, drying, extraction), data acquisition (analysis based on MS or NMR spectrometry), data processing including multivariate data analysis, and identification of metabolites [6, 7].

2 Materials

The materials and reagents described here are required for an NMR-based metabolomics. Prepare all solutions with deuterated solvents.

2.1 Sample Preparation Components (Harvesting and Drying)

1. Dewar barrels.
2. Liquid nitrogen (*see* **Note 1**).
3. Freezer (−80 °C) for sample storage.
4. Mortar and pestle.
5. Plastic tubes.
6. Freeze-dryer for sample drying.

2.2 Extraction Components (See Note 2)

1. Phosphate buffer (90 mM, pH 6.0): 1.232 g KH_2PO_4, 10 mg TMSP-d_4 in 100 mL D_2O (*see* **Note 3**).
2. 1.0 M NaOD.
3. 3-(trimethylsilyl)propionic-2,2,3,3-d_4 acid sodium salt (TSP), 99 at.% Deuterium oxide (D_2O) >99.9 at.% D.
4. Methanol-d_4 (CD_3OD) 99.8 %.
5. Sodium deuteroxide (NaOD) 99.5 % (40 % in D_2O).

2.3 Equipment

1. Ultrasonicator operating at a high frequency (≥2 MHz).
2. Microcentrifuge.
3. 2 mL Eppendorf tubes.
4. pH meter with electrode.
5. 5 mm NMR tubes.

3 Methods

3.1 Harvesting Plant Material

1. Prepare container filled with liquid nitrogen.
2. Isolate desired part (leaf, root, flower, etc.) from plant and transfer to plastic bag.
3. Keep plastic bag in the liquid nitrogen (*see* **Note 4**).

3.2 Sample Preparation

The next step after harvesting the plants is to prepare the plant material for the extraction procedures. The sample preparation steps include grounding, drying, and weighing.

1. Precool the pestle and mortar.
2. Place harvested plant material in the pestle and grind using mortar under liquid nitrogen.
3. Transfer powdered material to tubes (*see* **Note 5**).
4. Place them in the freeze drier for 1–2 days (*see* **Note 6**).
5. Weight samples precisely and transfer to tubes for extraction (*see* **Note** 7).

3.3 Extraction

3.3.1 Two-Phase Extraction Method for NMR Analysis (See Note 8)

1. Add 4 mL of chloroform, 4 mL of methanol:water (1:1) to 100 mg of dried plant material.
2. Vortex them for 30 s.
3. Ultrasonicate for 5 min and centrifuge at 14,000–17,000 × *g* for 20 min at 4 °C.
4. Take upper phase of methanol/water fraction (polar metabolites) and lower phase of chloroform/methanol fraction (non-polar metabolites).
5. Evaporate both fraction using rotary evaporator and add 1 mL of chloroform-*d* to chloroform–methanol fraction and 1 mL of methanol-d_4 and D_2O (KH_2PO_4 buffer, pH 6.0) to methanol–water fraction for NMR measurement.

3.3.2 Single Solvent Extraction Method for NMR Metabolomics (See Note 9)

1. Add 0.75 mL of methanol-d_4 and 0.75 mL of KH_2PO_4 buffer (pH 6.0) in D_2O containing 0.01 % trimethylsilylpropionic acid sodium salt (TMSP, w/w) solution to 50–100 mg of plant materials.
2. Vortex for 1 min at room temperature.

3. Ultrasonication for 10–20 min at room temperature.
4. Centrifuge at 14,000–17,000 × *g* for 15 min at room temperature until obtain a clear supernatant.
5. Transfer 800 μL supernatant to NMR tube (*see* **Note 10**).

3.4 Data Acquisition Using NMR Spectrometry (See Note 11)

1. Load NMR tubes inside the magnet of the NMR spectrometer.
2. Set the sample temperature to 298 K (25 °C) and leave for a few minutes for thermal equilibration.
3. Tune and match the NMR tube.
4. Lock the spectrometer frequency to the deuterium resonance arising from the NMR solvents.
5. Shim the sample using either manual or automated method.
6. Start acquisition with 64 or 128 scan numbers, which takes 5–10 min.

3.5 Data Processing and Multivariate Analysis

1. Make a Fourier transformation of FID with LB = 0.3 Hz.
2. Manually phase the spectra and correct the baseline.
3. Calibrate the spectra using solvent signal or internal standard (*see* **Note 12**) (Fig. 1).

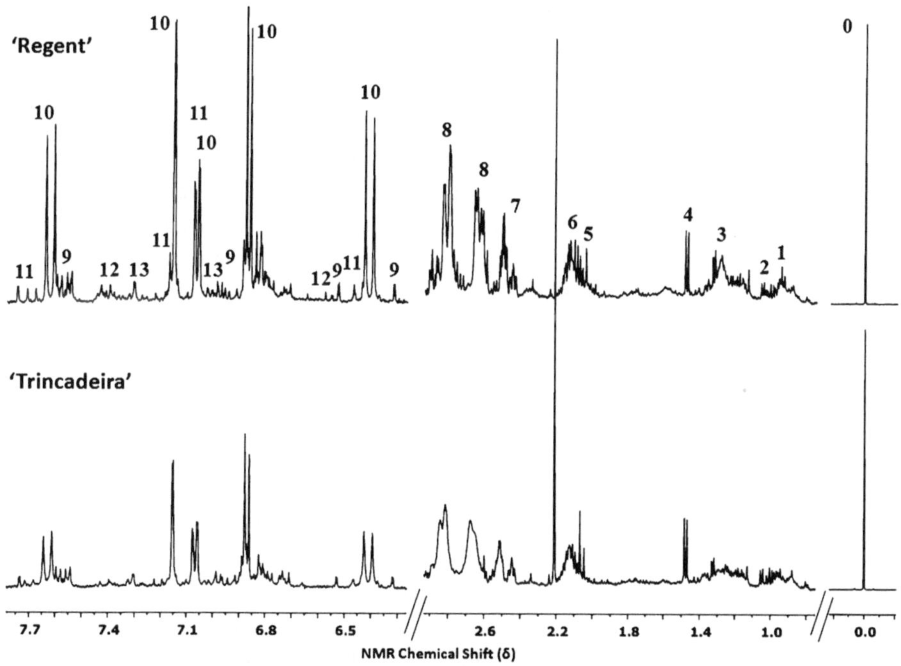

Fig. 1 Example of ^{1}H NMR spectra. Two different grape cultivars "Trincadeira" and "Regent" were analyzed after 48 h of pathogen inoculation. Numbers indicate major signals. *0*: TMSP (internal standard), *1*: leucine, *2*: valine, *3*: threonine, *4*: alanine, *5*: glutamate, *6*: proline and methionine, *7*: glutamine, *8*: malate, *9*:quercetin glucoside, *10*: caffeoyl moiety, *11*: feruloyl moiety, *12*: myricetin, *13*: Tyrosine (adapted from ref. [14] after permission)

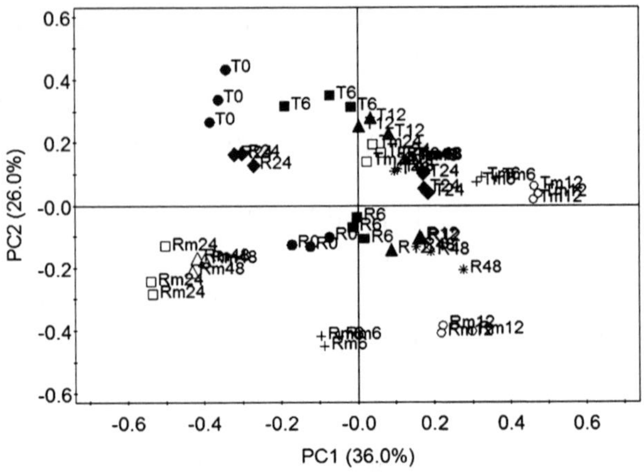

Fig. 2 Typical PCA score plot of two grape cultivars "Trincadeira" and "Regent." T and R represent each cultivar and number indicates time after inoculation with *Plasmoparaviticola* sporangia (adapted from ref. [14] after permission)

4. Convert NMR spectra to a suitable form for further multivariate analysis (*see* **Note 13**).
5. Perform multivariate analysis (PCA and PLS-DA) using SIMCA-P software or equivalent softwares (*see* **Note 14**) (Fig. 2).

3.6 Identification of Metabolites

1. Interpret the loading plots in case of PCA or PLS-DA to find the discriminating metabolites in the given samples.
2. Perform column chromatography if necessary to identify minor metabolites from the samples.
3. Analyze them using 2D NMR spectroscopy such as ^{1}H *J*-resolved spectroscopy, ^{1}H–^{1}H COSY (COrrelation SpectroscopY) or HMBC (Heteronuclear Multiple Bond Correlation).

3.7 Applications of Metabolomics in the Plant Biodiversity

NMR-based metabolomics has been successfully applied in many different fields such as chemotaxonomy, classification and quality control of (medicinal) plants as well as in the study of plant physiology and ecology. Using the protocols mentioned above, we have performed various studies and it has been well documented in several reviews [3, 5–8]. Some of the examples are as below.

3.8 Chemotaxonomy and Classification of Plants

Chemotaxonomy study or classification can be an important issue to differentiate plants species or to identify adulterants. Using NMR-based metabolomics, it was able to differentiate 11 species of Ilex which has been used throughout South America [9, 10].

3.9 Plant Physiology or Ecology

Plants responded chemically when they are under stress or infected with other organisms. It was well demonstrated in the study of Arabidopsis, Brassica, Senecio, and Grape [11–14].

Interaction with other organisms, Senecio with thrips (*Frankliniella occidentalis*) for instance, showed that changes in metabolism play an important role in the resistance of plants [13].

3.10 Quality Control of Medicinal Plants

Most of current quality controls are limited to detect one or two major active metabolites. Many medicinal plants, however, are used as crude extracts or plant powders, which means examine their metabolome is important to guarantee their pharmacological efficacy and safety. NMR-based metabolomics has been applied for the quality control of medicinal plants including Cannabis, Ephedra, Strychnos, and Ginseng [15–18].

4 Notes

1. Liquid nitrogen should be handled carefully and gloves and glasses should be used for protection. Dry ice can also be used to keep the sample material frozen during the harvesting.
2. There are different methods of extraction and the method to be chosen should be designed to detect as many metabolites as possible and meet the specific requirements of the analytical method to be used.
3. Adjustment of pH can be done after the stirring of the solution until total dissolution. Use 1.0 M NaOD.
4. To prevent the formation of new metabolites, as well as enzymatic and oxidative reactions the collection should be conducted very delicately and rapidly, and placing it in a liquid nitrogen tank or dry ice should immediately freeze the plant material. Harvested samples can be kept in a freezer (−80 °C) for weeks or months before extraction.
5. These samples can be kept in a freezer (−80 °C) for weeks or months before extraction.
6. In freeze-drying, frozen water in the sample sublimes under low pressure, passing directly from solid to gas phase. It is a rapid and mild method compared with other methods that use heat to evaporate and eliminate the water [19]. After drying, the samples these should be kept in a cold dark room to avoiding further enzymatic reactions.
7. The desirable amounts for NMR analysis is around 100–500 mg of fresh material, or 50–100 mg of freeze-dried, however less can be used depending on the sample types.
8. The extraction procedure should be repeated at least twice, to increase the yield extraction. Although this method of extraction allows to extract a broad range of polarities of compounds in both polar and nonpolar solvent, it is laborious, difficult to handle many samples, and some compounds can be lost during

the procedure due to degradation or loss occurring in the long preparation time.

(a) For the metabolomics analysis by NMR the buffered mixture methanol:water 1:1 coupled to sonication has been extensively used [6, 11–13] and has shown to extract a broad range of primary and secondary metabolites. By using these solvents in deuterated form, extracts can be measured directly after extraction. To avoid shifts due to differences in pH a buffered solvent (KH_2PO_4) should be used (pH 6.0).

9. Extract can be kept for few days in the cold dark room (0–4 °C) before NMR analysis. However, it is recommended to place at room temperature at least half an hour before NMR measurement to avoid poor shimming owing to the temperature difference of the samples.
10. All these steps are set up in the automation system, but it is recommended to do the first sample manually in order to obtain good resolved spectra.
11. We usually calibrate NMR spectra using TSP signal as δ 0.00. TSP is a reference for calibration of NMR shift and can also serve as a reference (internal standard) in the quantification of metabolites.
12. At this stage, signals of remaining solvents have to be removed for the statistical analysis. AMIX software (Bruker Biospin GmbH) is commonly used for converting spectra to an ASCII file. In this step, the peak is integrated into a small bin (bucket) the size of which is defined by the user. The size is preferably 0.04 ppm in order to avoid the effect of signal fluctuation due to pH or concentration. The other softwares available are MestReNova (v 7.02. program 2011 Mestrelab Research S.L), a MATLAB tool box (ProMetab for Bruker format), NMR suite (Chenomx) for Bruker, JEOL, or Varian format, and ACD NMR Manager (Advanced Chemistry Development, Toronto, Ontario, Canada).
13. Several commercial software programs for multivariate analysis are available such as AMIX-TOOLS (Bruker Biospin GmbH), Matlab with statistics toolbox (MathWorks, Natick, MA, USA), Minitab (Minitab, State College, PA, USA), Pirouette (Infometrix, Bothell, WA, USA), SIMCA-P (Umetrics, Umeå, Sweden), SPSS (SPSS, Chicago, IL, USA), or Unscrambler (CAMO software, Woodbridge, NJ, USA).
14. As the first step of identifying metabolites, several database can be used such as NMR chemical shift database for natural products (http://nmrshiftdb.pharmazie.uni-marburg/de/nmrshiftdb/), for pulse sequences of 2D-NMR and illustration of their mechanism (http://www.chem.queensu.ca/FACILITIES/NMR/nmr/webcourse/), for commercial ^{1}H

NMR Database of metabolites (ca. 250 metabolites) Chenomx Profiler (NMR suite) (*see* http://www.chenomx.com). Some books such as comprehensive NMR pulse sequences (S. Berger, S. Braun. 2004. in 200 and More NMR Experiments. Wiley-VCH) are useful. Other databases that might be potentially useful for metabolite identification using NMR spectra include the NMRShiftDB (www.ebi.ac.uk/NMRshiftdb/), the Spectral Database for Organic Compounds (SDBS: www.riodb01.ibase.aist.go.jp), and the BioMagResBank (www.bmrb.wisc.edu). The ^{1}H NMR data of common metabolites found in plant are listed in Table 1.

Table 1
Most common metabolites found in plants by nuclear magnetic resonance (NMR)

Group	Metabolites	Selected characteristic signals in NMR
Amino acid	Alanine	δ 1.48 (H-3, *d*, *J*=7.2 Hz)
	Asparagine	δ 2.8 (*m*), δ 2.97 (*m*)
	Glutamate	δ 2.07 (H-2, *m*), δ 2.36 (H-3, *m*)
	Glutamine	δ 2.12 (H-2, *m*), δ 2.48 (H-3, *m*)
	Leucine	δ 0.96 (H-5, *d*, *J*=8.0 Hz)
	Proline	δ 4.08 (H-2, *dd*, *J*=8.6, 6.4 Hz), δ 2.34 (H-3, *m*)
	Threonine	δ 1.32 (H-4, *d*, *J*=6.6 Hz)
	Tryptophan	δ 7.20 (H-5′, *t*, *J*=7.4 Hz), δ 7.29 (H-6′, *t*, *J*=7.5 Hz), δ 7.32 (H-2′, *s*) δ 7.54 (H-7′, *d*, *J*=8.1 Hz), δ 7.73 (H-4′, *d*, *J*=7.9 Hz)
	Tyrosine	δ 7.19 (H-2,H-6, *t*, *J*=8.5 Hz), δ 6.68 (H-3, H-5, *t*, *J*=7.5 Hz)
	Valine	δ 1.00 (H-4, *d*, *J*=7.8 Hz), δ 1.06 (H-5, *d*, *J*=7.8 Hz)
Carbohydrates	α-Glucose	δ 5.18 (H-1, *d*, *J*=3.8 Hz)
	β-Glicose	δ 4.58 (H-1, *d*, *J*=7.8 Hz)
	Fructose	δ 4.17 (H-1, *d*, *J*=9.0 Hz)
	Sucrose	δ 5.40 (H-1, *d*, *J*=3.8 Hz), δ 4.17 (H-1′, *d*, *J*=8.5 Hz)
	α-Rhamnose	δ 5.10 (H-1, *d*, *J*=2 Hz), δ 1.27 (H-6, *d*, *J*=6 Hz)
	β-Rhamnose	δ 4.85 (H-1′, br.*d*, *J*<2 Hz), δ 1.30 (H-6, *d*, *J*=5.5 Hz)
	L-Fucose	δ 4.20 (H-1, *d*, *J*=7.9 Hz)
Organic acid	Citric acid	δ 2.74 (H-2, *d*, *J*=17.6 Hz), δ 2.56 (H-2′, *d*, *J*=17.6 Hz)
	Formic acid	δ 8.46 (s)
	Fumaric acid	δ 6.56 (s)
	γ-Amino-butyrate (GABA)	δ 1.90 (H-3, *m*), δ 2.30 (H-2, *t*, *J*=7.2 Hz), δ 3.01 (H-4, *t*, *J*=7.5 Hz)
	Malic acid	δ 4.34 (H-2, *dd*, *J*=6.6,4.7 Hz), δ 2.74 (H-3, *dd*, *J*=16.6,4.7 Hz), δ 2.68 (H-2, *dd*, *J*=16.6, 6.6 Hz)
	Succinic acid	δ 2.56 (*s*)

(continued)

Table 1 (continued)

Group	Metabolites	Selected characteristic signals in NMR
Phenylpropanoids/ flavonoids	Kaempferol	δ 8.04 (H-2′, H-6′, *d*, *J*=8.6 Hz), δ 6.74 (H-3′, H-5′, *d*, *J*=8.6 Hz), δ 6.52 (H-8, *d*, *J*=2.0 Hz), δ 6.28 (H-6,*d*, *J*=2.0 Hz)
	(+)-Catechin	δ 2.49 (H-4, *dd*, *J*=16.1. 8.2 Hz), δ 2.83 (H-4, *dd*, *J*=16.0, 5.4 Hz), δ 4.04 (H-3, *m*), 4.55 (H-2, *d*, *J*=7.5 Hz), δ 5.91(H-6, *d*, *J*=2.2 Hz), δ 6.75 (H-8, *d*, *J*=8.0 Hz)
	(−)-Epicatechin	δ 2.72 (H-4, *dd*, *J*=16.8. 2.8 Hz), δ 2.85 (H-4, *dd*, *J*=16.7, 4.6 Hz), δ 5.91(H-6, *d*, *J*=10.0, 2.3 Hz), δ 6.96 (H-8, *d*, *J*=2.2 Hz)
	Quercetin	δ 7.71 (H-2′, *d*, *J*=2.0 Hz), δ 7.66 (H-6′, *dd*, *J*=8.6, 2.0 Hz), δ 6.99 (H-5′, *d*, *J*=8.6 Hz), δ 6.49 (H-8, *d*, *J*=1.8 Hz), δ 6.28 (H-6, *d*, *J*=1.8 Hz)
	Quercetin 3-*O* rhamnoside	δ 7.32 (H-2′, *d*, *J*=2.0 Hz), δ 7.27 (H-6′, *dd*, *J*=8.0, 2.0 Hz), δ 6.89 (H-5′, *d*, *J*=8.0 Hz), δ 6.97 (H-8, *d*, *J*=1.8 Hz), δ 6.27 (H-6, *d*, *J*=1.8 Hz)
	Gallic acid	δ 7.03 (s)
	p-Benzoic acid	δ 6.83 (H-3,H-5, d, *J*=8.7 Hz), δ 7.94 (H-2, H-4, d, *J*=8.8 Hz)
	Rutin	δ 7.66 (H-2′, *d*, *J*=2.0 Hz), δ 7.62 (H-6′, *dd*, *J*=8.6, 2.0 Hz), δ 6.87 (H-5′, *d*, *J*=8.6 Hz), δ 6.39 (H-8, *d*, *J*=1.8 Hz), δ 6.21 (H-6, *d*, *J*=1.8 Hz), δ 5.02 (H-1′, *d*, *J*=7.7 Hz), δ 4.54 (H-1′, *d*, *J*=1.5 Hz), δ 1.12 (H-6, *d*, *J*=6.22 Hz),
	p-Coumaric acid	δ 6.38 (H-8′, *d*, *J*=16.0 Hz), δ 6.84 (*d*, *J*=8.8), δ 7.50 (*d*, *J*=8.8), δ 7.59 (H-7′, *d*, *J*=16.0)
	Caffeic acid	δ 6.24 (H-8′, *d*, *J*=16.0 Hz), δ 6.87 (*d*, *J*=8.8), δ 7.02 (*dd*, *J*=8.8,2.0 Hz), δ 7.12 (H-7′, *d*, *J*=2.0), δ 7.52 (H-7′, *d*, *J*=16.0)
	Ferulic acid	δ 7.56 (H-7′, *d*, *J*=15.9 Hz), δ 7.19 (H-2′, *d*, *J*=2.1 Hz), δ 7.10 (H-6′, *dd*, *J*=8.4, 2.1 Hz), δ 6.88 (H-5′, *d*, *J*=8.4), δ 7.15 (*d*, *J*=2.8), δ 6.33 (H-8′, *d*, *J*=15.9),
	Chlorogenic acid	δ 7.60 (H-7′, *d*, *J*=15.9 Hz), δ 7.15 (H-2′, *d*, *J*=2.1 Hz), δ 7.05 (H-6′, *dd*, *J*=8.4, 2.1 Hz), δ 6.87 (*d*, *J*=8.4), δ 6.36 (H-8′, *d*, *J*=15.9), *d* 5.33 (H-3. t*d*, *J*=10.0, 4.8), δ 4.21 (H-5, br.*q*, *J*=3.1)
	Sinapic acid	δ 7.48 (H-7′, *d*, *J*=16.0 Hz), δ 6.93 (H-2, H-6, s, *J*=8.8), δ 6.37 (H-8, *d*, *J*=16.0 Hz)
Fatty acids	Common fatty acid	δ 0.88 (H-ω, *t*, *J*=7.5 Hz), δ 1.30 (CH_2, br.*s*)
	α-Linolenic acid	δ 0.97 (H-ω, *t*, *J*=7.5 Hz)
Other compounds	Choline	δ 3.24(*s*)
	Adenine	δ 8.2 (H-3, *s*), δ 8.21 (H-1, *s*)
	Inositol	δ 4.00 (H-2, *t*, *J*=2.8 Hz), δ 3.61 (H-6, H-6, *t*, *J*=9.9 Hz), δ 3.44 (H-1, H-3, *dd*, *J*=9.9,2.9 Hz), δ 3.24 (H-5, *t*, *J*=9.3 Hz)

Acknowledgments

This work was partly supported by National Institute of Horticultural and Herbal Science, Rural Development Administration of Korea (Project No: PJ008700).

References

1. Verpoorte R, Choi YH, Kim HK (2007) NMR-based metabolomics at work in phytochemistry. Phytochem Rev 6:3–14
2. Schripsema J (2010) Application of NMR in plant metabolomics: techniques, problems and prospects. Phytochem Anal 21:14–21
3. Verpoorte R, Choi YH, Mustafa NR, Kim HK (2008) Metabolomics: back to basics. Phytochem Rev 7:525–537
4. Ebbels TMD, Cavill R (2009) Bioinformatic methods in NMR-based metabolic profiling. Prog Nucl Magn Reson Spectrosc 55:361–374
5. Verpoorte R, Choi YH, Kim HK (2010) Metabolomics: will it stay? Phytochem Anal 21:2–3
6. Kim HK, Choi YH, Verpoorte R (2010) NMR-based metabolomic analysis of plants. Nat Protoc 5:536–549
7. Kim HK, Choi YH, Verpoorte R (2011) NMR-based plant metabolomics: where do we stand, where do we go? Trends Biotechnol 29:267–275
8. Van der Kooy F, Maltese F, Choi YH, Kim HK, Verpoorte R (2009) Quality control of herbal material and phytopharmaceuticals with MS and NMR based metabolic fingerprinting. Planta Med 75:763–775
9. Choi YH, Sertic S, Kim HK, Wilson EG, Michopoulosa F, Lefeber AWM, Erkelens C, Verpoorte R (2005) Classification of Ilex species based on metabolomic fingerprinting using nuclear magnetic resonance and multivariate data analysis. J Agric Food Chem 53: 1237–1245
10. Kim HK, Saifullah K, Wilson EG, Kricun SDP, Meissner A, Goraler S, Deelder AM, Choi YH, Verpoorte R (2010) Metabolic classification of South American *Ilex* species by NMR-based metabolomics. Phytochemistry 71:773–784
11. Hendrawati O, Yao Q, Choi YH, Kim HK, Linthorst HJM, Erkelens C, Lefeber AWM, Verpoorte R (2006) Metabolic differentiation of *Arabidopsis* treated with methyl jasmonate using nuclear magnetic resonance spectroscopy. Plant Sci 170:1118–1124
12. Jahangir M, Kim HK, Choi YH, Verpoorte R (2008) Metabolomic response of *Brassica rapa* submitted to pre-harvest bacterial contamination. Food Chem 107:362–368
13. Leiss KA, Choi YH, Abdel-Farid IB, Verpoorte R, Klinkhamer PGL (2009) NMR metabolomics of thrips (*Frankliniella occidentalis*) resistance in *Senecio* hybrids. J Chem Ecol 35: 219–229
14. Ali K, Maltese F, Figueiredo A, Rex M, Fortes AM, Zyprian E, Pais MS, Verpoorte R, Choi YH (2012) Alterations in grapevine leaf metabolism upon inoculation with *Plasmopara viticola* in different time-points. Plant Sci 191–192:100–107
15. Choi YH, Kim HK, Hazekamp A, Erkelens C, Lefeber AW, Verpoorte R (2004) Metabolomic differentiation of *Cannabis sativa* cultivars using ^{1}H NMR spectroscopy and principal component analysis. J Nat Prod 67:953–957
16. Kim HK, Choi YH, Erkelens C, Lefeber AWM, Verpoorte R (2005) Metabolic fingerprinting of *Ephedra* species using ^{1}H-NMR spectroscopy and principal component analysis. Chem Pharm Bull 53:105–109
17. Frédérich M et al (2004) Metabolomic analysis of *Strychnos nux-vomica*, *icaja* and *ignatii* extracts by ^{1}H nuclear magnetic resonance spectrometry and multivariate analysis techniques. Phytochemistry 65:1993–2001
18. Yang SY, Kim HK, Lefeber AWM, Erkelens C, Angelova C, Choi YH, Verpoorte R (2006) Application of two dimensional nuclear magnetic resonance spectroscopy to quality control of ginseng commercial products. Planta Med 72:364–369
19. Kim HK, Verpoorte R (2010) Sample preparation for plant metabolomics. Phytochem Anal 21:4–13

Chapter 10

Extraction Protocol for Nontargeted NMR and LC-MS Metabolomics-Based Analysis of Hard Coral and Their Algal Symbionts

Benjamin R. Gordon, William Leggat, and Cherie A. Motti

Abstract

Metabolomics and in particular, nontargeted metabolomics, has become a popular technique for the study of biological samples as it provides considerable amounts of information on extractable metabolites and is ideal for studying the metabolic response of an organism to stressors in its environment. One such organism, the symbiotic hard coral, presents its own complexity when considering a metabolomics approach in that it forms intricate associations with an array of symbiotic macro- and microbiota. While not discounting the importance of these many associations to the function of the coral holobiont, the coral-*Symbiodinium* relationship has been the most studied to date and as such, is the primary focus of this extraction protocol. This protocol provides details for the sample collection, extraction, and measurement of hard coral holobiont metabolites using both ^{1}H nuclear magnetic resonance (NMR) spectroscopy and liquid chromatography coupled with mass spectrometry (LC-MS). Using this nontargeted metabolomics approach, the holobiont metabolism can be investigated for perturbations resulting from either (1) natural or anthropogenic environmental challenges, (2) the controlled application of stressors, and (3) differences between phenotypes or species. Consequently, this protocol will benefit both environmental and natural products based research of hard coral and their algal symbionts. Every effort has been made to provide the reader with all the details required to perform this protocol, including many of the costly and time consuming "pitfalls" or "traps" that were discovered during its development. As a result, this protocol can be confidently accomplished by those with less experience in the extraction and analysis of symbiotic hard coral, requiring minimal user input whilst ensuring reproducible and reliable results using readily available lab ware and reagents.

Key words Metabolomics, *Symbiodinium*, Coral, LC-MS, NMR, Extraction

1 Introduction

The coral holobiont is a complex symbiotic organism (Fig. 1) that consists of the animal host and a plethora of intracellular and extracellular macro- and microbiota, such as fish, crustaceans, polychaetes, dinoflagellates, prokaryotes, viruses, fungi, archaea, and endolithic algae [1–3]. While many of the these have mutualistic associations with the coral host, the most well known and studied of them is that

Ute Roessner and Daniel Anthony Dias (eds.), *Metabolomics Tools for Natural Product Discovery: Methods and Protocols*, Methods in Molecular Biology, vol. 1055, DOI 10.1007/978-1-62703-577-4_10, © Springer Science+Business Media, LLC 2013

Fig. 1 A hard coral branch tip belonging to the genus *Acropora*. Areas of coral tissue that have a high abundance of the algal symbiont, *Symbiodinium spp.*, can be seen by the dark marbled patterns they form. Photo credit by João Paulo Krajewski

of the photosynthetic dinoflagellate, *Symbiodinium spp.* Members of this diverse genus can live freely in the water column or sediment [4] and also in symbiosis with a number of marine invertebrates, such as coral, giant clams, and anemones [5]. This symbiosis allows the animal host to meet part or all of its carbon requirements through autotrophy, thereby gaining a competitive advantage through increased fitness [6, 7]. Likewise, the relationship enables the bilateral exchange of metabolites, including metabolites that are not produced solely by either organism [8, 9].

The success of this symbiosis is based upon the invertebrate host supplying inorganic nutrients to the algal symbiont, which are returned as organic compounds and used to supplement the host's energy and nutritional demands [10]. In the coral symbiosis, *Symbiodinium* are found in a host-derived vacuole (symbiosome membrane) within the gastrodermal cell layer, which forms during the acquisition and division of the algal symbiont. The symbiosome membrane closely resembles those in legumes where the plant membrane encloses the symbiotic rhizobium cells [11, 12]. Consequently, the exchange of all organic and inorganic nutrients must proceed through this cell membrane and thus it is critical to the metabolic interaction between the symbiont and host.

Of the metabolites involved in the algal-invertebrate symbiosis not all are involved in nutritional roles. Compounds such as mycosporine-like amino acids (MAAs) play an important role in the protection of coral from ultraviolet light, in addition to acting as free radical scavengers [13]. Likewise, dimethylsulfoniopropionate and its breakdown products, dimethylsulfide and acrylate, are involved in free radical scavenging [14], osmotic- and cryoprotection [15], and the structuring of microbial communities [16–18]. While the biological role of some of these metabolites has either been identified or closely examined, there are still many metabolites of which the exact function remains unknown. For example, free amino acids and other small compounds may act as "host release factors" that stimulate the production and release of other metabolites from *Symbiodinium* to the host [19–21]. Compounds such as zooxanthellatoxins [22], zooxanthellamides [23], betaines, alkaloids, and ceramides [24] have all been isolated from either *Symbiodinium* or invertebrates in symbiosis with *Symbiodinium*, however, their biological role still remains unclear. On that note, the diversity of metabolites involved in the algal-invertebrate symbiosis, along with the theoretical prospect of changing growth conditions and host factors to manipulate metabolite production, highlights the potential role that metabolomics will have in elucidating metabolites of interest from both hard coral and their algal symbionts for not only the biochemical sciences but also for natural products research.

The extraction protocol presented here was developed with the following criteria in mind. Firstly, the protocol had to be user friendly and as such, should not require any further handling or input from the analyst than was absolutely necessary. The benefits of such an approach were twofold; in addition to reducing the working time, it also ensured that any interference from the analyst that may affect the integrity of a sample was kept to a minimum. The next criterion, maximizing the number of features detected by ^{1}H NMR and LC-MS, was based on the nontargeted metabolomics approach taken. As such, the choice of extraction solvent and analysis conditions played an integral role in fulfilling that requirement and are discussed later in more detail. The last criterion, but by no means the least, was reproducibility. Reproducibility is the key to a successful metabolomics study with the results of statistical analyses, and hence the conclusions that can be drawn from an experiment, being reliant on reproducible data. Consequently, not only were experimental conditions kept constant and similar between samples, sample collection and preparation was also regulated by maintaining a constant handling time and temperature. As such, aspects of the protocol that could utilize multiple techniques (i.e., sample concentration) for achieving the same outcome were thoroughly tested to ensure that the least disruptive technique was employed.

As mentioned previously, the choice of extraction solvent played a crucial role in the development of this protocol. For metabolomics, this choice depends largely on the metabolites of interest, the analytical platform being used and the hypothesis put forward by the researcher. If, for example, one was interested in employing a targeted metabolomics approach, such as the analysis of free amino acids, then in most cases an appropriate solvent system that targets those particular compounds is deemed most suitable. However, it is important to keep in mind that in symbiotic coral, the organism is composed of two distinctly different cell types, animal and algal cells, of which the latter have more robust cell walls than that of the coral host. Consequently, many high polarity solvent systems (i.e., those high in water content and suitable for free amino acid extraction) will not easily permeate the cell walls of the algal symbionts and would therefore require the employment of more time consuming mechanical disruption methods. With that in mind, a biphasic methanol–chloroform extraction that is capable of permeating the algal cell walls [25], while partitioning polar and nonpolar metabolites into two separate phases, would in most cases, be deemed most suitable for the targeted metabolomics analysis of symbiotic coral.

Considering the two distinct cell types of symbiotic coral, the fact that this protocol was developed for a nontargeted metabolomics approach and the array of solvent choices, we examined the effectiveness of five commonly used solvents employed in the extraction of compounds from both plant and animal material. Each extraction was carried out in the same manner as described in the methods using the symbiotic hard coral, *Acropora aspera*. The solvents tested for this protocol were 100 % methanol, 70 % aqueous methanol, 90 % aqueous acetone, 100 % water, and methanol:dichloromethane:water (MeOH:DCM:H_2O) biphasic solvent system. Of the five solvents, the 100 % methanol and 90 % aqueous acetone solvents produced the greatest number of features in both the ^{1}H NMR and LC-MS analyses. However, they also extracted large amounts of lipid, which is detrimental to reversed-phased LC-MS analysis and as such, were eliminated as viable extraction solvents. The 100 % water solvent was eliminated after microscopy analysis of the residual biological matter showed that the algal cells were not effectively lysed. Furthermore, in 100 % water enzymatic and metabolic activity was not halted, affecting the integrity of the sample. While the biphasic MeOH:DCM:H_2O extraction may be suitable for a targeted metabolomics analysis, we found a distinct disadvantage of this technique when applied to the nontargeted metabolomics analysis of symbiotic hard coral. Essentially, too many of the less polar metabolites partitioned into the organic phase of the solvent system, along with the undesirable lipids. This resulted in an aqueous phase that had very few features for both NMR and LC-MS analyses. Of the five extraction solvents examined, the 70 % aqueous methanol solvent displayed excellent reproducibility, extracted minimal amounts of lipid, effectively lysed the algal cells, complied with the proposed criteria, and hence, was considered the most suitable for a nontargeted metabolomics study of hard coral and their algal symbionts.

Metabolomics has become a popular technique for the study of biological samples, which can be attributed to the high number of quality studies published in a diverse number of fields such as environmental science, medicine, and pharmaceuticals. A quality metabolomics study depends largely on the hypotheses put forward by the researcher and a thorough understanding of the scientific principals involved. Consequently, each study needs to be scrutinized depending on the questions being asked. Here, we present an extraction protocol for a nontargeted metabolomics analysis of symbiotic hard coral using LC-MS and ^{1}H NMR. The protocol requires minimal user input and provides reproducible and reliable results using readily available lab ware and reagents. Every effort has been made to provide the reader with all the details required to perform the technique, including many of the costly and time consuming "pitfalls" or "traps" that were discovered during its development. As such, this protocol can be confidently accomplished by those with less experience in the extraction and analysis of symbiotic hard coral and their algal symbionts.

2 Materials

Extraction solvents should be prepared in ultra-clean glassware (*see* **Note 1**) at 25 °C using mass spectrometry (MS) grade solvents (*see* **Note 2**) and ultrapure (MilliQ) water having a resistance of 18 MΩ/cm. Extractions should be performed at approximately 4 °C or cooler (*see* **Note 3**) in ultra-clean glassware or disposable scintillation vials of known purity. High-quality centrifuge tubes are recommended for removing particulates from liquid extracts (*see* **Note 4**). To avoid unwanted contamination, all aspects of the extraction process should be performed while wearing appropriate gloves washed with MilliQ water.

2.1 Coral Collection and Quenching Materials

1. Liquid nitrogen.
2. Disposable foam Dewar.
3. 50 mL plastic centrifuge tubes (*see* **Note 4**).
4. Double-action, Stille-Liston bone cutters for cutting coral into nubbins.
5. Stainless steel forceps or clamps.
6. Liquid nitrogen or dry ice for temporary storage.
7. –80 °C freezer for long-term storage.
8. Nally bin with a tether attached.

2.2 Sample Extraction, Concentration, and Clarification Components

1. 70 % aqueous methanol extraction solvent: Using a clean glass measuring cylinder, measure and combine 70 parts of MS grade methanol to 30 parts of MilliQ water up to the required volume and store in an acid-washed glass Schott bottle or similar at −20 °C until required for use.
2. 15 mL centrifuge tubes (*see* **Note 4**).
3. 20 mL glass scintillation vials (*see* **Note 5**).
4. Dry ice or liquid nitrogen contained in two disposable foam Dewars.
5. Double-action, Stille-Liston bone cutters.
6. Glass Pasteur pipettes and silicon pipette bulb.
7. Freeze dryer.
8. Speed vacuum evaporator.
9. Sonic water bath maintained at 0–4 °C.
10. An appropriate labelling system (*see* **Note 6**).
11. An appropriate rack for upright handling and storage of glass vials in a −80 °C freezer.
12. Centrifuge with a carousel suitable for 15 mL centrifuge tubes.

2.3 LC-MS Materials and Conditions

1. MS certified sample vials (*see* **Note 7**).
2. 99 % formic acid in 1 mL glass ampules (*see* **Note 8**).
3. Superficially porous C_{18} or XB-C_{18} HPLC column and matching guard column.
4. MS grade acetonitrile (*see* **Note 2**).
5. 1,000 μL auto pipette.
6. LC-MS platform: Low resolution mass spectral data were measured on a Bruker Daltonics Esquire 3000 Plus mass spectrometer (ESI-MS) with an Apollo source connected to an Agilent 1100 HPLC system comprising degasser, binary pump, autosampler, and PDA. All LC-MS data was collected using Bruker Daltonics Esquire Control v5.3 and Hystar v3.1 operating on Windows XP Professional.

2.4 NMR Materials

1. NMR sample tubes, 509-UP-7 (*see* **Note 9**).
2. Deuterated methanol (CD_3OD, D 99.8 %) (*see* **Note 10**).
3. 1 mL graduated glass syringe.
4. Lint-free tissue.
5. 7 mL glass scintillation vials (*see* **Note 5**).
6. 1,000 μL auto pipette.
7. NMR spectrometer (^{1}H NMR spectra in this study were collected on a Bruker Avance 600 MHz NMR spectrometer complete with TXI cryoprobe operating at 600 MHz for ^{1}H in CD_3OD, δ_H 3.31 ppm).

3 Methods

3.1 Coral Collection and Metabolism Quenching

Harvesting coral and the quenching method used is dependent on the conditions at the collection site. It is recommended to harvest coral from a sheltered lagoon or reef flat where the collection can be undertaken safely in calm, shallow water that is approximately knee deep. As this method requires small amounts of liquid nitrogen to be taken to the site of collection, care should be taken to ensure the safety of personnel at all times. Generally, 2–3 L of liquid nitrogen in a disposable foam dewar is sufficient to harvest coral nubbins over a 30 min period. The foam dewar containing the liquid nitrogen can be placed in a Nally bin and floated on the water surface so that nubbins can be placed into the liquid nitrogen immediately after they are excised from the colony. This timing is critical as it is well documented in coral that physical interference can result in an immediate chemical response [26].

Collection of hard coral from deeper water, which will often involve the use of SCUBA and a boat, requires a different sampling technique. For this method, we recommend using a hammer and a cold chisel with a 25 mm cutting surface to remove all or part of a colony from the substrate. Whole colonies can then be dissected into smaller nubbins while still submerged just below the water surface and immediately placed into liquid nitrogen that is kept on the boat. Care should be taken to ensure that dive profiles do not involve multiple ascents and descents and as such, all colonies should be brought close to the surface in only one attempt per dive.

For metabolomics, diligent work and a reliable method will produce sound results that in some cases will not require the commonly used techniques of data scaling (i.e., normalization and transformation). In fact, as a means of verifying the robustness of this method and the importance of quickly and effectively quenching the coral metabolism, we compared five stressed coral nubbins (nubbins collected and snap frozen after 30 min of agitation in a bucket of seawater) with five non-stressed coral nubbins (nubbins snap frozen immediately according to this method) using ^{1}H NMR and principal components analysis (PCA). As expected, the results of the PCA show that the stressed nubbins displayed considerably greater variability (Fig. 2). Moreover, scaling of the data matrix rows and/or columns reduced the amount of variability explained by the PCA from 97.96 % in three PCs without scaling, to 97.16 % in four PCs with scaling. The fact that scaling reduced the ability of PCA to explain the variability associated with our data adds considerable weight to the reproducibility of this method.

1. Fill a disposable foam dewar with liquid nitrogen, store, transport, and handle with appropriate care.
2. Place the foam Dewar containing the liquid nitrogen, along with the bone cutters into a tethered Nally bin that is suitable for use as a stable, floating work platform.

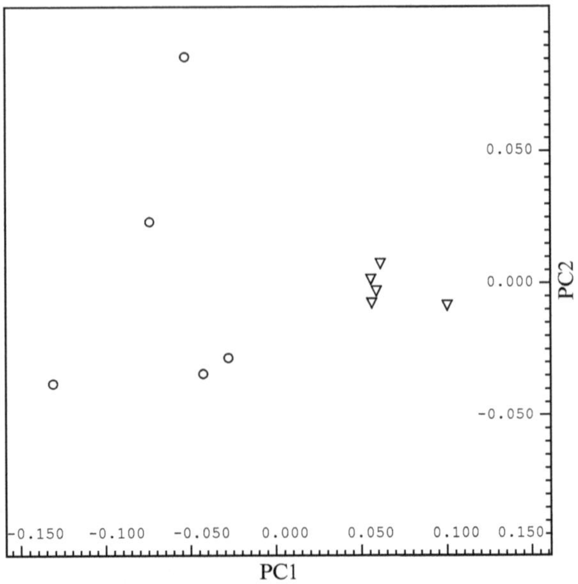

Fig. 2 PCA scores plot of 70 % methanol extracts of stressed (*open circles*) versus non-stressed (*open triangles*) *Acropora aspera* analyzed by ^{1}H NMR. The much tighter grouping of the non-stressed samples highlights the reproducibility obtained by this protocol and the importance of fast and effective quenching of the metabolism

3. Use the double-hinged bone cutters to cut the coral into nubbins of approximately 5 cm in length (*see* **Note 11**).
4. Immediately after cutting, place each nubbin into the foam Dewar containing liquid nitrogen to snap freeze. Nubbins can remain in the liquid nitrogen until ready for transfer into 50 mL centrifuge tubes.
5. Upon returning to shore, remove the coral nubbins from the liquid nitrogen using steel forceps and place immediately into labelled 50 mL centrifuge tubes (*see* **Note 12**).
6. Place the filled tubes back in liquid nitrogen or onto dry ice to keep frozen until ready for transfer into a −80 °C freezer for long-term storage.

3.2 Sample Extraction, Concentration, and Clarification

Given the desire to extract and analyze as many metabolites as possible whilst excluding as many lipid classes as possible for reasons mentioned previously, we recommend starting with the 70 % methanol extraction and then storing the sample at −80 °C in the dark. Using this approach, the lipids, chlorophylls, and other less polar classes of compounds can be examined at a later time by re-extracting the biological material in a less polar solvent. As such, long-term storage conditions are an important consideration if samples are to be analyzed at a later time and it is widely accepted that storing samples at or below −80 °C and in the dark is adequate for halting enzyme or metabolite activity and preserving the integrity of samples [27]. It should be noted, however, that storing samples above −25 °C is not adequate for performing the same task [26, 28, 29].

While the reproducibility of a method is influenced by the choice of extraction solvent, other factors such as sample concentration, extraction temperature, sample state before extraction (i.e., wet or dried sample), collection methods, sample storage, and sample handling (i.e., the amount of time kept at room temperature for analysis) can also have a significant effect on the outcome of a particular experiment. With regards to the extraction temperature, sample storage, and handling, it is largely a case of maintaining constant and similar conditions for all samples while reducing exposure to elevated temperatures that may cause sample degradation. However, where a number of choices are available, as is the case for sample concentration (speed vacuum, lyophilization, or nitrogen stream) and sample state prior to extraction (wet or lyophilized sample), it is appropriate to consider each option with respect to its potential to minimize sample degradation and data variability. With that in mind, we used ^{1}H NMR and PCA to examine the variability associated with three different solvent removal techniques (speed vacuum, lyophilization, and nitrogen stream at 25 °C) and the extraction of either lyophilized or wet coral nubbins. Of the three sample concentration techniques, drying under a stream of nitrogen gas at 25 °C introduced the greatest variability, while lyophilization and speed vacuum introduced similar levels of variability. As such, drying under a nitrogen stream at 25 °C was not considered an appropriate sample concentration technique. Furthermore, the analysis of extractions performed on wet and lyophilized samples showed that the extraction of wet sample was less reproducible, most likely due to the greater differences in water and salt composition between wet and dried samples. Consequently, we recommend that extractions are performed on lyophilized coral and that a combination of lyophilization and speed vacuum are used for sample concentration.

1. Fill two, appropriately sized, disposable foam Dewar's with dry ice or liquid nitrogen. Place the bone cutters and labelled 20 mL scintillation vials in the first foam Dewar to cool and place the 50 mL centrifuge tubes containing the coral nubbins collected in Subheading 3.1, **step 4** into the second Dewar.
2. Using the chilled bone cutters, cut each coral nubbin into several 1 cm^3 pieces so that each nubbin fits into the 20 mL glass scintillation vials (*see* **Note 13**). Remove any ice that may have dislodged from the nubbin when cutting (*see* **Note 14**). Return vials to the dry ice to keep frozen until lyophilization.
3. Undo each lid of the cold vials containing the frozen coral nubbins a quarter of a turn to ensure the atmosphere within

each vial can escape when subjected to the vacuum associated with the lyophilization. Vials should be placed upright in a suitably sized rack or tray and placed into a freeze drier for 24 h or until completely dry (*see* **Note 15**).

4. After lyophilization, fill each scintillation vial containing the coral nubbin pieces with cold (0–4 °C) 70 % methanol to approximately 75 % of its volume, ensuring that the coral nubbin is completely submersed in the extraction solvent.
5. Sonicate each vial for 5 min in a sonication bath chilled to 0–4 °C (*see* **Note 16**).
6. Decant the extract into 15 mL centrifuge tubes (*see* **Note 4**) and centrifuge at 5,800 × *g* for 5 min to settle any particulates still present in the extract solvent.
7. Using a glass Pasteur pipette remove the supernatant and transfer into clean, pre-weighed, 20 mL glass scintillation vials (*see* **Note 17**).
8. Remove the lids of each vial and place the vials containing the 70 % methanol extracts into a speed vacuum and dry at ambient temperature for ~1 h (*see* **Note 18**).
9. Remove the samples from the speed vacuum and freeze the remaining sample in liquid nitrogen or by placing in a −80 °C freezer for 1 h.
10. Undo the lids of each vial one-quarter of turn and place the frozen samples into a freeze drier and lyophilize to complete dryness.
11. Remove the samples from the freeze drier and tighten caps immediately to avoid any absorption of water.
12. Weigh each vial and calculate the extract weight by subtracting the empty weight of each vial from the final weight (*see* **Note 19**).
13. Resuspend each of the dried extracts in an appropriate amount of 70 % methanol, ensuring that each extract is made up to a uniform concentration (*see* **Note 20**).
14. Store samples at −80 °C until ready for analysis.

3.3 LC-MS Analysis

It is important to consider the optimization of the analytical platforms, which must be done on a *case-by-case* basis. While it is not pertinent to discuss in detail the optimization of our own LC-MS instrument and its associated peripheries, the relevant details are provided. In particular, it is worth mentioning some of the details related to our choice of column and mobile phase, along with the parameters and conditions used for our own LC-MS analysis. With regards to the chromatography, we tested the ability of three superficially porous columns to separate the 70 % methanol extract (20 mg/mL) of symbiotic hard coral using two different acidified organic mobile phases, 0.1 % formic acid in methanol, and 0.1 % formic acid in acetonitrile. Columns were rated based on

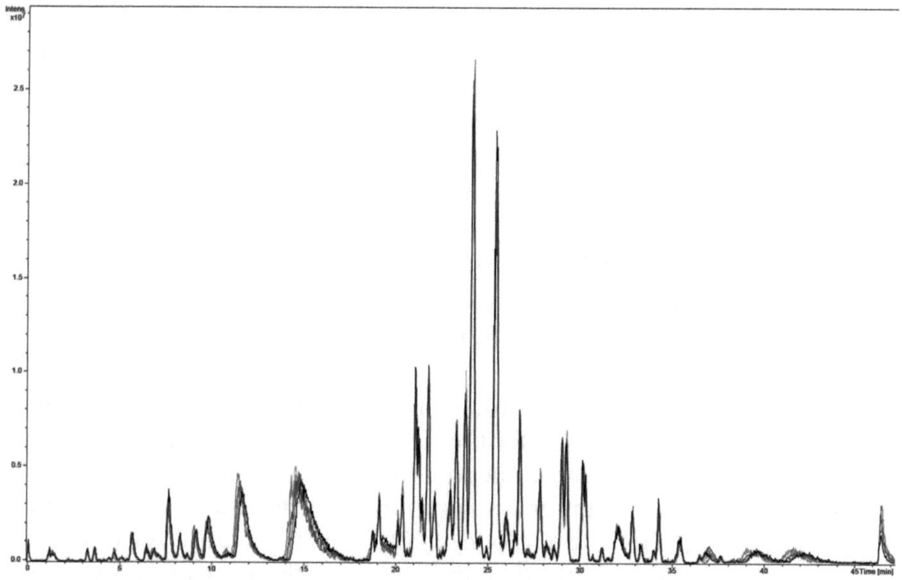

Fig. 3 Overlaid LC-MS base peak chromatograms of five replicate 70 % methanol extracts of *Acropora aspera*

peak separation, resolution, and the number of features within each chromatogram. Of the three different columns and two different mobile phases, the XB-C_{18} column and the 0.1 % formic acid in acetonitrile mobile phase gave reproducible chromatograms with the best resolution, peak separation, and the greatest number of features (Fig. 3).

1. Remove the samples prepared according to Subheading 3.2 from the −80 °C storage and store at 4 °C for 1 h to defrost.
2. Using a 1,000 μL autopipette and clean pipette tips, place 1 mL of each sample into new 1.5 mL centrifuge tubes and centrifuge each sample at 4 °C for 5 min at 21,900×*g* to ensure that no particulates remain in the sample (*see* **Note 21**).
3. Transfer 100 μL of each sample into a new reduced volume HPLC sample vial or one containing a reduced volume insert (*see* **Note 22**). Return the remaining sample back to −80 °C storage.
4. Store HPLC vials in the dark at −80 °C until ready for analysis.
5. Perform the LC-MS analysis. Conditions described in **steps 6** and 7 can be used as an initial starting point or guide (*see* **Note 23**).
6. Chromatography conditions: An XB-C_{18} superficially porous (Phenomenex, Kinetex 3×100 mm, 2.6 μm) column was mated with a 0.5 μm stainless steel filter and guard cartridge of the same stationary phase, 2 μL injection volume, mobile phase A (0.1 % formic acid in water), mobile phase B (0.1 % formic acid in acetonitrile), gradient elution from 50 % A to 100 % B

at a flow rate of 350 μL/min for 35 min. After 35 min 100 % B was maintained for 7.1 min at the same flow rate. At 42.1 min the mixture was changed to its initial setting (50 % A, 50 % B) and the column equilibrated until 47.1 min.

7. MS Acquisition parameters: ESI ion source, positive ion polarity, scan range from 50 to 1,000 *m/z*, capillary exit of 120 V, accumulation time of ~8,000 μs, Drying temperature of 350 °C, Nebulizer pressure of 32 psi, Drying gas flow of 8 L/min.
8. MS processing parameters: Base peak chromatograms were calculated with background removal.

3.4 ^{1}H NMR Analysis

It is not within the scope of this protocol to provide in detail the methods used for the ^{1}H NMR analysis of symbiotic hard coral. As such, we recommend using the well-described protocol written by Mark Viant as a solid foundation for the development of a high-quality metabolomics NMR analysis [25]. Although our NMR analysis is not described in detail, there are aspects worth mentioning. For example, and of particular importance, is that ^{1}H NMR is proven to be effective at identifying differences between coral that were either stressed or healthy at the time of snap freezing (Fig. 1) (*see* **Note 24**). However, upon examination of the 1D ^{1}H NMR spectra there were some noticeable, yet minor, shifts in signal position, which were attributed to the effects of high salt concentration and differences in the pH of samples. While most NMR metabolomics experiments employ the use of pH buffers to avoid such signal shifts, in this case, the shifts were small and hence the PCA was still very effective. As such, buffering of the sample pH using commonly employed buffers, such as sodium phosphate, were not included as part of this protocol (*see* **Note 25**) nor was a chemical shift standard (*see* **Note 26**).

1. Remove the samples prepared according to Subheading 3.2 from the −80 °C storage and store at room temperature (20–25 °C) for 30 min to thaw.
2. Using a 1,000 μL auto pipette and a clean tip for each sample, place 800 μL of each 70 % methanol extract into new, individually labelled, 7 mL glass scintillation vials.
3. Remove the lids of each vial and place the vials containing the extracts into a speed vacuum and dry at ambient temperature for ~10 min (*see* **Note 18**).
4. Remove the samples from the speed vacuum and freeze the remaining sample in liquid nitrogen or by placing in a −80 °C freezer for 1 h.
5. Undo the lids of each vial one-quarter of turn and place the frozen samples into a freeze drier and lyophilize to complete dryness.

6. Remove the samples from the freezer drier and tighten caps immediately to avoid any absorption of water.
7. Using a 1 mL graduated glass syringe, reconstitute each sample to its original 20 mg/mL concentration by adding 800 μL of deuterated methanol (CD_3OD, 99.8 %).
8. Place each vial containing the extracts in CD_3OD into a sonicating bath for 5 min to completely dissolve the extracts.
9. Using a 1,000 μL auto pipette and a clean tip for each sample, transfer the 800 μL of each sample into clean, individual and appropriately labelled 5 mm NMR tubes.
10. Cap each tube, seal with parafilm and store upright at −80 °C until ready for analysis.
11. Prior to analysis, thermally equilibrate samples by storing the tubes at room temperature (~25 °C) for 30 min (*see* **Note 27**).
12. Perform the NMR analyses using the acquisition and processing parameters in **steps 13** and **14** as a guide.
13. Acquisition parameters: Bruker pulse sequence zgesgp (1D excitation sculpting using 180° water-selective pulses) comprising [relaxation delay–180°–acquire]; 8.3 kHz spectral width; 3 s relaxation delay; typically four dummy scans followed by 32 transients (ns) are collected into 32 K data points; receiver gain was constant for all samples; temperature set at 298 K.
14. Processing parameters: Zero-filling was not applied; exponential line broadening of 0.3 Hz; Fourier transformation; manual phase correction (zero- and first-order corrections); baseline correction was not performed (*see* **Note 28**); calibrate the spectrum by setting CD_3OD peak to 3.31 ppm.

4 Notes

1. It is critical to ensure that all reusable glassware is free from contaminants that may affect the analysis. LC-MS using superficially porous columns is particularly susceptible to contaminants such as polyethylene glycol, slip agents, biocides, and plasticizers. They can be concentrated on reverse phase LC columns and eluted during a gradient. These contaminants are commonly found in detergents used for washing glassware and any plastic items they come in contact with (i.e., drying racks and gloves). They are often identifiable by broad peaks in the chromatogram with repeating mass units in the corresponding mass spectrum. As such, care should be taken to avoid any unnecessary exposure of glassware to plastic items. Gloves should be washed in ultrapure water to remove any slip agents before handling and all glassware of unknown purity should be acid-washed in a solution of 50 % nitric acid (HNO_3) for 30 min then rinsed at least three

times with MilliQ water. Glass Schott bottles, beakers, and measuring cylinders can be filled to the brim with 50 % HNO_3 and left for 30 min and then triple-rinsed with MilliQ water. It is not necessary to wash the outside of any glassware.

2. The use of high-quality MS grade solvents is highly recommended in order to eliminate the possibility of inferior solvents introducing spurious features to an analysis. This method recommends the use of Fisher Optima® LC-MS solvents (Thermo Fisher Scientific, Scoresby, VIC, Australia).

3. While the 70 % methanol extraction solvent effectively halts enzyme and metabolic activity, it is considered good practice to ensure that the temperature is kept as low as feasibly possible throughout the extraction and analysis to avoid unwanted changes in chemistry. Also, as temperature is not as critical for extraction efficiency as solvent choice [30], a temperature range of 0–4 °C is recommended for this protocol.

4. All plastic centrifuge tubes should be made of virgin polypropylene and free from slip agents, biocides, and plasticizers.

5. Glass scintillation vials should be of known purity and utilize an inert, contaminant and extractable free Teflon or foil-lined gasket. Vials having a plastic lid without such a gasket should be avoided.

6. This protocol recommends labels be made of plastic film and be solvent resistant. Labels made of paper or absorbent material can effect the calculation of extract yields as they contribute to a loss in mass during the solvent removal steps. Alternatively, permanent markers can be used to label vials; however, care should be taken to ensure that spilt solvents do not remove the markings.

7. During the development of this protocol, we found that the silicone/Teflon septa of standard autosampler vials contributed significant levels of plasticizer contamination to each analysis. As such, sample vials, lids, and septa used in LC-MS autosamplers should be certified as MS grade. We recommend the use of certified Verex™ vials, lids, and septa (Phenomenex, Lane Cove, NSW, Australia). For cases where reduced volume glass inserts are used, then it is acceptable to use standard vials with MS certified lids and septa as the glass inserts are of sufficient purity.

8. The cost of formic acid supplied in 1 mL glass ampoules can be very expensive and as such, it is not considered essential for this protocol, however, it does buy peace of mind for the analyst and is preferable. If using formic acid supplied in larger volumes, it is highly recommended to ensure the ongoing integrity of the reagent by exerting extra care when storing and taking aliquots from the storage container. As such, a number

of 1 mL aliquots should be prepared in individual clean glass vials (2 mL MS certified HPLC vials and lids are ideal) and stored at 4 °C for later use. In this way, it is a straightforward process of adding the premeasured 1 mL of formic acid to 1 L of mobile phase to give a final formic acid concentration of 0.1 %.

9. NMR sample tubes should be compatible with the field frequency of the users NMR and the volume of the sample must cover the entire active region of the shim coils (700 µL in a 5 mm NMR tube). Dirty tubes can be cleaned by filling them with 50 % HNO_3 and soaking for 30 min. The acid solution can then be removed and the tubes rinsed twice with ultra-pure water and again with acetone. Tubes can be dried by either air-drying or placing them under vacuum and should never be dried at high temperature. Always follow the manufactures instructions.
10. Volatile NMR solvents should be stored in a sealed container with a suitable desiccant to minimize the possibility of solvents absorbing water from the atmosphere.
11. When cutting the coral nubbins, it is preferable to cut the nubbin and allow it to fall into a waiting hand or container. The nubbin can then be quickly transferred directly into the liquid nitrogen and snap frozen with minimal handling. Care should be taken to avoid handling each nubbin more than required as this will damage the coral tissue, invoking the rapid production of coral mucous resulting in a change of chemistry.
12. It is acceptable to place replicate nubbins in a single 50 mL centrifuge tube after the nubbins have been snap frozen.
13. Once cut into smaller pieces, a nubbin should occupy approximately 30–50 % of the volume of each scintillation vial. The nubbin pieces should be completely submersed once the vial is filled to 75 % of its volume with the extraction solvent.
14. Most of the residual seawater that was frozen with the nubbin during the collection process will separate from the nubbins when cutting them into smaller pieces. It is important that this ice be discarded to reduce the amount of undesirable salts within the sample.
15. In any metabolomics study it is important to recognize the effects that certain techniques will have upon the sample. In the case of lyophilization, the reduced atmospheric pressure will remove many of the volatile compounds. If these volatile compounds are of interest to the analyst then drying under such conditions may need to be avoided and a more suitable technique examined.
16. At this point extracts should be orange in color if effective lysis of *Symbiodinium* cells has occurred.

17. New glass scintillation vials are weighed at this point in order to acquire the empty weight of each vial, which will be used to calculate the weight of the crude, dry extract.
18. The aim of this step is to remove ~90–95 % of the extraction solvent. Therefore, it is important to remove the vials from the speed vacuum concentrator before all the liquid has been evaporated so that the extracts remain cooled by the evaporation process, which commonly takes from 10 to 60 min depending on the volume of solvent. The speed vacuum concentrator should not be used with sample heating. Drying alcohol and water mixtures under vacuum can be time consuming without sample heating due to the nonideal interaction of water and methanol molecules at low concentrations [31]. As such, this protocol utilizes lyophilization as a second drying step. In this way, the water–methanol interaction is overcome, the sample does not need to be heated and thus the sample integrity is maintained.
19. By recording the weight of each extract, samples can be prepared at the same concentration for analysis. This has the effect of normalizing each sample.
20. For this method, a final concentration of 20 mg/mL is appropriate and will provide sufficient sample concentration for both ^{1}H NMR and LC-MS analyses. However, factors such as LC column loading and NMR sensitivity need to be accounted for when deciding on the appropriate concentration.
21. Syringe filtering was not employed in this method due to the contamination that the rubber syringe plungers introduce into the sample. Using glass syringes as an alternative was ruled out because cleaning them between samples was deemed too time consuming, in addition to the negative impact that additional sample handling could have on the sample. Alternatively, centrifugal solvent resistant filters could be used; however, they were not tested as part of this protocol.
22. In this instance, a 100 μL aliquot of the 20 mg/mL extract was transferred into each HPLC vial, hence the use of reduced volume inserts. This small amount is sufficient for multiple 2 μL injections and avoids the risk of compromizing the entire sample.
23. While the conditions described in **steps 6** and 7 will provide a good starting point, LC-MS conditions may still need to be optimized depending on the type of instrument and its associated peripheries.
24. In addition to the typical 1D ^{1}H NMR experiment we examined the effectiveness of the 2D JRES experiment [32], using a double spin-echo water suppression method [33], for the analysis of coral. The primary advantage of the JRES experiment for metabolomics lies in its ability to reduce the complexity of

the typical 1D ^{1}H NMR experiment by shifting the *J*-coupling into a second dimension. However, upon comparing the two experiments, we found that there was not a sufficient gain in spectral decongestion to warrant using the more time consuming JRES experiment.

25. The decision not to use a buffer was based on a number of reasons. Firstly, the observed signal shifts were small and within the tolerance of the 0.04 ppm bucket widths. For the small number of instances where shifts in NMR peaks did occur across the borders of a bucket, the loadings of the PCA reflected these shifts whereby multiple and consecutive bins contributed to the explanation of variability in the data. The final reason for avoiding the use of a pH buffer was salt concentration. We reasoned that adding a buffer to a marine sample, which already had a naturally high salt content, would contribute to the ionic strength of the sample, exacerbating the detrimental effects of high salt concentration [34]. That being the case, it is critical that the spectroscopist thoroughly examines and compares all NMR spectra for shifts in peak position to ascertain the need for pH buffering. When the use of a buffer is required, we recommend using a buffer with low ion mobility to preserve the sensitivity of the NMR probe, particularly when analyzing samples high in salt concentration using cryogenically cooled probes [35].
26. Chemical shift standards were not used in this protocol as all spectra were calibrated to the residual methanol signal at 3.31 ppm, thus avoiding the need to manipulate the sample. While chemical shift standards often perform the secondary function of an internal standard used for quantification, we recommend the use of the ERETIC experiment for this purpose [36, 37]. The advantage of the ERETIC experiment lies in its ability to add an electronic signal, at a chemical shift of one's choice, into previously acquired spectra. The electronic signal is calibrated and integrated to a known concentration of any analyte of choice and inserted into the previously acquired spectra where it can be used for the quantification of all other signals.
27. Warming the sample to 25 °C is necessary for cases where the NMR experiment is performed at 25 °C (298 K), thereby reducing the time of thermal equilibration within the instrument.
28. Manual baseline correction was not performed. In our case, all baselines were observed to be flat within the signal region, yet distortions did occur at the extremities (visible as upturned edges of the baseline). These distortions could not be removed using baseline correction without being detrimental to nearby regions. To rectify this, our sweep width was increased to 14 ppm (8.3 kHz) and the spectrum offset changed to 6 ppm.

In this way, the distortion at the extremities of the baseline was shifted to regions having no signals (approximately -1 and 13 ppm) and were then removed from the subsequent PCA by only integrating the signal region from 0 to 10 ppm.

Acknowledgments

This work was supported by the Australian Research Council, Centre of Excellence for Coral Reef Studies, and a Science for Management grant from the Great Barrier Reef Marine Park Authority (GBRMPA), Australia. We would also like to thank João Paulo Krajewski for kindly allowing us to use his photograph and Mark Viant and Christian Ludwig for their kind assistance with the DSE-JRES pulse program.

References

1. Rohwer F, Seguritan V, Azam F, Knowlton N (2002) Diversity and distribution of coral-associated bacteria. Mar Ecol Prog Ser 243:1–10
2. Stella JS, Pratchett MS, Hutchings PA, Jones GP (2011) Coral-associated invertebrates: diversity, ecological importance and vulnerability to disturbance. Oceanogr Mar Biol Annu Rev 49:43–104
3. Bourne DG, Garren M, Work TM, Rosenberg E, Smith GW, Harvell CD (2009) Microbial disease and the coral holobiont. Trends Microbiol 17:554–562
4. Adams LM, Cumbo V, Takabayashi M (2009) Exposure to sediment enhances primary acquisition of *Symbiodinium* by asymbiotic coral larvae. Mar Ecol Prog Ser 377:149–156
5. Stat M, Carter D, Hoegh-Guldberg O (2006) The evolutionary history of *Symbiodinium* and scleractinian hosts—symbiosis, diversity, and the effect of climate change. Perspect Plant Ecol Evol Systemat 8(1):23–43
6. Muscatine L, Porter JW (1977) Reef corals: mutualistic symbioses adapted to nutrient-poor environments. Bioscience 27:454–460
7. Trench RK (1979) The cell biology of plant-animal symbiosis. Annu Rev Plant Physiol Plant Mol Biol 30:485–531
8. Lewis DH, Smith DC (1971) The autotrophic nutrition of symbiotic marine coelenterates with special reference to hermatypic corals. I. Movement of photosynthetic products between the symbionts. Proc R Soc Lond Ser B Biol Sci 178:111–129
9. Gordon BR, Leggat W (2010) *Symbiodinium*-invertebrate symbioses and the role of metabolomics. Mar Drugs 8:2546–2568
10. Yellowlees D, Rees TAV, Leggat W (2008) Metabolic interactions between algal symbionts and invertebrate hosts. Plant Cell Environ 31:679–694
11. Roth E, Jeon K, Stacey G (1988) Homology in endosymbiotic systems: the term 'symbiosome'. Molecular genetics of plant microbe interactions. Proceedings of the 4th international symposium on molecular genetics of plant–microbe interaction. pp 220–225
12. Rands ML, Loughman BC, Douglas AE (1993) The symbiotic interface in an alga invertebrate symbiosis. Proc R Soc Lond Ser B Biol Sci 253:161–165
13. Dunlap WC, Yamamoto Y (1995) Small-molecular antioxidants in marine organisms: antioxidant activity of mycosporine-glycine. Compar Biochem Physiol 112B:106–114
14. Sunda W, Kieber D, Kiene R, Huntsman S (2002) An antioxidant function for DMSP and DMS in marine algae. Nature 418:317–320
15. Trevena AJ, Jones GB, Wright SW, van den Enden RL (2000) Profiles of DMSP, algal pigments, nutrients and salinity in pack ice from eastern Antarctica. J Sea Res 43:265–273
16. Miller TR, Hnilicka K, Dziedzic A, Desplats P, Belas R (2004) Chemotaxis of *Silicibacter* sp. strain TM1040 toward dinoflagellate products. Appl Environ Microbiol 70:4692–4701
17. Sjoblad RD, Mitchell R (1979) Chemotactic responses of *Vibrio alginolyticus* to algal extracellular products. Canadian J Microbiol 25:964–967
18. Raina J-B, Dinsdale EA, Willis BL, Bourne DG (2010) Do the organic sulfur compounds DMSP and DMS drive coral microbial associations? Trends Microbiol 18:101–108
19. Muscatine L (1967) Glycerol excretion by symbiotic algae from corals and *Tridacna* and its control by the host. Science 156:516–519

20. Trench RK (1971) The physiology and biochemistry of zooxanthellae symbiotic with marine coelenterates. III. The effect of homogenates of host tissues on the excretion of photosynthetic products *in vitro* by zooxanthellae from two marine coelenterates. Proc R Soc Lond Ser B Biol Sci 177:251–264
21. Gates RD, Hoegh-Guldberg O, McFall-Ngai MJ, Bil KY, Muscatine L (1995) Free amino acids exhibit anthozoan "host factor" activity: they induce the release of photosynthate from symbiotic dinoflagellates *in vitro*. Proc Natl Acad Sci USA 92:7430–7434
22. Nakamura H, Asari T, Ohizumi Y, Kobayashi J, Yamasu T, Murai A (1993) Isolation of zooxanthellatoxins, novel vasoconstrictive substances from the zooxanthella *Symbiodinium* sp. Toxicon 31:371–376
23. Onodera K, Nakamura H, Oba Y, Ojika M (2003) Zooxanthellamide A, a novel polyhydroxy metabolite from a marine dinoflagellate of *Symbiodinium* sp. Tetrahedron 59:1067–1071
24. Nakamura H, Kawase Y, Maruyama K, Murai A (1998) Studies on polyketide metabolites of a symbiotic dinoflagellate, *Symbiodinium* sp.: a new C30 marine alkaloid, zooxanthellamine, a plausible precursor for *zoanthid* alkaloids. Bull Chem Soc Jpn 71:781–787
25. Viant MR (2008) Environmental metabolomics using ^{1}H-NMR spectroscopy. Methods Mol Biol (Totowa, NJ, United States) 410 (Environmental Genomics):137–150
26. Tapiolas D, Motti C, Holloway P, Boyle S (2010) High levels of acrylate in the Great Barrier Reef coral *Acropora millepora*. Coral Reefs 29:621–625
27. Fiehn O (2002) Metabolomics—the link between genotypes and phenotypes. Plant Mol Biol 48:155–171
28. Lauridsen M, Hansen SH, Jaroszewski JW, Cornett C (2007) Human urine as test material in ^{1}H NMR-based metabonomics: recommendations for sample preparation and storage. Anal Chem 79:1181–1186
29. Ettinger-Epstein P, Motti CA, De Nys R, Wright AD, Battershill CN, Tapiolas DM (2007) Acetylated sesterterpenes from the Great Barrier Reef sponge *Luffariella variabilis*. J Nat Prod 70:648–651
30. Beltran A, Suarez M, Rodriguez MA, Vinaixa M, Samino S, Arola L, Correig X, Yanes O (2012) Assessment of compatibility between extraction methods for NMR and LC/MS-based metabolomics. Anal Chem 84:5838–5844
31. Wakisaka A, Abdoul-Carime H, Yamamoto Y, Kiyozumi Y (1998) Non-ideality of binary mixtures water-methanol and water-acetonitrile from the viewpoint of clustering structure. J Chem Soc Faraday Trans 94:369–374
32. Aue WP, Karhan J, Ernst RR (1976) Homonuclear broad band decoupling and two-dimensional J-resolved NMR spectroscopy. J Chem Phys 64:4226–4227
33. Thrippleton MJ, Edden RAE, Keeler J (2005) Suppression of strong coupling artefacts in J-spectra. J Magn Reson 174:97–109
34. Weljie AM, Newton J, Mercier P, Carlson E, Slupsky CM (2006) Targeted profiling: quantitative analysis of ^{1}H NMR metabolomics data. Anal Chem 78:4430–4442
35. Kelly AE, Ou HD, Withers R, Dotsch V (2002) Low-conductivity buffers for high-sensitivity NMR measurements. J Am Chem Soc 124:12013–12019
36. Akoka S, Barantin L, Trierweiler M (1999) Concentration measurement by proton NMR using the ERETIC method. Anal Chem 71:2554–2557
37. Tapiolas DM, Raina J-B, Lutz A, Willis BL, Motti CA (2013) Direct measurement of dimethylsulfoniopropionate (DMSP) in reef-building corals using quantitative nuclear magnetic resonance (qNMR) spectroscopy. J Exp Mar Biol Ecol 443:85–89

Chapter 11

Determination of Absolute Configuration Using Single Crystal X-Ray Diffraction

Abigail L. Albright and Jonathan M. White

Abstract

Single crystal X-ray crystallography is the most powerful structural method for the determination of the 3D structures of molecules. While the results of a routine diffraction experiment readily provide unambiguous determination of the relative configuration of all stereogenic centers in the molecule, determination of absolute configuration is more challenging. This chapter provides some helpful tips towards increasing the chances of success in the determination of the absolute configuration of a chiral, enantiomerically pure natural product using X-ray crystallography.

Key words X-ray crystallography, Absolute configuration determination, Light atom structure, Chiral reference

1 Introduction

Perhaps the single most challenging task facing the research chemists today is the efficient, cost-effective synthesis and characterization of enantio-pure compounds. The paramount importance of chirality is based on the fact that the physiological and/or pharmacological activity of a compound is often determined by its interaction with chiral-based receptors in the body. It is now a standard requirement for all new drugs to be characterized as single enantiomers [1, 2] and the FDA has not approved the use of a drug in its racemic form in over a decade. As illustrated in Fig. 1 the total number of pharmaceuticals approved worldwide as racemates, single enantiomers, or achiral compounds, respectively, and aptly depicts the sharp decline in the prevalence of racemic drugs. Historical precedent—founded on such sobering events as the thalidomide tragedy—and modern day drug development, in which numerous enantiomeric pharmaceuticals have been shown to display drastically different physiological activity (Dexfenfluramine, Dextromethorphan, Ritalin), speak to the need for improved asymmetric methodology [3–5].

Ute Roessner and Daniel Anthony Dias (eds.), *Metabolomics Tools for Natural Product Discovery: Methods and Protocols*, Methods in Molecular Biology, vol. 1055, DOI 10.1007/978-1-62703-577-4_11, © Springer Science+Business Media, LLC 2013

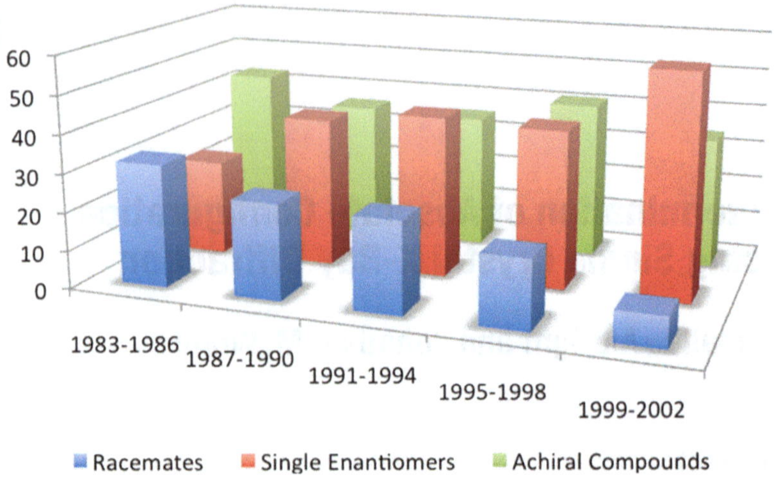

Fig. 1 Number of drugs approved worldwide as racemates, single enantiomers, or achiral compounds [3]

From the simple yet arduous task carried out by Louis Pasteur in the separation of the visually distinct enantiomers of tartaric acid, to the advent of advanced chiroptical methods such as vibrational and electronic circular dichroism (CD), the determination of absolute configuration can be achieved in numerous and varied ways [6]. However, it is the direct method of X-ray diffraction, which presents the scientific community with the most definitive proof of structure. This cornerstone methodology was pioneered by Bijvoet in the early twentieth century [7, 8] and, with the advent of the Flack and Hooft parameters, can be used to differentiate between enantiomers to a high degree of precision and accuracy [9]. Today it is often paired with CD [10], multidimensional NMR, and chiral derivatization [11] to provide rigorous determination of absolute configuration.

Indeed, we have been unable to find a single case in the literature in which absolute configuration was incorrectly assigned by XRD analysis, where this was the objective of the experiment. In the famous case of diazonamide A, in which researchers were met with conflicting data and a seemingly intractable absolute configuration determination, it was not the X-ray data but rather its misinterpretation which was the source of error [12].

The continued development of X-ray crystallography hardware, particularly with sensitive charge-coupled device (CCD) area detectors and intense, stable X-ray sources, has made the collection of high quality data with high redundancy a readily achieved objective. The differentiation between enantiomers using X-ray crystallography relies on the accurate measurement of the intensity differences of the so-called Friedel pairs. Reflections of X-rays

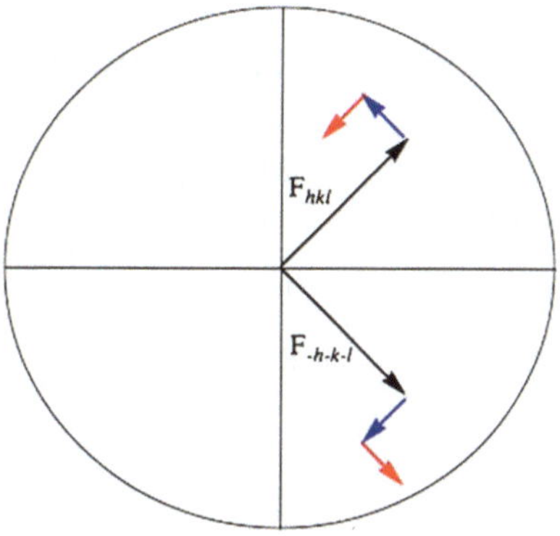

Fig. 2 Graphical depiction of maximized amplitudinal and phasic differences resulting from the anomalous dispersion in a Friedel pair [67]

by crystals occur from many thousands of planes in the crystal, each of which is assigned a Miller index (hkl). Friedel pairs (hkl and -h–k–l) represent the measurement of X-ray diffraction from opposite sides of a given set of planes, related by inversion through the origin.

The Friedel pairs have equal intensity in a crystal having a centrosymmetric space group, however small differences arise in the non-centrosymmetric space groups due to anomalous dispersion effects. The magnitude of anomalous dispersion, and hence the ability to differentiate these differences, depends upon several factors, including the X-ray wavelength, the atomic numbers of the atoms composing the crystal, and the quality of the X-ray diffraction data. Anomalous dispersion increases with increasing atomic number, and hence it is an advantage to have a heavy atom in the crystal to maximize these differences (Fig. 2).

Chiral, enantiomerically pure molecules crystallize in non-centrosymmetric space groups, which do not contain improper symmetry elements: The so-called Schonke Space Groups. There are 65 of these space groups, which are given in Table 1.

The strategies that can be employed in the determination of absolute configuration using X-ray crystallography fall into two general categories:

1. The direct approach, where the absolute configuration is determined by the diffraction experiment alone, and relies upon accurate determination of the small differences in scattering intensities of Friedel opposites which arise due to anomalous dispersion effects (resonant scattering effects).
2. Indirect approaches, where a chiral reference compound with known absolute configuration is incorporated into the structure, either as a counter ion, a chiral auxiliary, a solvate, or as a co-crystal.

Table 1
The 65 Schonke space groups for chiral, enantiomerically pure compound

Crystal class	Space group symbol	Space group number	Crystal class	Space group symbol	Space group number
Triclinic	P1	1	Trigonal	P3	143
Monoclinic	P2	3		$P3_1$	144
	$P2_1$	4		$P3_2$	145
	C2	5		R3	146
Orthorhombic	P222	16		P312	149
	$P222_1$	17		P321	150
	$P2_12_12$	18		$P3_112$	151
	$P2_12_12_1$	19		$P3_121$	152
	$C222_1$	20		$P3_212$	153
	C222	21		$P3_221$	154
	F222	22		R32	155
	I222	23	Hexagonal	P6	168
	$I2_12_12_1$	24		$P6_1$	169
Tetragonal	P4	75		$P6_5$	170
	$P4_1$	76		$P6_2$	171
	$P4_2$	77		$P6_4$	172
	$P4_3$	78		$P6_3$	173
	I4	79		P622	177
	$I4_1$	80		$P6_122$	178
	P422	89		$P6_522$	179
	$P42_12$	90		$P6_222$	180
	$P4_122$	91		$P6_422$	181
	$P4_12_12$	92		$P6_322$	182
	$P4_222$	93	Cubic	P23	195
	$P4_22_12$	94		F23	196
	$P4_322$	95		I23	197
	$P4_32_12$	96		$P2_13$	198
	I422	97		$I2_13$	199
	$I4_122$	98		P432	207

(continued)

Table 1 (continued)

Crystal class	Space group symbol	Space group number	Crystal class	Space group symbol	Space group number
				$P4_232$	208
				F432	209
				$F4_132$	210
				I432	211
				$P4_332$	212
				$P4_132$	213
				$I4_132$	214

1.1 The Direct Determination of Absolute Configuration Using X-Ray Crystallography

Forethought and effective planning of the X-ray diffraction experiment are critical to determining absolute configuration in an efficient and reliable manner. One useful indicator to determine the likelihood of a reliable determination of absolute configuration using the effects of anomalous dispersion is the Friedif value, described by Flack and Bernardinelli [13]. The Friedif value is calculated from both the chemical composition of the compound and from the wavelength of the X-radiation. A simple spreadsheet application has been made available for this purpose. For compounds having Friedif values greater than 80 the determination of the absolute configuration is fairly routine, provided that good quality X-ray diffraction data is available.

Light atom structures having Friedif values significantly less than 80 provide a major challenge for the direct determination of absolute configuration because the effects of anomalous dispersion are relatively small; in other words, the difference in the intensities of the Friedel opposites are minute. However, there now exist sound statistical techniques for the analyses of the differences of the many thousands of Friedel pairs collected during a data collection, including the Flack parameter [14], the *Q*-value method [15], the Hooft parameter [16], and more recently, Leverage analysis [17]. Notably, these methods all rely on the availability of high quality X-ray data (although an alternative statistical treatment for poor quality data sets has been suggested [18], this is not routinely available). Sophisticated refinement methods, including charge-density deformation methods, can further improve the certainty of the Flack and Hooft parameters. Absolute configuration of molecules, with Friedif values as low as 30–40 can now be determined

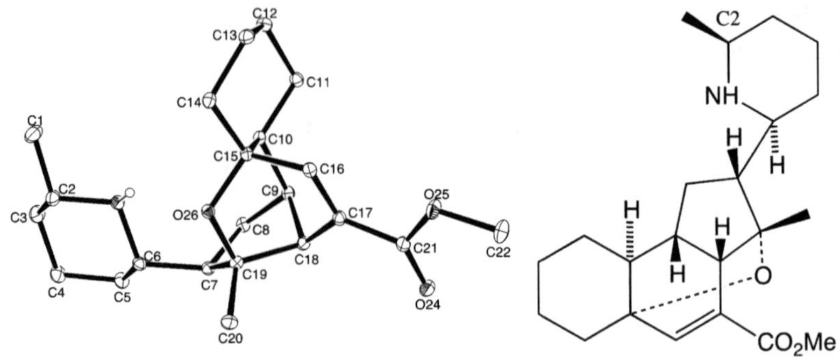

Fig. 3 Thermal ellipsoid plot and chemical structure diagram for compound **1**, a member of the himbacine class of alkaloids

with reasonable certainty. There are a number of excellent reviews and discussions detailing the best approaches to the determination of absolute configuration in these cases [19–24].

Because the direct approach necessitates a high quality crystal, the greatest challenge of the overall analysis lies is in the hands of the chemist (*see* **Note 1**). Cu-Kα X-radiation is strongly recommended for diffraction experiments used to determine the absolute configuration of structures containing only light atoms and the X-ray data should be collected at low temperature with high redundancy, as this significantly improves the quality of the data, particularly at high theta angles. The data collection should be set up to ensure the inclusion of all accessible Friedel pairs to high resolution (minimum 0.833 Å for Cu radiation).

Our recent analysis of the himbacine alkaloid (**1**) shown in Fig. 3 is a sound example of the successful direct absolute configuration determination of a light-atom structure. This alkaloid, which has the chemical formula $C_{22}H_{35}NO_2$, has a Friedif value of 5 for Mo-Kα radiation and 26 for Cu-Kα radiation; the data were therefore collected using the Cu-Kα radiation source. The X-ray data collection was carried out at 130 K employing an Oxford Cryostream low temperature device. The experiment was set up to ensure the collection of all Friedel opposites and high redundancy of data. Some experimental details are shown in Table 2 [25].

The structure was solved and refined using the Shelx programs as implemented within the WinGX suite of programs. The Flack parameter converged to 0.04 (13). The standard uncertainty on the Flack parameter is 0.13, which given that the sample was known to be enantiomerically pure, is strongly indicative that the absolute configuration is correct [19–24]. The Hooft parameter, which is obtained from a careful analysis of the Bijvoet differences using Bayesian statistics, was determined to be 0.03 with a standard uncertainty 0.05. The Hooft parameter, which is similar to the

Table 2
Selected crystal and refinement details for 1

Crystal system	Orthorhombic
Space group	$P\ 2_1 2_1 2_1$
Unit cell dimensions	$a = 7.42370(10)$ $b = 11.01590(10)$ $c = 24.4602(4)$
Crystal size	$0.38 \times 0.21 \times 0.10\ mm^3$
Theta range for data collection	3.61–73.31°
Reflections collected	28,593
Independent reflections	3,984 [*R*(int) = 0.0317]
Completeness to theta = 73.31°	99.50 %
Data/restraints/parameters	3,984/0/368
Goodness-of-fit on F2	1.074
Final *R* indices [I > 2sigma(I)]	R1 = 0.0300, wR2 = 0.0808
R indices (all data)	R1 = 0.0317, wR2 = 0.0818
Flack parameter	0.04(13)

Flack parameter, is generally determined with greater certainty. The analysis of the Bijvoet pairs and determination of the Hooft parameter is implemented within the Platon suite of programs (*see* **Note 2**). A screen capture of this calculation is given in Fig. 4. The absolute configuration determined for **1** in this experiment also corresponds to that obtained for all members of this class of alkaloids, namely the *S* configuration at C2.

1.2 Heavy Atom Derivatives

To increase the chances of a successful outcome in the direct determination of absolute configuration, introduction of a heavy atom into the structure is recommended. For example, the first determination of the absolute configuration of tartaric acid was achieved as its sodium rubidium salt in the famous paper by Bijvoet [7, 8]. If the compound can be derivatized readily as an ester or amide, then a *p*-bromobenzoate derivative may be employed [26]. Such derivatization may also have the advantage of converting a poorly crystalline substance into a more readily crystallized derivative. Compounds containing amino groups may be determined as a heavy atom containing salt; for example, determination of absolute configuration of the chiral base brucine, the antihcolinergic agent (+) dexbenzetimide [27], latifoline [28], and the chiral auxiliary

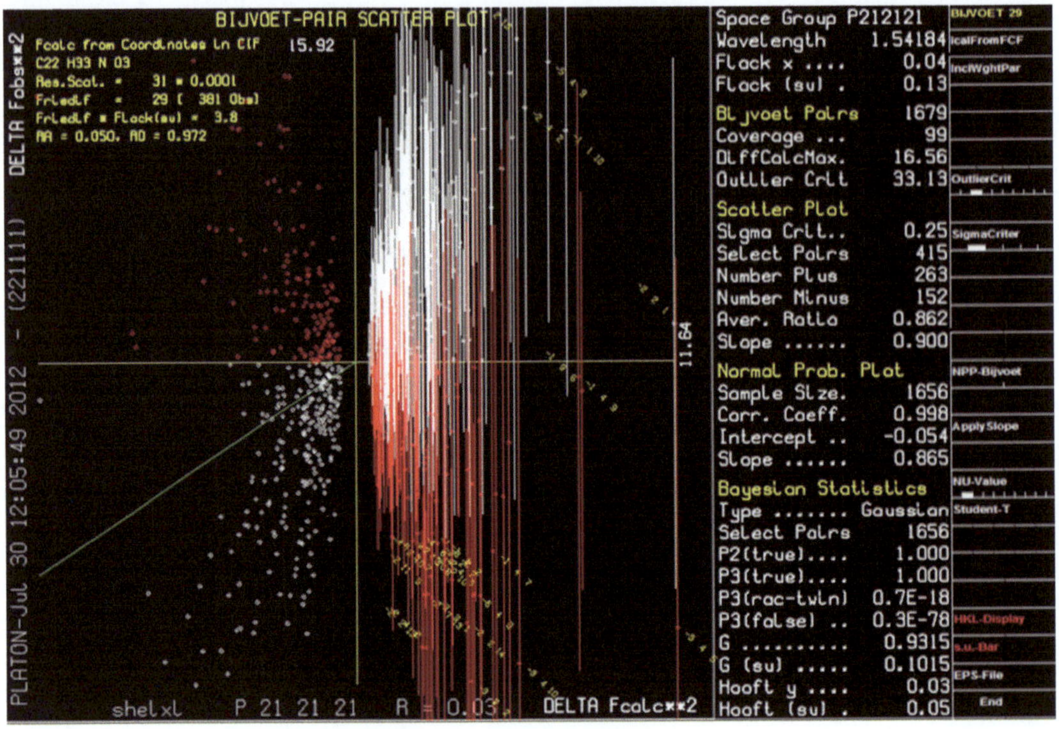

Fig. 4 Analysis of the Bijvoet pairs for **1** using the program Platon

1-(α-aminobenzyl)-2-naphthol, or Betti's base [29], were all carried out on the corresponding hydrobromide salts.

A heavy atom may also be introduced as part of a solvate of crystallization, e.g., chloroform or dichloromethane. Furthermore, recent advances in crystal engineering have led to the use of multi-component molecular co-crystals for heavy atom incorporation in molecules lacking acidic or basic functionality. For example, halogenated phenol derivatives have been successfully co-crystallized with pregnenolone and cholesterol. Well-defined co-crystals have even been reported for a great number of light atom organic liquids [30]. With development of exciting new crystal engineering technology and the availability of a wide variety of acids including various substituted chloro-, bromo-, and iodo-substituted carboxylic acids and sulphonic acids, the possibilities for incorporation of a heavy atom into the structure are seemingly endless.

1.3 Indirect Methods for the Determination of Absolute Configuration: Incorporation of a Chiral Reference

The incorporation of a component of known absolute configuration into a target reduces the diffraction experiment to a simple determination of the 3D structure of the compound. The relative configurations of all chiral centers are established and it is therefore simply a matter of ensuring the stereocenter of the chiral reference is set to the correct absolute configuration. The experiment can give conclusive results even if the crystal quality is poor and the data is of low quality. The only caveat to this approach is that the chiral reference employed must have been correctly determined in a previous experiment.

Just as with heavy atom incorporation, there are a multitude of ways in which a chiral reference may be incorporated into a structure, coupled in many cases with the serendipitous transformation of an oily compound into a highly crystalline derivative. These methods may be generally divided according to the method by which an element of known absolute configuration is incorporated into the crystal: Namely, as a counterion in a salt, a covalently bonded moiety or a co-crystallized element coordinated to the target molecule by hydrogen bonds or other relatively weak intermolecular interactions. Salt formation is of course reliant on some acidic or basic functionality in the target molecule, and as this is a common structural feature, there are a multitude of examples in which a chiral counterion has been successfully employed in the indirect determination of absolute configuration by XRD. In addition to serving as proof of absolute configuration, diastereomeric salts find extensive application in the resolution of enantiomers.

1.4 Ionically Bound Chiral Internal References: Diastereomeric Salts

If the target compound contains a basic functionality like an amine, then the use of tartrate as chiral anion is an excellent place to start. Tartaric acid has held a prominent place in the study and determination of absolute configuration since it was first encountered by Pasteur in 1847, and because of its ease of use, availability and affordability, its use as a resolving agent and a chiral reference via the formation of tartrate salts today is no exception. The absolute configuration of many molecules has been determined by X-ray analysis of their corresponding tartrate salts, including (−)-Crispine A [31], Irindalone [32], cisapride [33], and ketamine [34]. There is also precedence for the use of dibenzoyl tartrate as a chiral anion [35]. In addition to organic molecules containing basic functionality, tartrate anion has also been employed as a chiral reference in the analysis of Co(III) salts [36]. Other chiral anions in the literature include mandelate, applied in the absolute configuration determination of chiral pyrollidines [37], and camphor-10-sulfonate, which was applied in the absolute configuration determination of chiral telluronium salts [38] and azetidines [39].

Strychnine and brucine are far and away the most common of basic chiral references employed in the indirect determination of absolute configuration of unknown compounds containing an acidic moiety. VANOL [40], bilirubin [41], and saphenamycin [42] were all characterized as brucinium salts when their absolute configuration was first determined unequivocally. Prominent indirect determinations utilizing the strychnium salts are that of bromochlorofluoroacetic acid [43] and in the neutron structure of (+)-neopentyl-1-d alcohol [44], although the use of strychnine as a chiral reference is somewhat limited by that fact that it is quite poisonous.

Because it is a fact often overlooked, it is worth stating the obvious: No chiral reference is ubiquitous, and any compound of known absolute configuration is a potential candidate for an internal reference. For example, 1-phenylethylamine [45] and *cis*-2-(benzylamino)cyclohexylmethanol [46] have been reported in the literature as chiral cations of known configuration in the analysis of novel molecules containing carboxylic acid functionalities. And the absolute configuration of Troger's base was determined in reference to an axially chiral phosphate anion [47]. The ever-increasing number of chiral auxiliaries appearing in the literature should be concurrently viewed as a growing pool of internal chiral references for the determination of absolute configuration.

1.5 Covalently Bound Chiral Internal References

The field of covalently bound internal chiral references is dominated by the chiral phthalic acids developed by Harada et al. [11, 48, 49]. This class of compounds, notably camphorsultam dichlorophthalic acid (CSDP acid), has been applied to the absolute configuration determination of a remarkable scope of carboxyl- and alcohol-containing molecules, including some with very unique chirality elements such as paracyclophanes and conformationally chiral, congested alkenes. CSDP acid serves as an excellent resolving agent, in addition to being highly crystalline. It therefore allows for the isolation of enantiomerically pure compounds (via HPLC or recrystallization) concurrently with the determination of absolute configuration (via XRD). Since these topics have been extensively reviewed [11, 48, 49], they have only been summarized briefly herein.

One particularly striking example of the utility of an internal reference in X-ray analysis was recently put forth. C_{60}-Fullerenes are notoriously reluctant to crystallize, so the absolute configuration determination of chiral fullerene derivatives has therefore been reliant on the use of CD. But when several conflicting reports of absolute configuration appeared in the literature, Toriumi et al. set about the acquisition of the crystal structure [50]. (*2R*, *3R*)-(−)-butanediol was incorporated into the structure and some crystals were grown, albeit of very low quality. Yet despite the high *R*-value of the data (*see* **Note 3**), the X-ray analysis clearly showed the relative stereochemistry of the structure and the disputed absolute configuration of the chiral fullerene was determined with certainty.

1.6 Co-crystallized Chiral Internal References

A discussion of internal chiral references for XRD analysis is certainly lacking if the emerging field of co-crystallization goes unmentioned. The formation of diastereomeric salts is dependent on having a functional group that can be manipulated by pH, while one must be mindful that the introduction of a covalently bound chiral reference does not affect the absolute configuration of the stereocenters one seeks to determine. The use of co-crystals conveniently circumnavigates both of these potentially intractable issues. While still in the early stages of development [51, 52], the precedence of co-crystallization as a means of introducing a chiral reference has been established: the absolute configuration of a number of 3-arylbutanoic acids was recently determined by X-ray analysis of their co-crystals with isonicotinamide [53]. Remarkably, diastereomeric co-crystals have even been used for absolute configuration determination by powder diffractometry [54].

2 Software

Although the chemist may be loath to admit it, software is indeed the cornerstone of absolute configuration determination by XRD. A brief summary of the software packages available for absolute configuration determination follows:

1. Shelx [55].
   (a) A software suite for the solution of single crystal diffraction data. It employs the Flack Parameter for determination of absolute configuration, in which a crystal is treated as if it is twinned by inversion [13]. Flack parameter value of 0 indicates the absolute configuration is correct while a value of 1 indicates it should be inverted.
2. Platon [56].
   (b) The determination of absolute configuration using Bayesian statistics, the derived Hooft parameter has a similar meaning as the Flack parameter, but is generally better determined.
3. Crystals [57].
   (c) A software suite employing several means for determining absolute configuration, including the Flack and Hooft parameters, and Quotient restraints methods for the determination of absolute configuration.

To the ever-expanding toolbox of the today's synthetic organic chemist must be added the use of both direct and indirect methods for the determination of absolute configuration by X-ray diffraction. The last 60 years have beheld a revelation in the speed, reliability, and accuracy with which the 3D structure of a molecule might be determined by the use of X-rays, and more recent developments in software and technique have greatly expanded the scope of molecules, which might be analyzed by this technique. X-ray diffraction represents the most powerful and definitive method available to the scientific community for the determination of absolute configuration.

3 Notes

1. There are a number of helpful publications [58–62] and links on the Internet [63–66] describing a range of standard and specialized crystallization techniques.
2. A screencast showing how to perform an absolute structure determination using Bayesian statistics using Platon is available at the following link: www.youtube.com/watch?v=0m68d6hXIDk.
3. The *R*-value is used to quantify the quality of the data set from an X-ray diffraction experiment by measuring how well it agrees with the crystallographic model. It is analogous to standard deviation; the larger the *R*-value, the poorer quality the data.

References

1. Food and Drug Administration (1992) Policy statement for the development of new stereoisomeric drugs. US Food and Drug Administration regulatory guidance. www.fda.gov/cder/guidance/stereo.html
2. Gawley RE, Aubé J (1996) Principles of asymmetric synthesis. Pergamon, Oxford, UK
3. Rouhi AM (2003) Chiral at work: drug developers can learn much from recent successful and failed chiral switches. Chem Eng News 81:56–61
4. Shimazawa R, Nagai N, Toyoshima S, Okuda H (2008) Present status of new chiral drug development and review in Japan. J Health Sci 54:23–29
5. Caner H, Groner E, Levy L, Agranat I (2004) Trends in the development of chiral drugs. Drug Discov Today 9:105–110
6. Allenmark S, Gawronski J (2008) Determination of absolute configuration—an overview related to this special issue. Chirality 20:606–608
7. Bijvoet JM, Peerdeman AF, Van Bommel AJ (1951) Determination of the absolute configuration of optically active compounds by means of X-rays. Nature 168:271–272
8. Bijvoet JM, Wiebenga EH (1972) Structure determination by X-ray diffraction. Woolters-Wardhoff, Groningen
9. Flack HD, Bernardinelli G (1999) Absolute structure and absolute configuration. Acta Crytallogr A 55:908–915
10. Douglas MJ, Rutkowske RD, Miller LAD (2007) Strategies for successfully applying vibrational circular dichroism in a pharmaceutical research environment. Am Pharm Rev 10:118–123
11. Harada N (2008) Determination of Absolute Configurations by X-ray Crystallography and ^{1}H NMR Anisotropy. Chirality 5:691–723
12. Li J, Jeong S, Esser L, Harran PG (2001) Total synthesis of nominal diazonamides—part 1: convergent preparation of the structure proposed for (−)-diazonamide A. Angew Chem Int Ed 40:4765–4769
13. Flack HD, Bernardinelli G (2008) Applications and properties of the Bijvoet intensity ratio. Acta Crytallogr A A64:484–493
14. Flack HD (1983) On enantiomorph-polarity estimation. Acta Crystallogr A39:876–881
15. Parsons S, Flack H (2004) Precise absolute-structure determination in light-atom crystals. Acta Crystallogr A60:61
16. Hooft RWW, Straver LH, Spek AL (2008) Determination of absolute structure using Bayesian statistics on Bijvoet differences. J Appl Cryst 41:96–103
17. Parsons S, Wagner T, Presly O, Wood PA, Cooper RL (2012) Applications of leverage analysis in structure refinement. J Appl Cryst 45:417–429
18. Hooft RWW, Straver LH, Spek AL (2010) Using the *t*-distribution to improve the absolute structure assignment with likelihood calculations. J Appl Cryst 43:665–668
19. Flack HD, Bernardinelli G (2008) The use of X-ray crystallography to determine absolute configuration. Chirality 20:681–690
20. Thompson AL, Watkin DJ (2009) X-ray crystallography and chirality: understanding the limitations. Tetrahe Asymm 20:712–717
21. Deschamps JR (2010) X-ray crystallography of chemical compounds. Life Sci 86:585–589
22. Flack HD, Bernardinelli G (2000) Reporting and evaluating absolute-structure and absolute-configuration determinations. J Appl Cryst 33:1143–1148
23. Flack HD (2008) The use of X-ray crystallography to determine absolute configuration. Acta Chim Slov 55:689–691

24. Dittrich B, Strümpel M, Koritsánszky T, Schäfer M, Spackman MA (2006) Invarioms for improved absolute structure determination of light-atom crystal structures. Acta Crystallogr A62:217–223
25. Bradford TA, Willis AC, White JM, Herlt AJ, Taylor WC, Mander LN (2011) The structures of four new himbacine-like Galbulimima alkaloids. Tetrahe Lett 52:188–191
26. Hirayama N, Shirahata K (1987) Structural studies of mitomycins. Acta Crystallogr B43:555–559
27. Spek AL, Peerdeman AF, Van Wijngaarden I, Soudijn W (1971) The absolute configuration and crystal structure of the anticholinergic drug dexbenzetimide. Nature 232:575
28. Culvenor CCJ, Mackay MF (1992) The absolute structure of latifoline. Aust J Chem 45:451–456
29. Cardellicchio C, Ciccarella G, Naso F, Schingaro E, Scordari F (1998) The Betti base: absolute configuration and routes to a family of related chiral nonracemic bases. Tetrahe Asymm 9:3667–3675
30. Desiraju GR, Bhatt PM (2008) Co-crystal formation and the determination of absolute configuration. CrystEngComm 10:1747–1749
31. Louafi F, Moreau J, Shahane S, Golhen S, Roisnel T, Sinbandhit S, Hurvois JP (2011) Electrochemical synthesis and chemistry of chiral 1-cyanotetrahydroisoquinolines. an approach to the asymmetric syntheses of the alkaloid (−)-crispine A and its natural (+)-antipode. J Org Chem 23:9720–9732
32. Jensen B (1988) Structure of the (+)-tartrate of the selective 5-HT_2 antagonist irindalone. Acta Cryst C44:1602–1605
33. Peeters OM, Blaton NM, De Ranter CJ (1997) Absolute configuration of the double salt of *cis*-4-amino-5-chloro-*N*-{1-[3-(4-fluorophenoxy)propyl]-3-methoxypiperidin-1-ium-4-yl}-2-methoxybenzamide tartrate (cisapride tartrate). Acta Cryst C53:597–599
34. Ratti-Moberg E, Groth P, Aasen A, Arne J (1991) The absolute configuration of ketamine – a general anaesthetic. The crystal structure of the (R, R)-tartrate salt of (−)-(S)-ketamine. Acta Chem Scand 45:108–110
35. Kobayashi Y, Kinbara K, Sato M, Saigo K (2005) Synthesis, absolute configuration, and application of enantiopure *trans-1-aminobenz*[*f*]indan-2-ol. Chirality 17: 108–112
36. Bernal I, Korp J, Creaser II (1984) The absolute configuration of Λ-[Co(sen)] [(R, R)(+) tart] C1.6H2O and of Δ-[Co(sen)] [(R, R)(+) tart] C1.4. 5H2O (sen is 5-Methyl-5-(4-amino-2-azabutyl)-3,7-diazanonane-1,9-diamine): a comment on the mode of chiral resolutions by (R, R)(+)Tartrate on the antipodal cations of cobalt(III) salts. Aust J Chem 37: 2365–2369
37. Dahlen E, Hjalmarsson M, Norin T, Csoregh I, Ertan A (1991) Synthesis and resolution of cis-2-(1-Hydroxy-1-methylethyl)-5- methyl-pyrrolidine and X-ray crystal structure of its (S)-mandelate. Salt. Acta Chem Scand 45: 200–205
38. Shimizu T, Urakubo T, Kamigata N (1996) Synthesis and absolute configuration of optically active telluronium salts. Chem Lett 4: 297–298
39. Frigola J, Vano D, Torrens A, Gomez-Gomar A, Ortega E, Garcia-Granda S (1995) 7-azetidinylquinolones as antibacterial agents. 3. Synthesis, properties and structure-activity relationships of the stereoisomers containing a 7-(3-Amino-2-methyl-1-azetidinyl) moiety. J Med Chem 38:1203–1215
40. Polavarapu PL, Petrovic AG, Vick SE, Wulff WD, Ren H, Ding Z, Staples RJ (2009) Absolute configuration of 3,3′-Diphenyl-[2,2′-binaphthalene]-1,1′ -diol revisited. J Org Chem 74:5451–5457
41. Boiadjiev SE, Person RV, Puzicha G, Knobler C, Maverick E, Trueblood KN, Lightner DA (1992) Absolute configuration of bilirubin conformational enantiomers. J Am Chem Soc 114:10123–10133
42. Laursen JB, Jorgensen CG, Nielson J (2003) First synthesis of racemic saphenamycin and its enantiomers. Investigation of biological activity. Bioorg Med Chem 11:723–731
43. Costante J, Ehlinger N, Perrin M, Collet A (1996) Absolute configuration of bromochlorofluoroacetic acid. Enantiomer 1:377–386
44. Yuan HS, Stevens RC, Bau R, Mosher HS, Koetzle TF (1994) Determination of the absolute configuration of (+)-neopentyl-1-d alcohol by neutron and X-ray diffraction analysis. Proc Natl Acad Sci USA 91:12872–12876
45. Kremsner J, Wallfisch BC, Belaj F, Uray G, Kappe CO, Wentrup C, Kollenz G (2008) Tetra-*tert*-butyltrioxabicyclo[3.3.1]nonadienedicarboxylic acid: optical resolution, absolute configuration and application in chiral discrimination. Eur J Org Chem 19: 3382–3388
46. Wang Z, Hirose T, Shitara H, Goto M, Nohira H (2005) Structure and chiral recognition ability of endo-3-benzamidonorborn-5-ene-2-carboxylic acid. Bull Chem Soc Jap 78:880–885
47. Mason SF, Vane GW, Schofield K, Wells RJ, Whitehurst JS (1967) The circular dichroism and absolute configuration of Tröger's base. J Chem Soc B, 553
48. Harada N (2006) Chiral auxiliaries powerful for both enantiomer resolution and determination of absolute configuration by X-ray crystallography. Top Stereochem 25:177–203

49. Toyota S (1999) Camphorsultam – an excellent chiral reagent for enantiomer resolution and determination of absolute stereochemistry. Enantiomer 4:25–32
50. Kuwahara S, Obata K, Yoshida K, Matsumoto T, Harada N, Yasuda N, Ozawa Y, Toriumi K (2005) Conclusive determination of the absolute configuration of chiral C60-fullerene *cis*-3 bisadducts by X-ray crystallography and circular dichroism. Angew Chem Int Ed 44: 2262–2265
51. Friscic T, Jones W (2010) Benefits of cocrystallisation in pharmaceutical materials science: an update. J Pharm Pharmacol 62:1547–1559
52. Atkinson MBJ, Mariappan SVS, Bucar DK, Baltrusaitis J, Friscic T, Sinada NG, MacGilivray LR (2011) Crystal engineering rescues a solution organic synthesis in a cocrystallization that confirms the configuration of a molecular ladder. Proc Natl Acad Sci USA 108:10974–10979
53. Eccles KS, Deasy RE, Fabian L, Maguire AR, Lawrence SE (2011) The use of co-crystals for the determination of absolute stereochemistry: an alternative to salt formation. J Org Chem 76:1159–1162
54. Bock DA, Lehmann CW (2012) Chirality determination from X-ray powder data—diastereomeric co-crystals of mandelic acid and proline amide. CrystEngComm 14:1534–1537
55. Sheldrick GM (2008) A short history of shelx. Acta Cryst A64:112–122
56. Spek AL (2009) Structure validation in chemical crystallography. Acta Cryst D65:148–155
57. Thompson AL, Watkin DJ (2011) CRYSTALS enhancements: absolute structure determination. J Appl Cryst 44:1017–1022
58. Etter MC, Jahn DA, Donahue BS (1986) Growth and characterization of small molecule organic crystals. J Cryst Grow 76:645–655
59. Jones PG (1981) Crystal growing. Chem Br 17:222–225
60. Sluis P, Hezemans AM, Kroon J (1989) Crystallization of low-molecular-weight organic compounds for X-ray crystallography. J Appl Cryst 22:340–344
61. Holden A, Singer P (1960) Crystals and crystal growing. Anchor Books-Doubleday, New York, NY
62. Laudise RA (1970) The growth of single crystals, Solid state physical electronics series. Prentice-Hall, Inc., Englewood cliffs, NJ
63. http://www2.chemistry.msu.edu/facilities/crystallography/xtalgrow.pdf
64. Hampton research catalog, http://www.hamptonresearch.com/
65. Boyle PD (2006) http://www.udel.edu/chem/xray/Crystal_Growing_Guide.html
66. Lachicotte RJ How to grow X-ray quality crystals, http://chem.chem.rochester.edu/~nvd/crystalgrowth.html
67. Merritt E (1995) X-ray anomalous scattering, http://skuld.bmsc.washington.edu/scatter/AS_index.html

Chapter 12

Natural Product Chemistry in Action: The Synthesis of Melatonin Metabolites K_1 and K_2

Helmut M. Hügel and Oliver A.H. Jones

Abstract

Natural product chemistry often yields new compounds with great potential for economic and/or health benefits. However, most natural compounds must be artificially synthesized on an industrial scale to generate enough active ingredients to be commercially viable. Thus chemical synthesis is an essential tool of natural product chemistry. Chemical synthesis may be defined as the purposeful execution of a series of chemical reactions in order to obtain a product(s) of interest or to demonstrate important methodologies. Organic synthesis is (unsurprisingly) specifically concerned with the construction of organic compounds, which are often of an exceedingly high level of structural complexity. Each step of any form of chemical synthesis involves a chemical reaction, and the reagents and conditions for each of these need to be designed to give a good yield and a pure product. Here we illustrate this methodology using a simple protocol for the synthesis of the metabolites of the natural product melatonin, the kynurenamines *N1-acetyl-N2*-formyl-5-methoxykynurenine (K_1) and *N*-acetyl-5-methoxykynurenamine (K_2). The four key synthetic transformations involve (a) conversion of 4-methoxy aniline into the *tert*-butyl (4-methoxyphenyl) carbamate, (b) *ortho*-lithiation-iodination with *tert*-butyllithium-1,2-iodoethane, (c) use of the Sonogashira reaction with *N*-acetyl propargylamine, and (d) sequential alkyne hydration/formylation to give (K_1) or alkyne to give (K_2).

Key words Pineal hormone, Brain chemistry, Kynurenamine synthesis, *ortho*-Lithiation, Sonogashira reaction

1 Introduction

The main value of natural product screening in the present day is for lead discovery. The last decade has seen an explosive growth of the isolation and structural elucidation of natural products, particularly from marine organisms. Many newly discovered natural compounds show tremendous promise in human health and medicine [1] but are also highly structurally complex and require the use of highly specialized analytical techniques to elucidate their structure. The creation of useful end products often requires the synthesis of intricate molecules in significantly larger amounts than found in nature [2, 3]. Organic chemistry

Ute Roessner and Daniel Anthony Dias (eds.), *Metabolomics Tools for Natural Product Discovery: Methods and Protocols*, Methods in Molecular Biology, vol. 1055, DOI 10.1007/978-1-62703-577-4_12, © Springer Science+Business Media, LLC 2013

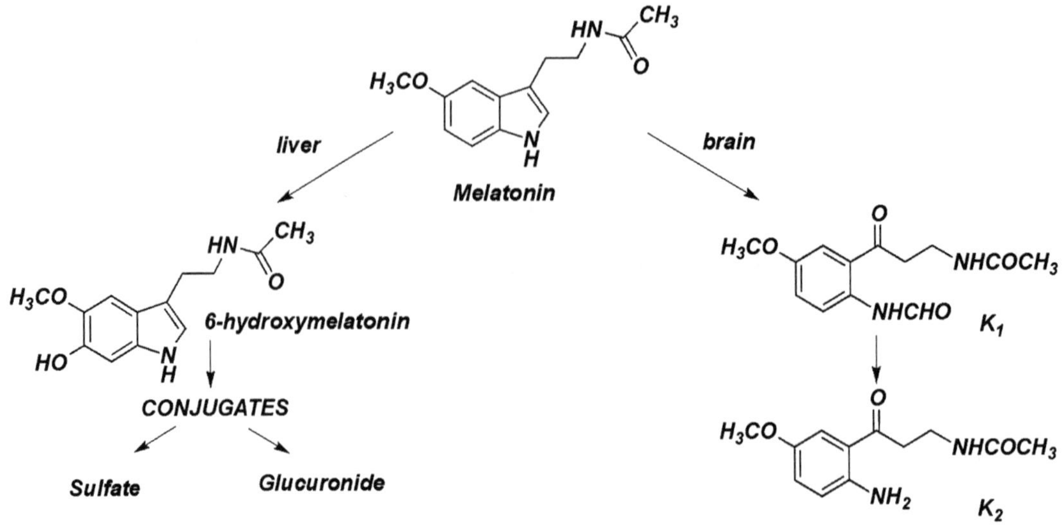

Fig. 1 Melatonin metabolism

has responded by the development of combinatorial methods of synthesis, enabling the economic preparation of large numbers of useful compounds. As such, although the nature of its major contributions has varied over time, organic synthesis has always been a vital part of the highly integrated and multidisciplinary process of natural products chemistry and this is unlikely to change in the foreseeable future.

One relevant example of a natural product used widely for beneficial health effects is Melatonin (*N*-acetyl-5-methoxytryptamine), a brain hormone produced in the pineal gland, which plays a significant role in regulating and controlling circadian and seasonal sleep rhythms. Melatonin is biosynthesized from tryptophan and has been implicated in pathologies related to circadian rhythm disorders, potent antioxidant protection. It is widely used as a supplement to treat a wide variety of conditions such as jet lag, seasonal affective disorder, depression, and delayed sleep phase syndrome [4, 5]. Circulating melatonin is metabolized in the liver by 6-hydroxylation followed by sulfation and glucuronidation to give the urinary metabolite 6-sulfatoxymelatonin as the major component (*see* Fig. 1). In the brain, melatonin is metabolized by the action of the indolamine-2,3-dioxygenase to give *N*1-acetyl-*N*2-formyl-5-methoxykynurenine K_1 [6, 7]. This metabolite is further transformed into *N*-acetyl-5-methoxykynurenamine (K_2). K_2 shows the highest activity against neural nitric oxide synthase inhibitors with an inhibition percentage of 65 % at a 1 mM concentration and may serve as a scaffold for the preparation of new lead compounds with potential therapeutic applications as neuroprotective agents [8]. Here we detail and present the stepwise methodology on how the two major metabolites of melatonin can be efficiently prepared. This protocol may be used to gain familiar-

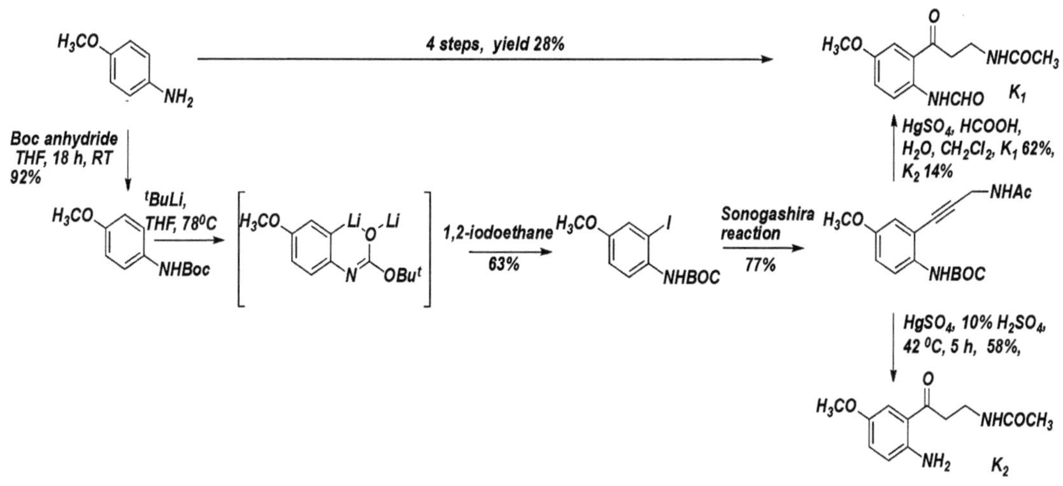

Fig. 2 Synthesis of melatonin metabolites K_1 and K_2

ity with the methods and procedures of organic synthesis prior to attempting more detailed reactions [9, 10]. The interested reader is also directed to http://www.orgsyn.org/ where detailed and carefully checked procedures for the synthesis of a range of organic compounds can be found.

2 Materials

All solutions and reactions should be prepared using analytical grade reagents and if necessary ultrapure water (18 MΩ). The purity of each compound may be checked using the NMR data provided at the end of the chapter. A detailed overview of the reaction sequence is provided in Fig. 2.

1. 1,2-Iodoethane ($C_2H_5I_2$).
2. 4-Methoxyaniline (C_7H_9NO).
3. Anhydrous magnesium sulfate ($MgSO_4$).
4. Anhydrous Tetrahydrofuran (THF).
5. Bis(triphenylphosphine)palladium(II) dichloride ($PdCl_2[PPh_3]_2$).
6. Brine solution (saturated NaCl).
7. Copper (I) iodide (CuI).
8. Di-*tert*-butyl dicarbonate (Boc_2O).
9. Dichloromethane (CH_2Cl_2).
10. Diethyl ether (Et_2O).
11. Distilled water (H_2O).
12. Ethyl acetate (EtOAc).
13. Formic acid (HCO_2H).

14. Magnetic stirrer (flea).
15. Mercuric sulfate ($HgSO_4$).
16. Methanol (MeOH).
17. *N*-acetyl propargylamine (C_5H_7NO).
18. *n*-Hexane (C_6H_{14}).
19. pH paper (or probe).
20. Round bottom flasks (100 mL).
21. Saturated aqueous sodium thiosulfate solution ($Na_2S_2O_3$).
22. Silica gel (for flash chromatography).
23. Sodium bicarbonate ($NaHCO_3$).
24. Sodium chloride (NaCl).
25. Schlenk flasks (with PTFE or Teflon septa).
26. Sulfuric acid (H_2SO_4) (at 10 % concentration).
27. *tert*-Butyl-lithium (*t*-$CH_3)_3CLi$ (in pentane).
28. Triethylamine (Et_3N).

3 Methods

Conduct all chemical synthesis reactions and manipulations in a well ventilated fumehood. Use a vacuum manifold in a fumehood for Schlenk glassware manipulations. All reagents may be prepared and stored at room temperature unless otherwise indicated. National and international waste disposal regulations should be followed at all times.

3.1 Synthesis Steps for the Preparation of tert-Butyl (4-Methoxyphenyl) Carbamate

1. Prepare a solution of 4-methoxyaniline (1.847 g, 15 mmol) and di-*tert*-butyl dicarbonate (5.1 mL, 22 mmol) in 15 mL of dry THF (15 mL).
2. Stir at room temperature for 18 h using a magnetic stirrer.
3. Evaporate solvent using rotary evaporator under reduced pressure.
4. Add 20 mL of ultrapure H_2O to dried extract from **step 2**.
5. Extract mixture with three washes of 30 mL of CH_2Cl_2.
6. Wash the CH_2Cl_2 extract with 20 mL of a saturated aqueous NaCl solution and then dry over $MgSO_4$.
7. Purify the residue of **step 6** via silica gel (70 g) column chromatography using an ethyl acetate–*n*-hexane mixture (ratio of 1:5) as eluent, to give colorless needle like crystals.
8. Recrystallize the end product from **step 7** from the Et_2O-n-hexane (*see* **Note 1**).

3.2 Synthesis of tert-Butyl (2-Iodo-4-Methoxyphenyl) Carbamate

1. Using the product of Section 3.1 in a Schlenk flask prepare a solution of *tert*-butyl (4-methoxyphenyl) carbamate (10.0 g, 44.82 mmol) in dry THF (112 mL).
2. Ensure the seal on the Schlenk flask is secure, immerse in a dry ice-acetone cooling bath. Using a N_2 gas flushed syringe, add a solution of *tert*-butyl lithium in pentane (68.56 mL, 116.53 mmol) under N_2 at −78 °C.
3. Stir and allow the solution to react for 15 min.
4. Warm solution to −20 °C and stir for 2.5 h.
5. Add a solution of 1,2-iodoethane (18.95 g, 67.23 mmol) in dry THF (40 mL).
6. Stir the reaction mixture at ambient temperature overnight.
7. Quench reaction with saturated aqueous $Na_2S_2O_3$ solution (170 mL).
8. Extract the reaction mixture with 270 mL (3 × 90 mL) of Et_2O.
9. Wash the organic phase with 160 mL of brine (2 × 80 mL) and dry over $MgSO_4$.
10. Filter and concentrate the final compound on a rotary evaporator.
11. Carry out flash chromatography using *n*-hexane–ethyl acetate in a 10:1 ratio to give the pure product as colorless prisms (*see* **Note 2**).

3.3 Synthesis of tert-Butyl(2-[3′-N-Acetylaminopropargyl]-4-Methoxy Phenyl) Carbamate

1. Prepare a solution of the product of Section 3.2 (1.75 g, 5 mmol), $PdCl_2[PPh_3]_2$ (140.3 mg, 0.2 mmol) and CuI (85 mg, 0.45 mmol), in dry Et_3N (9.12 mL).
2. At room temperature and under N_2 slowly (over 30 min) add *N*-acetyl propargylamine (0.65 g, 6.65 mmol) to the solution prepared in **step 1**.
3. Stir reaction mixture at room temperature for 1 h.
4. Partition the mixture between Et_2O (25 mL) and brine (7 mL).
5. Separate and dry the organic layer over $MgSO_4$.
6. Filter and evaporate to dryness to give the crude product.
7. Carry out flash chromatography using ethyl acetate–*n*-hexane in a 2:1 ratio to give pure product as white crystals (*see* **Note 3**).

3.4 Synthesis of K_1

1. In a 100 mL round bottom flask mix mercuric sulfate (0.36 g, 1.13 mmol), distilled H_2O (2.75 mL), formic acid (19.25 mL) and DCM (11 mL), using a magnetic stirrer, until the mixture has dissolved.
2. To this solution, add the product of Subheading 3.3 (0.36 g, 1.12 mmol) over 30 min.
3. Stir the flask contents for 4 h at 42 °C immersed in a silica oil bath.

4. Add solid $NaHCO_3$ and adjust solution to pH 7–8.
5. Freeze dry the reaction mixture, then extract with 3×50 mL washes of EtOAc
6. Wash organic phase with 100 mL of H_2O and 100 mL of brine.
7. Dry the organic phase over $MgSO_4$.
8. Filter and concentrate to give the crude compound.
9. Carry out flash chromatography using EtOAc-MeOH, in a 30:1 ratio to give the pure product as white needles (*see* **Note 4**).

3.5 Synthesis of K_2

1. In a 100 mL round bottom flask mix mercuric sulfate (0.36 g, 1.13 mmol), 10 % H_2SO_4 (25 mL) and CH_3OH (10 mL), using a magnetic stirrer, until the mixture has dissolved.
2. To this solution, add the product of Subheading 3.3 (0.36 g, 1.12 mmol) over 30 min.
3. Stir the flask contents for 4 h at 42 °C immersed in a silica oil bath.
4. Add solid $NaHCO_3$ and adjust solution to pH 7–8.
5. Freeze dried reaction mixture then extract with 3×50 mL washes of EtOAc.
6. Wash organic phase with 100 mL of H_2O and 100 ml of brine.
7. Dry reaction mixture over $MgSO_4$.
8. Filter and concentrate to give the crude compound.
9. Carry out flash chromatography using EtOAc-MeOH, in a 30:1 ratio to give the pure product as white needles (*see* **Note 5**).

4 Notes and Product Physical and Spectroscopic Data

The purity of each compound may be checked with the NMR, Infra Red (IR) and Mass spectral (MS) data given below.

1. The approximate yield of this reaction is 3.072 g (92 %). The melting point is 92–94 °C;^{1}H NMR ($CDCl_3$/TMS) δ (ppm) 1.51 (9H, s), 3.78 (3H, s), 6.32 (^{1}H, br), 6.83 (2H, d, J=9.2 Hz), 7.26 (2H, d, J=8.8 Hz); IR ν ($CHCl_3$) cm^{-1} 3,450, 1,730; High Resolution MS data for M^+ 223 m/z=223.1207.
2. The approximate yield of this reaction is 9.97 g (63.74 %). The melting point is 49–51 °C; ^{1}H NMR ($CDCl_3$/TMS) δ (ppm) 7.75 (d, J=9.1 Hz), 7.23 (d, J=2.9 Hz), 6.85 (dd, J=2.9, 9.0 Hz), 6.55 (br, 1H), 3.76 (s, 3H), 1.53 (s, 9H); ^{13}C NMR (90 MHz, $CDCl_3$) δ 155.9, 153.0, 132.3, 123.6, 114.8, 80.7, 55.6, 28.3; IR (KBr): 3,346 (s), 2,978 (m), 1,700 (s), 1,515 (s), 1,163 (s); High Resolution MS data for M^+ 349 m/z=349.0173.

3. The approximate yield of this reaction is 122 mg (76.7 %). The melting point is 114–118 °C; ^{1}H NMR ($CDCl_3$) δ 7.97 (d, J=9.89 Hz, 1H), 6.96 (br, 1H), 6.96–6.83 (mt, 2H), 6.05 (br, 1H), 4.33 (d, J=5.11 Hz, 2H), 3.75 (s, 3H), 2.05 (s, 3H), 1.53 (s, 9H); ^{13}C NMR (90 MHz, $CDCl_3$) δ 199.91, 154.32, 152.69, 133.13, 119.68, 116.28, 115.83, 111.91, 91.42, 80.55, 78.50, 55.36, 29.91, 28.20, 22.79; IR (KBr): 3,328(s), 1,698(s), 1,652(s), 1,525(s), 1,292(s), 1,163(s); High resolution MS (ES, Na) for M^+Na^+ 341.1473 m/z=341.3564.
4. The approximate yield of this reaction is 165 mg (62.0 %) of K_1 (melting point 142–118 °C) with 42.1 mg (14.1 %) of K_2; ^{1}H NMR ($CDCl_3$) δ 11.15 (br, 1H), 8.53 (d, J=9.16Hz, 1H), 8.31 (d, J=1.65 Hz, 1H), 7.26 (d, J=2.75 Hz, 1H), 7.03 (dd, J=2.75, 9.16 Hz), 6.99 (dd, J=2.75.9.16 Hz), 6.2 (br, 1H), 3.72 (s, 3H), 3.6–3.4 (m, 2H), 3.2–3.1 (m, 2H), 1.86 (s,3H); ^{13}C NMR (90 MHz, $CDCl_3$) δ 203.20, 170.242, 159.35, 154.83, 133.10, 123.07, 120.66, 118.07, 115.46, 55.64, 39.64, 34.42, 23.19. IR (KBr): 3,331(m), 1,687 (s), 1,671 (s), 1,649 (s), 1,540 (s), 1,195 (s); High resolution MS (ES, Na) for M^+Na^+ 287.1004 m/z=287.2662.
5. The approximate yield of this reaction is 154 mg (58.0 %) of K_2 (melting point 86–88 °C); ^{1}H NMR (200 MHz, $CDCl_3$) δ 7.13 (d, J=2.93 Hz, 1H), 6.99 (dd, J=2.93, 8.9 Hz, 1H), 6.95 (dd, J=2.93, 8.9 Hz, 1H), 6.3 (br, 1H), 5.8–5.0 (br, 2H); ^{13}C NMR ($CDCl_3$) δ 200.87, 170.10, 150.14, 145.04, 123.94, 118.83, 117.32, 112.95, 55.91, 38.72, 34.49, 23.30; IR (KBr): 3,446 (s), 3,339 (s), 2,937 (s), 1,731 (s), 1,651 (s), 1,557 (s), 1,237 (s); High resolution MS (ES, Na) for M^+Na^+ m/z 259.1054=259.2623.

Acknowledgments

The authors would like to thank the staff and students of the organic synthesis group at RMIT University for their ongoing help and support.

References

1. Holloway GA, Hügel HM, Rizzacasa MA (2003) Formal total synthesis of salicylihalamides A and B. J Org Chem 68:2200–2204
2. Tsegay S, Hügel H, Rizzacasa M (2009) Formal total synthesis of (+)-citrafungin A. Aust J Chem 62:676–682
3. Urban S, Separovic F (2005) Developments in hyphenated spectroscopic methods in natural product profiling. Front Drug Design Discov 1:113–166
4. Kopin IJ, Pare CMB, Axelrod J, Weissback H (1961) The fate of melatonin in animals. J Biol Chem 236:3072–3075
5. Kennaway DJ, Hügel HM, Clarke S, Tjandra A, Johnson DW, Royles P, Webb HA, Carbone F (1988) Structure–activity studies of melatonin analogues in prepubertal male rats. Aust J Biol Sci 41:393–400
6. Fujiwara M, Shibata M, Watanabe Y, Nukiwa T, Hirata F, Mizuno N, Hayaishi O (1978)

Indoleamine 2,3-dioxygenase. Formation of L-kynurenine from L-tryptophan in cultured rabbit pineal gland. J Biol Chem 253:6081–6087

7. Hirata F, Hayaishi O, Tokuyama T, Seno S (1974) *In vitro* and *in vivo* formation of two new metabolites of melatonin. J Biol Chem 249:1311–1313
8. Entrena A, Camacho ME, Carrion MD, Lopez-Cara LC, Velasco G, Len J, Escames G, Acuna-Castroviejo D, Tapias V, Gallo MA, Vivo A, Espinosa A (2005) Kynurenamines as neural nitric oxide synthase inhibitors. J Med Chem 48:8174–8181
9. Amiet G, Hügel HM, Nurlawis F (2002) The synthesis of the kynurenamines K_1 and K_2, metabolites of melatonin. Synlett 3:495–497
10. Hügel HM, Nurlawis F (2003) Short syntheses of melatonin. Heterocycles 60:2349–2354

Chapter 13

Discovery, Biosynthesis, and Rational Engineering of Novel Enterocin and Wailupemycin Polyketide Analogues

John A. Kalaitzis

Abstract

The marine actinomycete *Streptomyces maritimus* produces a structurally diverse set of unusual polyketide natural products including the major metabolite enterocin. Investigations of enterocin biosynthesis revealed that the unique carbon skeleton is derived from an aromatic polyketide pathway which is genetically coded by the 21.3 kb *enc* gene cluster in *S. maritimus*. Characterization of the *enc* biosynthesis gene cluster and subsequent manipulation of it via heterologous expression and/or mutagenesis enabled the discovery of other *enc*-based metabolites that were produced in only very minor amounts in the wild type. Also described are techniques used to harness the enterocin biosynthetic machinery in order to generate unnatural *enc*-derived polyketide analogues. This review focuses upon the molecular methods used in combination with classical natural products detection and isolation techniques to access minor metabolites of the *S. maritimus* secondary metabolome.

Key words Enterocin, Wailupemycin, Polyketide, Biosynthesis, Type II polyketide synthase, Biosynthesis gene cluster, Mutasynthesis, Heterologous expression, Enzymatic synthesis

1 Introduction

Microorganisms are a proven source of both novel and bioactive secondary metabolites (natural products) [1]. The ability to culture such organisms and therefore scale-up production levels of secondary metabolites to isolable quantities makes them attractive targets for natural product discovery programs despite the common notion that <1% of microorganisms can be cultured under laboratory or artificial conditions. Given this, the role of metabolomics, or more precisely secondary metabolomics, is an important component of the novel natural products discovery process for a number of reasons. Most important is the role metabolomics plays in surveying the hidden secondary metabolome of the microbe of interest. In order to discover new compounds it is essential that leads provided by the cultivable minority of microbes are taken full advantage of.

Ute Roessner and Daniel Anthony Dias (eds.), *Metabolomics Tools for Natural Product Discovery: Methods and Protocols*, Methods in Molecular Biology, vol. 1055, DOI 10.1007/978-1-62703-577-4_13,

Many natural product molecules from microbial sources are structurally complex rendering them synthetically intractable, and thus the cultured organism is often the best and only source of the compound. This is important, for example if the natural product happens to show significant bioactivity, and thus promise as a drug candidate, as greater amounts of the compound would be required for clinical trials.

Complete genome sequencing of microbes has revealed that the potential for novel natural product discovery from a microorganism's associated secondary metabolome [2–6] is greater than what was anticipated prior to the genomics-driven discovery era, and thus the development and application of metabolomics tools to streamline the discovery process is not only crucial but now commonplace. No single analytical tool is currently available to survey an entire secondary metabolome and so complementary methods need to be employed in order to obtain as near as complete picture as possible.

To access minor secondary metabolites, natural product researchers have a number of new tools available to them which complement the well-established chromatographic, mass spectrometry, and NMR methods. Integration of the so-called omics technologies to the traditional discovery platform has provided great insight into a microbe's secondary metabolome and its natural product biosynthetic potential [7, 8]. In combination with MS and NMR methods, "omics" technologies now form part of the initial dereplication process. Dereplication is vital for identifying potentially novel structures for isolation early in the discovery process rather than expending efforts isolating known natural products or simple analogues thereof.

Advances in DNA sequencing technologies together with molecular biology tools have enabled a firm understanding of the genetic basis for the biosynthesis of complex bacterial natural products [9–11]. Bioinformatics based genome mining tools are now used to predict partial or even near complete structures of secondary metabolites from genome sequence data alone. Bioinformatics analyses such as these, give insight into the potential or hidden secondary metabolome of a single microbe prior to isolation. This information can also assist in tailoring culture conditions for production of a specific metabolite as well directing the isolation of it [12]. Comparison of natural product biosynthesis-coding gene sequences with online databases can also be considered a form of dereplication and provide the foundation for the metabolomic survey [13].

With many turning their attention to the more challenging task of revealing the hidden secondary metabolome, recent research efforts have been focussed upon a single biosynthesis gene cluster, or the complement of gene clusters revealed through genome sequencing. This approach involves linking a biosynthesis gene cluster to its product or vice versa, and exploiting the biosynthetic

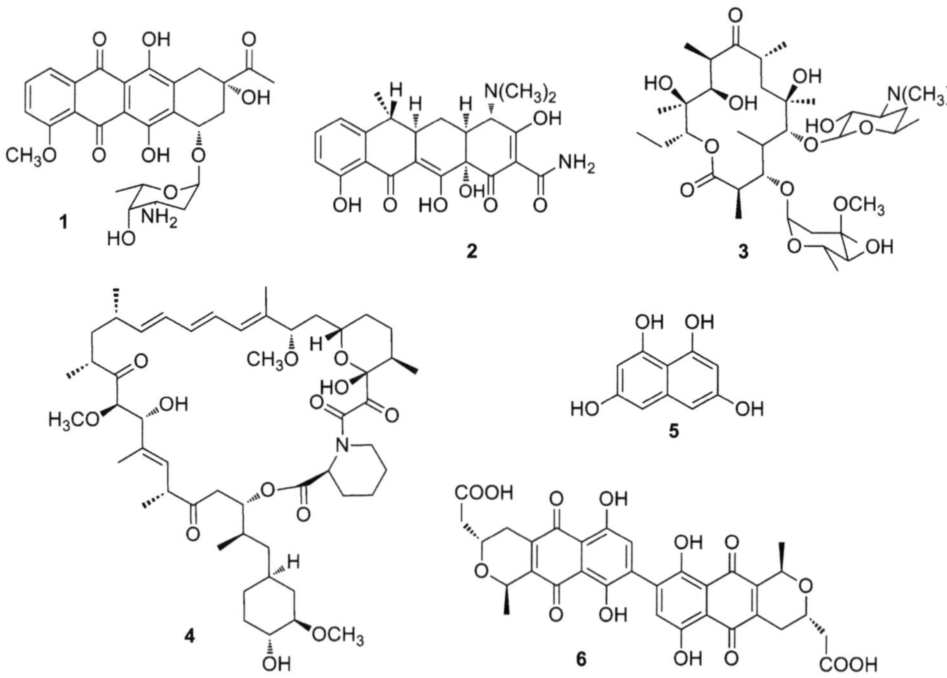

Fig. 1 A selection of polyketide natural products

capability of the identified cluster(s) [14]. A major advantage of harnessing a single biosynthesis gene cluster for the purpose of discovering new metabolites is the removal of major and/or interfering metabolites thus allowing access to minor metabolites produced by the parent microorganism.

This chapter focuses on the enterocin biosynthesis gene cluster (*enc*), and the assembly of the structurally unique polyketide natural products it codes for [15, 16]. The discovery of some minor polyketides, which were identified from extracts derived from heterologous expression systems and biosynthesis gene knockout mutants, is detailed along with an overview of the molecular methods used to engineer libraries of unnatural analogues. While this chapter highlights aspects of enterocin and polyketide biosynthesis, the relevant references cited throughout the text should be consulted for in-depth discussions. Before proceeding to the enterocin biosynthesis gene cluster it is necessary to briefly introduce bacterial polyketide biosynthesis and the enterocin family of polyketides.

2 Bacterial Polyketide Biosynthesis

Polyketide natural products (Fig. 1) possess a broad spectrum of biological activities and command an important role in human health with natural products such as the anticancer agent daunorubicin (**1**) and the antibiotic tetracycline (**2**) in clinical use. Actinomycetes, which include the common genus *Streptomyces*, are a proven source of similar aromatic polyketides as well as large macrocyclic polyketides such as erythromycin (**3**) and rapamycin (**4**) and small entities such as 1,3,6,8-tetrahydroxynaphthalene (THN) (**5**) [17, 18].

Nature has evolved a universal biosynthetic mechanism for the production of polyketides [18]. Polyketides are biosynthesized by microorganisms from small carboxylic acid building blocks and are assembled by repetitive Claisen condensations of an activated carboxylic acid starter unit with malonyl-CoA derived extender units to form a highly labile poly-β-ketide. The construction of the final polyketide product is catalyzed by a discrete set of biosynthetic enzymes collectively known as a polyketide synthase (PKS) and genes encoding the PKS, are typically clustered on the microbe's genome hence the common term "biosynthesis gene cluster."

Bacterial PKSs are classified into three groups on the basis of molecular genetics, biochemistry, and PKS structure. In the very simplest of terms, PKSs are categorized as either type I, II, or III and a number of in-depth reviews have been published on each of the three types [18–22]. Type I and III PKSs will not be dealt with in this chapter, though it is important to note that bacterial type I PKSs are responsible for the biosynthesis of large macrocyclic type polyketides such as erythromycin (**3**) and rapamycin (**4**) while bacterial type III PKSs yield small aromatic polyketides such as THN (**5**) [19]. Prototypical type II PKS products include tetracycline and the blue *Streptomyces coelicolor* pigment actinorhodin (**6**) [23]. The vast majority of type II PKS products are polycyclic aromatic polyketides with only a few exceptions, one of them being enterocin (**7**).

3 Type II Polyketide Synthases

Polycyclic aromatic polyketides, although numerous, belong to in general six major structure types, namely, the angucyclines, anthracyclines, aureolic acids, benzoisochromanequinones, tetracenomycins, and tetracyclines. Biosynthesis gene clusters for each of these structure types have been cloned and sequenced in actinomycetes and confirmed as type II PKS products. Type II PKSs are comprised of sets of iteratively used individual proteins that generate poly-β-keto intermediates and catalyze cyclodehydrations to yield multicyclic aromatic natural products such as the clinically important tetracenomycins, doxorubicin and tetracyclines. A type II PKS contains a minimal PKS which consists of two β-ketoacyl:ACP synthase subunits KS_α and KS_β, an acyl carrier protein (ACP) and the malonyl-CoA:ACP transacylase (MAT), and it governs the construction of the unmodified polyketide chain. The minimal PKS alone can function to assemble cyclized, unmodified polyketides and thus is the smallest set of proteins required for biosynthesis. Further associated proteins function to modify the polyketide and thus contribute directly to structural diversity. The prototypical actinorhodin (*act*) PKS has long been investigated in an effort to fully understand polyketide assembly in actinomycetes [23]. As with many type II PKSs, the *act* PKS shares architectural similarities with the enterocin (*enc*) PKS. Minor differences between the two biosynthesis gene clusters accounts for the divergence of pathways in operation in the producing organisms and thus the complement of polyketides isolated from the respective wild-type strains. Rather than describe in great detail type II PKSs, facets of these will be presented in context to the *enc* PKS in later sections.

4 The Enterocin Family of Polyketide Natural Products

The antibiotic polyketide enterocin (7) was first reported from two strains of *Streptomyces* spp. isolated from soil collected in Osaka, Japan [24, 25] and shortly thereafter (named as vulgamycin) from a strain of *S. hygroscopicus* [26]. Early biosynthesis feeding studies in *S. hygroscopicus* revealed that the enterocin molecule was derived from benzoate, methionine and seven acetates, and the cage-like core likely resulted from a Favorskii type rearrangement [26]. Later, chemical investigations of two specimens of marine ascidians of the genus *Didemnum* collected in Western Australia yielded enterocin (7) together with 5-deoxyenterocin (**8**), and the esters enterocin-5-behenate (**9**) and enterocin-5-arachidate (**10**) [27]. Related polyketides were reported from a marine derived actinomycete *Streptomyces maritimus* originally designated BD-26T(20), which was isolated from sediments collected from Wailupe Beach Park, Oahu, Hawaii [28]. When cultured in a seawater based media, *S. maritimus* produced the novel compounds 3-*epi*-5-deoxyenterocin (**11**), wailupemycins A-C (**12–14**) together with the major metabolites (7) and (**8**) (Fig. 2) [28]. This discovery was made around the same time as genetics approaches were starting to be used more often in biosynthesis studies. The relatively quick growth rate of *S. maritimus*, the structural diversity of the enterocins and wailupemycins, together with the unusual cage like core structure possessed by these metabolites, rendered this organism and its secondary metabolites an attractive target for further biosynthesis investigations.

5 The Genetic Basis for Enterocin Biosynthesis

The α-pyrone moiety, common to all of the *S. maritimus* derived structure variants and some other previously confirmed type II PKS products [29, 30] indicated a strong likelihood that these too were assembled via a type II PKS pathway. With this in mind, the search

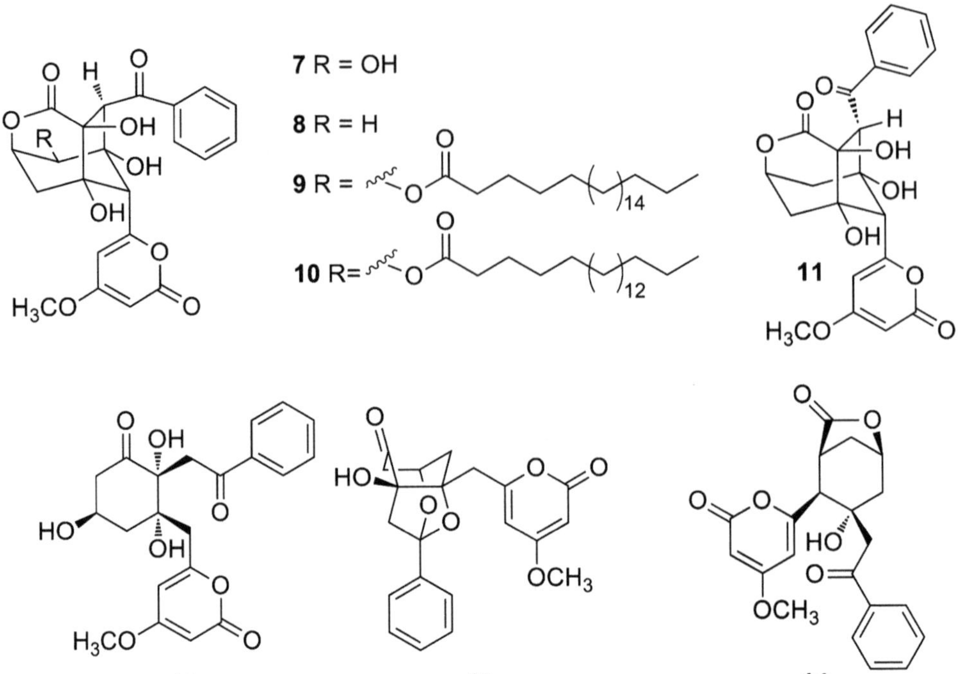

12 13 14

Fig. 2 Enterocin and related metabolites isolated directly from natural sources

for the enterocin polyketide synthase in *S. maritimus* began in earnest. An *S. maritimus* cosmid library constructed in SuperCosI was screened for *act*I (actinorhodin minimal PKS) gene homologues with an *act*I-ORF1 probe and revealed several candidate type II PKS biosynthesis gene sets [15]. Sequencing of one of the clusters, which additionally hybridized to an *act*III (actinorhodin ketoreductase) probe, revealed the presence of genes proposed to be involved in the biosynthesis of the benzoate derived starter unit, all but confirming the proposed candidate gene cluster [15]. Ultimately, unequivocal proof that this cluster was indeed coding for the assembly of the enterocin family of molecules was achieved through heterologous expression in the engineered host *S. lividans* K4-114 (*see* Subheading 6) [15].

6 The Enterocin Biosynthesis Gene Cluster (*enc*)

Discovery, sequencing and functional annotation of the 21.3 kb enterocin (*enc*) biosynthesis gene cluster in *S. maritimus* represented the first characterized natural product biosynthesis cluster from the marine environment [16]. The centrally located *enc* minimal PKS is flanked by a number of accessory enzymes involved in oxidation, reduction and cyclization of the poly-β-ketide together with those involved in starter unit biosynthesis and attachment (Fig. 3) [16].

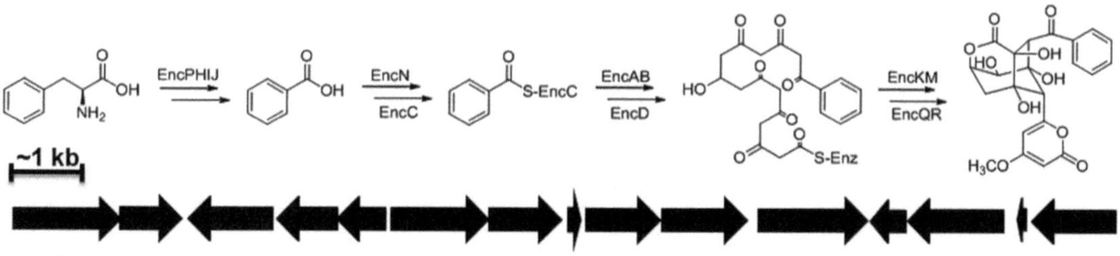

Fig. 3 Abbreviated biosynthesis of enterocin from L-phenylalanine, and partial map of the enterocin (*enc*) biosynthesis gene cluster in *S. maritimus*. The proposed functions of the respective gene products are: EncA-EncB (ketosynthase αβ subunits); EncC (acyl carrier protein); EncD (ketoreductase); EncH (cinnamate:CoA ligase), EncI (cinnamoyl-CoA hydratase); EncJ (β-oxoacyl-CoA thiolase); EncK (*O*-methyl transferase); EncL (type II thioesterase); EncM (FAD dependent oxygenase); EncN (benzoate:ACP ligase); EncO (unknown); EncP (phenylalanine ammonia lyase); EncQ (ferredoxin); EncR (cytochrome P450 hydroxylase)

Bioinformatics based interrogation of the sequenced gene cluster supported conclusions drawn from prior biosynthesis studies and provided evidence for proposed assembly routes. Specifically, the enterocin family of molecules is derived from a type II PKS, whose assembly is primed by benzoate which was proposed to be derived from phenylalanine. The observed structural diversity relating to the rearranged carbon backbones of the enterocins is attributed to the lack of cyclases and aromatases in the *enc* gene cluster, components which are present in other type II PKSs [16]. Furthermore, an FAD dependent oxygenase was predicted to play a role in the uncommon Favorskii rearrangement first proposed by Seto and coworkers [26]. With this information at hand the challenge was to then functionally characterize the discrete genes comprising the *enc* cluster and further elaborate the biosynthesis pathway.

The cosmid clone pJP15F11, a derivative of an *E. coli–Streptomyces* shuttle cosmid, containing the entire *enc* gene cluster was heterologously expressed in the genetically engineered host strain *S. lividans* K4-114 [31, 32]. Analysis of the resultant extract by LC-MS revealed the major metabolites 7 and **8** as well as a number of uncharacterized minor compounds [15]. Confirmation that these novel compounds were indeed products of the *enc* cluster was provided by supplementing the transconjugant *S. lividans* K4-114/pJP15F11 with benzoic acid-d_5. The resultant extract was analyzed for deuterium enrichment by LC-MS in order to confirm benzoate incorporation, and a total of ten novel benzoate derived metabolites were detected. The yet to be characterized benzoate derived metabolites had molecular masses of either 414, 364, or 346 g/mol, corresponding to polyketides with molecular formulas of $C_{21}H_{18}O_9$, $C_{21}H_{16}O_6$ and $C_{21}H_{14}O_5$ respectively [15]. One of the molecules was subsequently isolated from a 2 L culture (5 mg) of *S. maritimus* fermented in the saltwater based medium A1. Structure elucidation

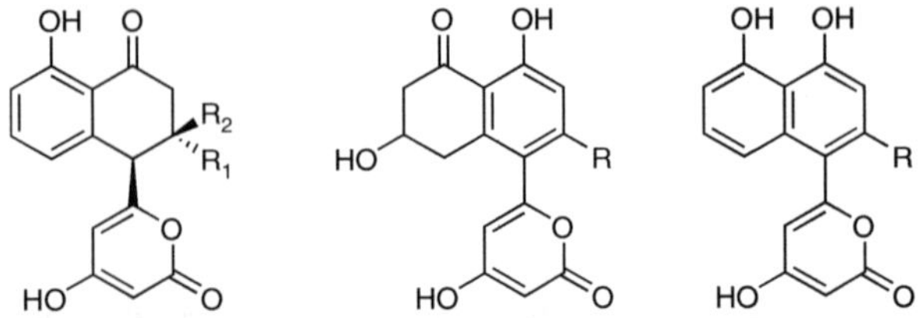

15 R_1=Ph R_2=OH
16 R_1=CH_3 R_2=OH
17 R_1=OH R_2=Ph

18 R=Ph
20 R=CH_3
23 R=n-Pr

19 R=Ph
21 R=CH_3
22 R=n-Pr

Fig. 4 Wailupemycins D–G and related compounds

of the novel compound, wailupemycin D (**15**), was achieved using common NMR and MS techniques, and was found to contain the characteristic α-pyrone further extending upon the known natural chemical diversity encoded by the *enc* cluster [15]. Wailupemycin D (**15**) is a structural analogue of EM18 (**16**), a shunt product of the *act* minimal PKS [33], with the only difference being the substitution of the phenyl group in **15** by a methyl in **16** (Fig. 4). In Section 8, the characterization of the wailupemycin D (**15**) analogues, wailupemycins E-G (**17–19**), minor products of both the wild type and *S. lividans* K4-114/pJP15F11 transconjugant, will be detailed.

7 Functional Characterization of the Enterocin Biosynthesis Gene Cluster

Genetic characterization of the enterocin type II PKS by way of establishing an individual gene's function was initially pursued via the construction of deletion, or gene inactivated, mutants of *S. maritimus.* Genes comprising the enterocin biosynthesis cluster were systematically deleted in an effort to abolish polyketide production and then often complemented with a biosynthetic precursor or the wild type biosynthesis gene to restore function and thus enterocin biosynthesis. Analysis of organic extracts derived from the various mutants by LC-MS or HPLC and comparison with an *S. maritimus* extract usually confirmed or dispelled a proposed biosynthetic hypothesis. Rather than discuss all mutants individually, some key outcomes relevant to this chapter are discussed in greater detail in the following sections.

8 The Favorskii-Like Rearrangement

In order to better understand the biosynthetic Favorskii-like rearrangement, the gene coding the proposed "favorskiiase" *encM*, was inactivated by single crossover homologous recombination [34].

The resultant *encM* knockout strain, *S. maritimus* KM was grown on solid A1 media, extracted and analyzed by LC-MS. As expected, production of the enterocins was completely abolished; however, minor metabolites derived from the pathway which were previously observed from the wild type strain and the *S. lividans* K4-114/pJP15F11 transconjugant were present in levels which allowed them to be isolated and characterized [34]. Wailupemycin D (**15**) was reisolated along with its diastereomer wailupemycin E (**17**) and another isomer wailupemycin F (**18**). The dehydration product wailupemycin G (**19**) (346 g/mol), was also purified from the extract and it is notable that **18** and **19** are structural analogues of the *act* PKS products mutactin (**20**) [35] and dehydromutactin (**21**) [36] respectively, differing only at position C15, as in the case of the analogous pair wailupemycin D (**15**) and EM18 (**16**) (Fig. 4) [34].

An almost identical extract was derived from a knockout mutant where the methyl transferase coding gene *encK* was deleted. Based on this data it was proposed that EncK and EncM jointly catalyze the Favorskii-like rearrangement as well as the pyrone methylation in enterocin biosynthesis [34]. Later, Graziani and coworkers reported the structures of related polyketides phaeochromycins A-E from the soil actinomycete *Streptomyces phaeochromogenes* LL-P018 [37]. Like dehydromutactin (**21**) and mutactin (**20**), phaeochromycin A (**22**) and phaeochromycin B (**23**) are analogues of wailupemycin polyketides, the only difference being the attachment of a butyrate derived *n*-propyl group rather than the benzoate derived phenyl ring (Fig. 4). This discovery further exemplifies the diverse biosynthetic capability of the actinomycetes and the utility of culturing microbes for novel natural products.

9 Starter Unit Biosynthesis

The function of each of the biosynthesis enzymes proposed to be involved in starter unit assembly was partially assigned through analysis of the cluster's gene sequence [16, 38]. Functional characterization was undertaken mainly by mutagenesis in which each proposed starter unit biosynthesis gene (*encH, I, J, P*, and *N*) was inactivated by single and/or double crossover homologous recombination [39, 40]. These studies confirmed that the benzoate derived starter unit is arrived at by one of two pathways, either endogenously from phenylalanine via a plant like pathway, or exogenously from the free carboxylate. A key step in the endogenous pathway is the deamination of L-phenylalanine to *trans*-cinnamic acid which is facilitated by the phenylalanine ammonia lyase (PAL) EncP (Fig. 5) [41]. Inactivation of *encP* in *S. maritimus* resulted in the complete abolishment of benzoate primed polyketides; however, on supplementation with cinnamic or benzoic acid to the mutant (*S. maritimus* XP), wild type levels of enterocin and wailupemycins

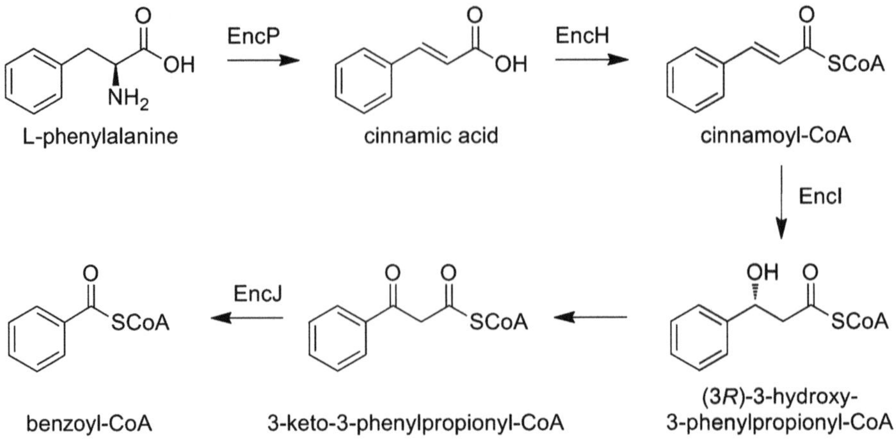

Fig. 5 Plant like biosynthetic route to the enterocin starter unit benzoyl-CoA in *S. maritimus*

D–G were successfully restored [39]. Supplementation with benzoates to restore polyketide production by the mutant *S. maritimus* XP via the exogenous pathway, was facilitated by EncN [42].

10 Exploiting Starter Unit Biosynthesis to Generate Novel Compounds via Mutasynthesis

Early work on enterocin biosynthesis by Seto and coworkers showed that supplementation feeding experiments with fluorinated benzoates in the enterocin producer *S. hygroscopicus* No. A-5294 resulted in the biosynthesis of minor amounts of fluorinated enterocins [26]. By exploiting the exogenous route to benzoate primed polyketides in the mutant *S. maritimus* XP, a series of *p-fluorobenzoate* derived enterocins and wailupemycins (**24–29**) (Fig. 6) were prepared by mutasynthesis (Fig. 7, Table 1) [43]. The beauty of mutasynthesis as opposed to precursor directed biosynthesis is that it results in the sole production of the artificial metabolites with all natural metabolites (background) eliminated. This allows the identification of newly mutasynthesized metabolites and the generation of isolable amounts of compound. LC-UV-MS proved to be the ideal analytical tool to (a) determine the success of the mutasynthesis experiments, and (b) identify the artificial analogues from the resultant organic extract. Typically, artificial analogues of the natural products possessed similar retention times and displayed comparable UV spectra. Also, given the nature of mutasynthesis, the expected molecular ions can be calculated and the anticipated metabolites can be easily correlated with retention times and UV data.

A series of carboxylic acids (~1 mg) were individually administered to 25 mL cultures of *S. maritimus* XP grown on solid A1. In total 20 carboxylates including some cinnamates were tested, the

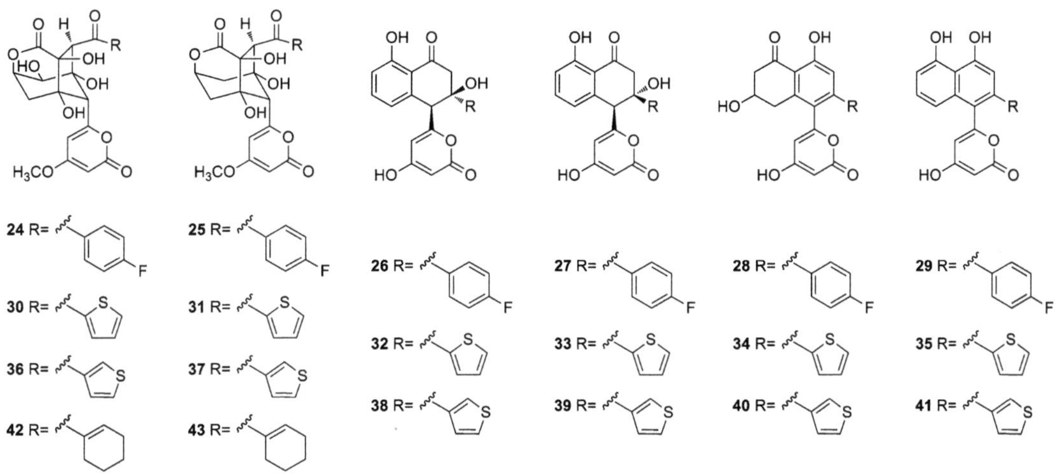

Fig. 6 Enterocin and wailupemycin analogues generated through mutasynthesis

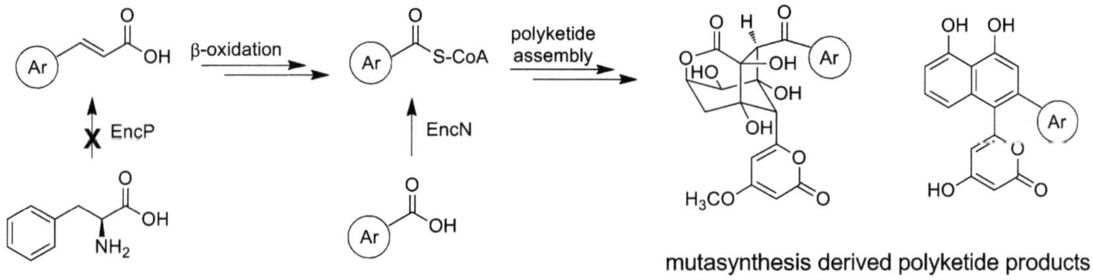

Fig. 7 Mutasynthesis of enterocins and wailupemycins in *S. maritimus* XP

Table 1
Engineered *enc*-based polyketides

Starter unit	Mutasynthesis	Enzymatic synthesis	
	S. maritimus XP	EncA/B, -C, -D, -N	EncA/B, -C, -D, -N, -M
p-fluorobenzoate	24–29	28–29	28–29, 55, 61
2-thiophene carboxylate	30–35	34–35	56, 34–35
3-thiophene carboxylate	36–41	40–41	57, 40–41
cyclohex-1-enecarboxylate	42–43	–	–
p-chlorobenzoate	–	51–52	51–52, 60, 62
3-hydroxybenzoate	–	45–46	58, 45–46
4-hydroxybenzoate	–	47–48	59, 47–48
3,4-dihydroxybenzoate	–	49–50	49–50
nicotinate	–	53–54	53–54

majority of which were not incorporated into polyketide products by the mutant. Analysis of culture extracts revealed the successful mutasynthesis of 2- and 3-thiophenecarboxylic acid derived enterocins and wailupemycins (**30–41**) (Fig. 6). Supplementation with cyclohex-1-enecarboxylic acid resulted in the biosynthesis of enterocin and 5-deoxyenterocin analogues (**42–43**); however, wailupemycin analogues were not detected (Table 1) [43]. These results indicated that EncN possessed relaxed substrate specificity and this was further probed and evaluated in vitro. Competitive assays were used to determine the in vitro conversion of various carboxylates (39 in total) to their corresponding CoA thioesters and analysis by LC-MS revealed that *p*-fluorobenzoate, cyclohex-1-enecarboxylate, 2- and 3-thiophenecarboxylates displayed the highest relative activities in comparison to the natural benzoate substrate, mirroring the in vivo observations [43]. This was the first reported exploitation of a type II PKS for the purposes of mutasynthesis.

11 Exploring Enterocin and Wailupemycin Biosynthesis In Vivo

Using the heterologous expression system described earlier, combinations of enterocin biosynthesis genes were expressed in *S. lividans* K4-114 cultured on solid R2YE supplemented with benzoate (Table 2). Comparison of culture extracts with authentic standards by HPLC confirmed biosynthesis proposals that were based on bioinformatics analyses of the *enc* cluster gene sequence, and data obtained from mutagenesis experiments. Most notably, the data revealed that the enterocin ketoreductase (EncD) is absolutely required for the production of benzoate primed polyketides, while EncL, originally assigned as a benzoyl-CoA:ACP acyltransferase and later formally characterized as a type II thioesterase, was not [44, 45].

Table 2
Heterologous expression of *enc* genes and polyketide products

Culture conditions	Expression plasmid (*enc* genes)	Major products	Reference
Solid R2YE + benzoate	pCH16 (ABCLMN)	–	[44]
	pCH16 (ABCLMN)/pCH20 (D)	**15–16, 18–19**	[44]
	pMP6 (ABCLMN/D)	**15–16, 18–19**	[44]
	pJK494 (ABCN/D)	**18–19**	[45]
	pBM43 (ABCLMN/DK)	**8, 18–19**	[46]
Liquid R2YE + benzoate	pMP6 (ABCLMN/D)	**15–16, 18–19, 44**	[46]
	pBM43 (ABCLMN/DK)	**8, 15–16, 18–19, 44**	[46]
Liquid R2YE + 44	pBM43 (ABCLMN/DK)	**8**	[46]

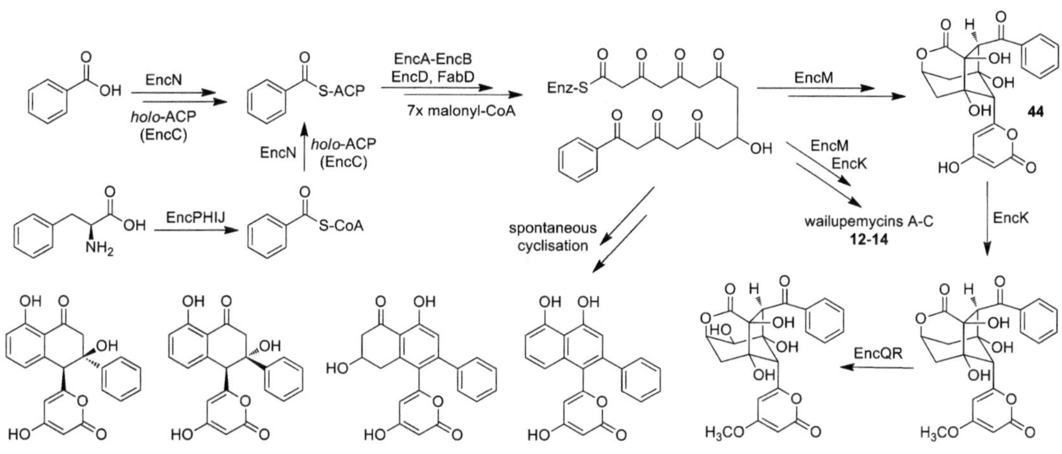

Fig. 8 Total biosynthesis of *enc*-based polyketides

An expression plasmid based on pRM5 [36], harboring the gene set *encABCLMN/encDK* was introduced into the host *S. lividans* K4-114. When the transformant was fermented in liquid, rather than on solid R2YE media, and supplemented with benzoic acid, not only were 5-deoxyenterocin (**8**) and wailupemycins D–G (**15**, **17**–**19**) produced, as observed when grown on solid media, but also the new natural product desmethyl-5-deoxyenterocin (**44**) (Fig. 8) [46]. The isolation and structure elucidation of **44** (414 g/mol), which was previously detected only by LC-MS provides a model example of accessing the hidden metabolome of a strain using innovative techniques. While it is only produced in very minor amounts by the wild type, almost 10 mg was isolated from a 1 L fermentation of the transformant [46]. Heterologous expression of a similar gene set lacking the methyltransferase coding gene *encK*, produced even greater relative amounts of the newly characterized metabolite. In contrast to the previous in vivo mutagenesis results, these data suggested that EncM alone catalyzes the Favorskii rearrangement and that EncK acts as a methyltransferase with rearranged polyketide substrates (Fig. 8) [46]. Interestingly, complementation of aureothin biosynthesis genes in vivo with *encK* resulted in the production of an *O*-methylated pyrone aureothin analogue supporting the functional assignment of EncK as a dedicated methyltransferase [47]. Taken together all of these observations provided the knowledge base for the subsequent ex vivo reconstitution of the enterocin pathway.

12 Enzymatic Synthesis of Enterocin Polyketides

Possessing an intimate knowledge of enterocin biosynthesis and the enzymes, cofactors and substrates involved in its assembly, Moore and coworkers set out to reconstitute the biosynthetic pathway ex vivo [48]. The multienzyme total synthesis was achieved with benzoate and malonate substrates in a single vessel and represented the first total in vitro assembly of a type II PKS natural product using purified enzymes (Fig. 8) [48].

In the absence of benzoate, an assay containing the minimal set of enterocin biosynthesis enzymes resulted in the production of a series of all malonate derived polyketides never detected in vivo. LC-MS of an extract derived from a [2-^{13}C] malonyl-CoA supplemented assay revealed the presence of three yet to be characterized nonaketides. A mass shift of 9 amu from 342 and 298 to 351 and 307 (indicating the incorporation of nine acetates) confirmed the nonaketides and HR FT-MS allowed the determination of the molecular formulas $C_{18}H_{14}O_7$ and $C_{17}H_{14}O_5$ [48]. While the structures of the three nonaketides are yet to be confirmed, this finding serves to highlight the utility of examining a single biosynthesis gene cluster in order to reveal even more of the organism's hidden secondary metabolome. Identification of the proposed nonaketides also suggested that the size of the primer unit may dictate polyketide chain length.

13 Enzymatic Synthesis of Unnatural Enterocins and Wailupemycins

Using the same biosynthetic machinery, an ex vivo library of unnatural enterocin and wailupemycin analogues were prepared by exploiting, once again, the relaxed substrate specificity of the benzoate:ACP ligase EncN (Fig. 9, Table 1) [49].

In the presence of the natural primer benzoate, incubation of enterocin minimal PKS enzymes, the ketoreductase EncD, the *S. glaucescens* malonyl-CoA:ACP transacylase (*Sg*FabD) and EncN together with malonyl-CoA and the appropriate cofactors, yielded the spontaneous cyclization products wailupemycins D–G (**15**, **17–19**) [48]. Alternative incubation with *p*-fluorobenzoate

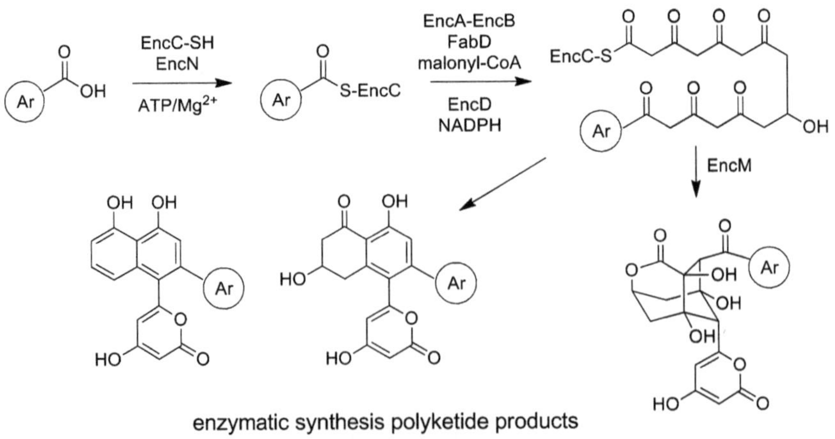

Fig. 9 Enzymatic synthesis of enterocins and wailupemycins

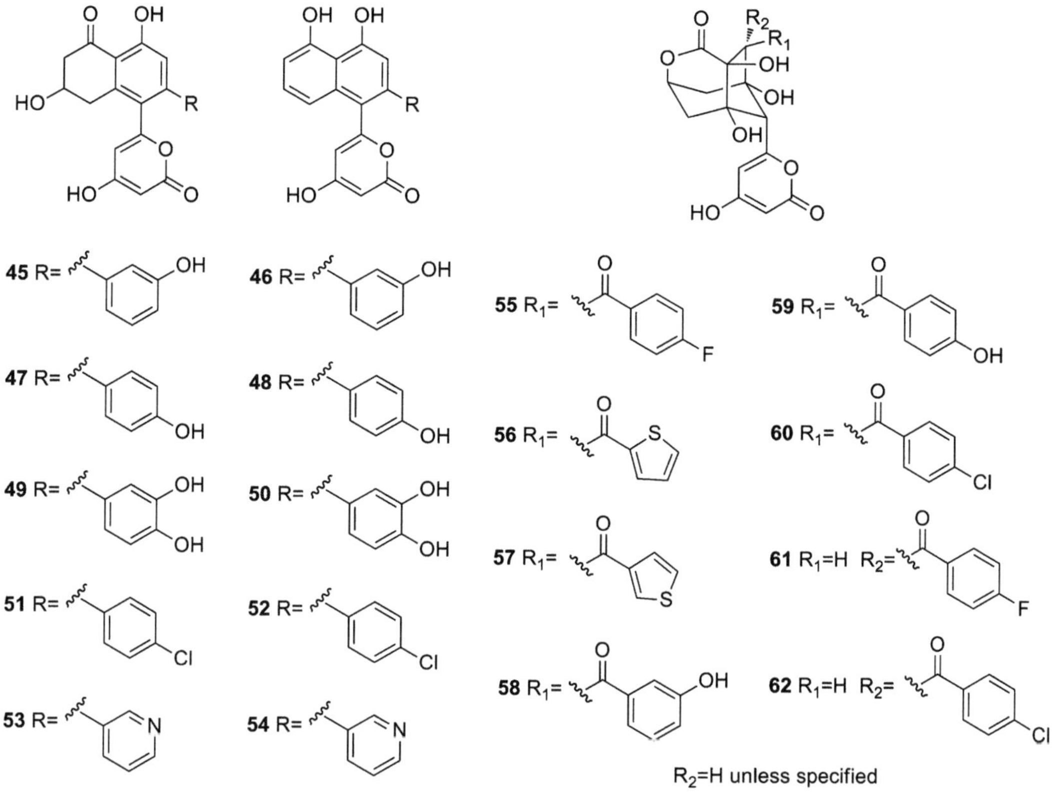

Fig. 10 Further enterocin and wailupemycin analogues generated through enzymatic synthesis

and 2- and 3-thiophene carboxylates gave wailupemycin F analogues (**28**, **34**, **40**) and wailupemycin G analogues (**29**, **35**, **41**) which were identical by LC-MS to standards generated by in vivo mutasynthesis [49]. Further molecules generated in vitro that were not detected from mutasynthesis experiments included 3- and 4-hydroxybenzoate, 3,4-dihydroxybenzoate, *p*-chlorobenzoate, and nicotinate derived wailupemycin F and G analogues (**45–54**, Fig. 10) [49]. It is likely that wailupemycin D (**15**) and E (**17**) analogues were also produced; however, these were not detected.

A subsequent set of enzymatic assays were performed in which the FAD-dependent favorskiiase EncM was added to the minimal set of biosynthesis enzymes. As expected a set of desmethyl-5-deoxyenterocin analogues (**55–60**) (Fig. 10, Table 1) were able to be generated from the same starters as successfully employed above with the exception of 3,4-dihydroxybenzoate and nicotinate [49]. Also detected by LC-MS were additional analogues (**61–62**) corresponding to the C-3 epimers of the halogenated **55** and **60**, analogues of **11**, a minor metabolite of the wild type strain [49]. These results further emphasized the usefulness of natural biosynthesis enzymes for the production of synthetically intractable molecules such as the enterocins albeit on a micro scale.

Furthermore, using this ex vivo enzymatic approach, additional analogues were generated that were not detected from mutasynthesis studies, thus expanding upon the functional nature of the enterocin biosynthesis gene cluster.

14 Concluding Remarks

The marine microbe *Streptomyces maritimus* produces a diverse range of polyketides including the unusual type II PKS product enterocin. Prior to any biosynthetic investigations of *S. maritimus*, six metabolites 7, **8**, and **11–14**, were thought to be produced from a single, yet to be described, polyketide synthase. Through characterization of the enterocin PKS mainly via mutagenesis, a further five natural products **15**, **17–19** and **44** produced by the wild type strain in only minor amounts, were identified. Armed with a thorough knowledge of the enterocin biosynthesis pathway and its underlying biosynthetic genetic machinery, the hidden *S. maritimus* secondary metabolome became readily accessible. Overviewed here were the molecular methods used to further understand the biosynthesis of the enterocins and wailupemycins at the genetic level. The modular and seemingly simple architecture of the enterocin biosynthesis gene cluster prompted it's manipulation with the aim of generating novel chemistries. As exemplified by studies of the *enc* cluster, the efficacy of harnessing biosynthesis genes for the generation of unnatural natural product analogues proved successful. The identification of novel metabolites produced in microgram quantities through enzymatic and mutasynthesis, relied heavily upon robust metabolomic techniques, predominantly LC-MS, and comparison of UV spectra with their respective natural analogues. In short, 20 natural product analogues were generated through mutasynthesis and at least another 20 via enzymatic synthesis illustrating this organism's extremely impressive genetic biosynthetic capability (Fig. 11).

Underpinned by the knowledge derived from the many natural product biosynthesis clusters characterized during the late 1990s through the early part of this century, chemical structures of microbial secondary metabolites can be predicted with some degree of confidence from genome sequence data thus providing the basis for a predicted secondary metabolome. When coupled with highly sensitive mass spectrometry techniques, this or similar genome mining approaches provide a powerful means of streamlining the natural product discovery process while helping to ensure that the biosynthetic capabilities of prolific microorganisms are fully capitalized on.

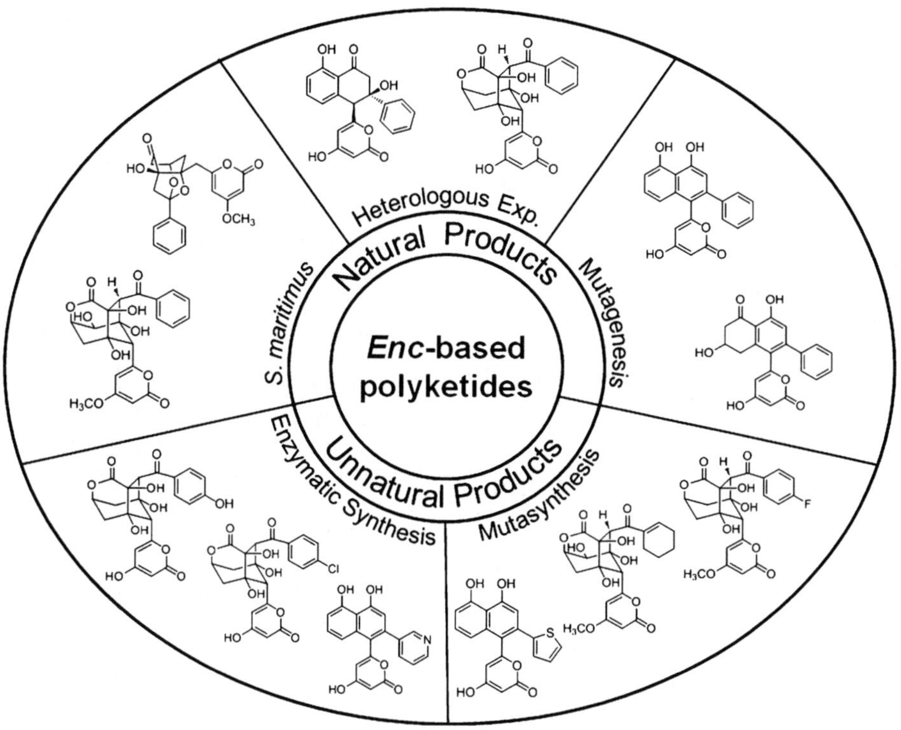

Fig. 11 Diversity and sources of natural and unnatural *enc*-based polyketides

Acknowledgments

I wish to thank Professor Bradley S. Moore (Scripps Institution of Oceanography, UCSD) for providing me with the opportunity to work on the enterocin project, and also for his advice, support, and mentorship. I also thank many colleagues and collaborators who are referenced throughout this paper for their vast contributions.

References

1. Demain AL, Sanchez S (2009) Microbial drug discovery: 80 years of progress. J Antibiot 62:5–16
2. Omura S, Ikeda H, Ishikawa J, Hanamoto A, Takahashi C, Shinose M, Takahashi Y, Horikawa H, Nakazawa H, Osonoe T, Kikuchi H, Shiba T, Sakaki Y, Hattori M (2001) Genome sequence of an industrial microorganism *Streptomyces avermitilis*: deducing the ability of producing secondary metabolites. Proc Natl Acad Sci USA 98:12215–12220
3. Bentley SD, Chater KF, Cerdeno-Tarraga AM, Challis GL, Thomson NR, James KD, Harris DE, Quail MA, Kieser H, Harper D, Bateman A, Brown S, Chandra G, Chen CW, Collins M, Cronin A, Fraser A, Goble A, Hidalgo J, Hornsby T, Howarth S, Huang CH, Kieser T, Larke L, Murphy L, Oliver K, O'Neil S, Rabbinowitsch E, Rajandream MA, Rutherford K, Rutter S, Seeger K, Saunders D, Sharp S, Squares R, Squares S, Taylor K, Warren T, Wietzorrek A, Woodward J, Barrell BG, Parkhill J, Hopwood DA (2002) Complete genome sequence of the model actinomycete *Streptomyces coelicolor* A3(2). Nature 417:141–147

4. Ikeda H, Ishikawa J, Hanamoto A, Shinose M, Kikuchi H, Shiba T, Sakaki Y, Hattori M, Omura S (2003) Complete genome sequence and comparative analysis of the industrial microorganism *Streptomyces avermitilis*. Nat Biotechnol 21:526–531
5. Udwary DW, Zeigler L, Asolkar RN, Singan V, Lapidus A, Fenical W, Jensen PR, Moore BS (2007) Genome sequencing reveals complex secondary metabolome in the marine actinomycete *Salinispora tropica*. Proc Natl Acad Sci USA 104:10376–10381
6. Krug D, Zurek G, Revermann O, Vos M, Velicer GJ, Muller R (2008) Discovering the hidden secondary metabolome of *Myxococcus xanthus*: a study of intraspecific diversity. Appl Environ Microbiol 74:3058–3068
7. Winter JM, Behnken S, Hertweck C (2011) Genomics-inspired discovery of natural products. Curr Opin Chem Biol 15:22–31
8. Donadio S, Monciardini P, Sosio M (2007) Polyketide synthases and nonribosomal peptide synthetases: the emerging view from bacterial genomics. Nat Prod Rep 24:1073–1109
9. Fischbach MA, Walsh CT (2006) Assembly-line enzymology for polyketide and nonribosomal peptide antibiotics: logic, machinery, and mechanisms. Chem Rev 106:3468–3496
10. Nett M, Ikeda H, Moore BS (2009) Genomic basis for natural product biosynthetic diversity in the actinomycetes. Nat Prod Rep 26:1362–1384
11. Van Lanen SG, Shen B (2006) Microbial genomics for the improvement of natural product discovery. Curr Opin Microbiol 9:252–260
12. Lautru S, Deeth RJ, Bailey LM, Challis GL (2005) Discovery of a new peptide natural product by *Streptomyces coelicolor* genome mining. Nat Chem Biol 1:265–269
13. Kalaitzis JA, Lauro FM, Neilan BA (2009) Mining cyanobacterial genomes for genes encoding complex biosynthetic pathways. Nat Prod Rep 26:1447–1465
14. Walsh CT, Fischbach MA (2010) Natural products version 2.0: connecting genes to molecules. J Am Chem Soc 132:2469–2493
15. Piel J, Hoang K, Moore BS (2000) Natural metabolic diversity encoded by the enterocin biosynthesis gene cluster. J Am Chem Soc 122:5415–5416
16. Piel J, Hertweck C, Shipley PR, Hunt DM, Newman MS, Moore BS (2000) Cloning, sequencing and analysis of the enterocin biosynthesis gene cluster from the marine isolate '*Streptomyces maritimus*': evidence for the derailment of an aromatic polyketide synthase. Chem Biol 7:943–955
17. Staunton J, Weissman KJ (2001) Polyketide biosynthesis: a millennium review. Nat Prod Rep 18:380–416
18. Hertweck C (2009) The biosynthetic logic of polyketide diversity. Angew Chem Int Ed 48:4688–4716
19. Austin MB, Noel JP (2003) The chalcone synthase superfamily of type III polyketide synthases. Nat Prod Rep 20:79–110
20. Khosla C, Kapur S, Cane DE (2009) Revisiting the modularity of modular polyketide synthases. Curr Opin Chem Biol 13:135–143
21. Shen B (2003) Polyketide biosynthesis beyond the type I, II and III polyketide synthase paradigms. Curr Opin Chem Biol 7:285–295
22. Van Lanen SG, Shen B (2008) Advances in polyketide synthase structure and function. Curr Opin Drug Disc 11:186–195
23. Hopwood DA (1997) Genetic contributions to understanding polyketide synthases. Chem Rev 97:2465–2497
24. Miyairi N, Sakai HI, Konomi T, Imanaka H (1976) Enterocin, a new antibiotic - taxonomy, isolation and characterization. J Antibiot 29:227–235
25. Tokuma Y, Miyairi N, Morimoto Y (1976) Structure of enterocin - x-ray analysis of *m*-bromobenzoyl enterocin dihydrate. J Antibiot 29:1114–1116
26. Seto H, Sato T, Urano S, Uzawa J, Yonehara H (1976) Utilization of C-13-C-13 coupling in structural and biosynthetic studies XII. Structure and biosynthesis of vulgamycin. Tetrahedron Lett 17:4367–4370
27. Kang H, Jensen PR, Fenical W (1996) Isolation of microbial antibiotics from a marine ascidian of the genus *Didemnum*. J Org Chem 61:1543–1546
28. Sitachitta N, Gadepalli M, Davidson BS (1996) New alpha-pyrone-containing metabolites from a marine-derived actinomycete. Tetrahedron 52:8073–8080
29. Yu TW, Shen YM, McDaniel R, Floss HG, Khosla C, Hopwood DA, Moore BS (1998) Engineered biosynthesis of novel polyketides from *Streptomyces* spore pigment polyketide synthases. J Am Chem Soc 120:7749–7759
30. Shen YM, Yoon P, Yu TW, Floss HG, Hopwood D, Moore BS (1999) Ectopic expression of the minimal *whiE* polyketide synthase generates a library of aromatic polyketides of diverse sizes and shapes. Proc Natl Acad Sci USA 96:3622–3627

31. Ziermann R, Betlach MC (1999) Recombinant polyketide synthesis in *Streptomyces*: engineering of improved host strains. Biotechniques 26:106–109
32. Ziermann R, Betlach MC (2000) A two-vector system for the production of recombinant polyketides in *Streptomyces*. J Ind Microbiol Biot 24:46–50
33. Alvarez MA, Fu H, Khosla C, Hopwood DA, Bailey JE (1996) Engineered biosynthesis of novel polyketides: properties of the *whiE* aromatase/cyclase. Nat Biotechnol 14:335–338
34. Xiang L, Kalaitzis JA, Nilsen G, Chen L, Moore BS (2002) Mutational analysis of the enterocin Favorskii biosynthetic rearrangement. Org Lett 4:957–960
35. Zhang HL, He XG, Adefarati A, Gallucci J, Cole SP, Beale JM, Keller PJ, Chang CJ, Floss HG (1990) Mutactin, a novel polyketide from *Streptomyces coelicolor* - structure and biosynthetic relationship to actinorhodin. J Org Chem 55:1682–1684
36. McDaniel R, Ebert Khosla S, Fu H, Hopwood DA, Khosla C (1994) Engineered biosynthesis of novel polyketides - influence of a downstream enzyme on the catalytic specificity of a minimal aromatic polyketide synthase. Proc Natl Acad Sci USA 91:11542–11546
37. Graziani EI, Ritacco FV, Bernan VS, Telliez JB (2005) Phaeochromycins A-E, anti-inflammatory polyketides isolated from the soil actinomycete *Streptomyces phaeochromogenes* LL-P018. J Nat Prod 68:1262–1265
38. Hertweck C, Moore BS (2000) A plant-like biosynthesis of benzoyl-CoA in the marine bacterium '*Streptomyces maritimus*'. Tetrahedron 56:9115–9120
39. Xiang L, Moore BS (2002) Inactivation, complementation, and heterologous expression of *encP*, a novel bacterial phenylalanine ammonia lyase gene. J Biol Chem 277:32505–32509
40. Xiang L, Moore BS (2003) Characterization of benzoyl coenzyme A biosynthesis genes in the enterocin-producing bacterium *Streptomyces maritimus*. J Bacteriol 185:399–404
41. Xiang L, Moore BS (2005) Biochemical characterization of a prokaryotic phenylalanine ammonia lyase. J Bacteriol 187:4286–4289
42. Izumikawa M, Cheng Q, Moore BS (2006) Priming type II polyketide synthases via a type II nonribosomal peptide synthetase mechanism. J Am Chem Soc 128:1428–1429
43. Kalaitzis JA, Izumikawa M, Xiang L, Hertweck C, Moore BS (2003) Mutasynthesis of enterocin and wailupemycin analogues. J Am Chem Soc 125:9290–9291
44. Hertweck C, Xiang L, Kalaitzis JA, Cheng Q, Palzer M, Moore BS (2004) Context-dependent behavior of the enterocin iterative polyketide synthase: a new model for ketoreduction. Chem Biol 11:461–468
45. Kalaitzis JA, Cheng Q, Meluzzi D, Xiang L, Izumikawa M, Dorrestein PC, Moore BS (2011) Policing starter unit selection of the enterocin type II polyketide synthase by the type II thioesterase EncL. Bioorgan Med Chem 19:6633–6638
46. Xiang L, Kalaitzis JA, Moore BS (2004) EncM, a versatile enterocin biosynthetic enzyme involved in Favorskii oxidative rearrangement, aldol condensation, and heterocycle-forming reactions. Proc Natl Acad Sci USA 101:15609–15614
47. Werneburg M, Busch B, He J, Richter MEA, Xiang L, Moore BS, Roth M, Dahse HM, Hertweck C (2010) Exploiting enzymatic promiscuity to engineer a focused library of highly selective antifungal and antiproliferative aureothin analogues. J Am Chem Soc 132:10407–10413
48. Cheng Q, Xiang L, Izumikawa M, Meluzzi D, Moore BS (2007) Enzymatic total synthesis of enterocin polyketides. Nat Chem Biol 3:557–558
49. Kalaitzis JA, Cheng Q, Thomas PM, Kelleher NL, Moore BS (2009) *In vitro* biosynthesis of unnatural enterocin and wailupemycin polyketides. J Nat Prod 72:469–472

Chapter 14

Bioassays for Anticancer Activities

Janice McCauley, Ana Zivanovic, and Danielle Skropeta

Abstract

The MTT/MTS in vitro cell proliferation assay is one of the most widely used assays for evaluating preliminary anticancer activity of both synthetic derivatives and natural products and natural product extracts. The highly reliable, colorimetric based assay is readily performed on a wide range of cell lines. This assay gives an indication of whole cell cytotoxicity; however, to determine the exact molecular target further assays need to be performed. Of these, kinase inhibition assays are also one of the most widespread enzyme inhibition screening assays performed. Kinases are enzymes that play a key role in a number of physiological processes and their inhibitors have been found to exhibit anticancer activity against various human cancer cell lines. Herein, we describe the methods for performing both in vitro MTT/MTS cytotoxicity and kinase enzyme inhibition assays. These are two of the most useful anticancer screening techniques available that are relatively economical and can be easily and routinely performed in the laboratory to characterize anticancer activity. Both assays are highly versatile and can be modified to test against targeted disease processes by using specific kinase enzymes or cell lines.

Key words MTS/MTT assays, Cytotoxicity, Anticancer activity, Human cancer cell lines, Enzyme inhibition, Kinases

1 Introduction

A goal of natural product chemistry and organic synthesis laboratories is ultimately drug discovery. An important aspect of the drug development process is testing both natural products and synthesized compounds for bioactivities that are involved in targeted diseases processes. Cancer is a general term to define a number of diseases that are characterized by the uncontrolled proliferation of cells resulting from the disruption or dysfunction of regulatory signaling pathways that are normally under tight control [1, 2]. Cancer can spread rapidly and invade other tissues and organs and different cancers are recognized to have unique characteristics or markers [3].

To assess for preliminary anticancer activity in terms of cell viability, the 3-(4,5-dimethylthiazol-2-yl)-2,5-diphenyltetrazolium

Ute Roessner and Daniel Anthony Dias (eds.), *Metabolomics Tools for Natural Product Discovery: Methods and Protocols*, Methods in Molecular Biology, vol. 1055, DOI 10.1007/978-1-62703-577-4_14, © Springer Science+Business Media, LLC 2013

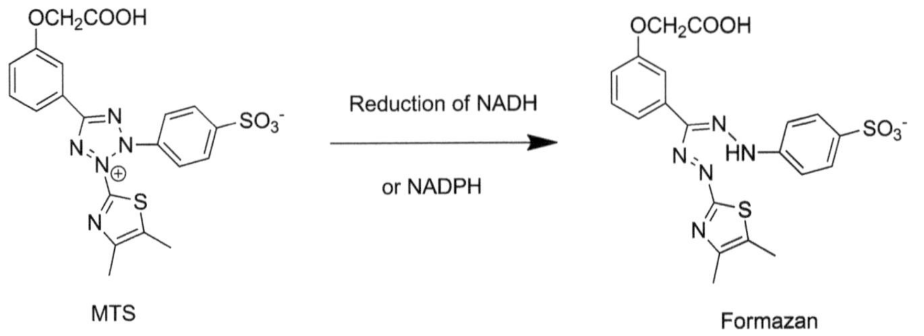

Fig. 1 The reduction of the MTS tetrazolium salt to the red formazan product by viable cells

bromide (MTT) and 3-(4,5-dimethylthiazol-2-yl)-5-(3-carboxymethoxyphenyl)-2-(4-sulfophenyl)-2H-tetrazolium inner salt (MTS) in vitro cytotoxicity assays are considered two of the most economic, reliable and convenient methods. This is based on their ease of use, accuracy and rapid indication of toxicity [4], as well as their sensitivity and specificity [5]. Both assays are in vitro whole cell toxicity assays that employ colorimetric methods for determining the number of viable cells based on mitochondrial dehydrogenase activity measurement and differ only in the reagent employed. In the MTT assay, 3-(4,5-dimethylthiazol-2-yl)-2,5-diphenyltetrazolium bromide is bioreduced by dehydrogenase inside living cells to form a colored formazan dye, while in the MTS assay, a similar bioconversion takes places utilizing 3-(4,5-dimethylthiazol-2-yl)-5-(3-carboxymethoxyphenyl)-2-(4-sulfophenyl)-2H-tetrazolium, inner salt and an electron coupling reagent (phenazine ethosulfate) PES (Fig. 1).

The MTT assay requires the addition of solubilizing agents to dissolve the insoluble formazan product formed, while the MTS assay generates a water-soluble formazan product, thus simplifying the assay. The number of viable cells is measured through colorimetry and works on the principle that the mitochondrial dehydrogenase enzymes which produces NADH or NADHP, reduces the colorless tetrazolium salt into a colored aqueous soluble formazan product by the mitochondrial activity of viable cells at 37 °C (Fig. 1). The quantity of the colored product is directly proportional to the number of live cells in the culture since the MTT/MTS reagent can only be reduced to formazan by metabolically active cells.

This is a useful assay to characterize potential anticancer agents and can be performed routinely and easily in the laboratory without the need to forego intellectual property. The MTT and MTS assays assess for toxicity to the particular cell under investigation (not anticancer activity per se). Therefore, most researchers screen for cytotoxicity against either murine or human cancer cell lines, as well as against a normal cell line such as peripheral blood lymphocytes. A selectivity index of the compound for cancer cells over normal cells can then be determined. It is also advisable to confirm MTS/MTT results with qualitative observations under the microscope of the cell morphology both before and after the assay. This can often assist in identifying potential modes of actions and deciding which further assays should be performed such as caspase activation, assessing stage of cell cycle arrest and microtubule stabilization or destabilization [6].

In addition to performing MTT or MTS assays against specific cell types, many researchers worldwide also submit their compounds to screening by the National Cancer Institute (NCI, http://www.cancer.gov), which is part of the US Government National Institutes of Health (NIH). The NCI offers a rapid in vitro primary anticancer drug screen to support cancer researchers worldwide [7]. The screen, which is performed at no cost to the researcher other than shipping of their sample, consists of a panel of 60 different human tumor cell lines from several cancer types including leukemia, melanoma and cancers of the lung, colon, brain, ovary, breast, prostate, and kidney. The NCI60 screen tests the degree of growth, inhibition, or cytotoxicity of a compound against each cell line over a range of concentrations to generate a characteristic profile or fingerprint of cellular response [8]. A computer program (COMPARE) is used to assess the pattern of response across the cell lines, which can point towards a likely mechanism of action or identify compounds with unique modes of activity and/or selectivities for specific cell types. Operational details are provided by the NCI Developmental Therapeutics Program (DTP) and can be found at http://dtp.nci.nih.gov and for information regarding the submission of compounds for testing in the NCI screens *see* http://dtp.nci.nih.gov/docs/misc/common_files/submit_compounds.html, [9]. Only pure compounds are screened in the assay and the supplier of the sample is required to provide the molecular structure of any compounds submitted, which are then reviewed and considered for testing.

The above assays are whole cell assays; however, often a researcher will want to determine activity against a specific molecular or cellular target in order to confirm the mechanism of action and to assess for selectivity towards different targets and any off-target effects. This can be achieved using in vitro inhibition assays against either purified enzymes or cell-free extracts enriched in the enzyme target of interest. One such anticancer screening method that can be easily and routinely performed in the laboratory in low-, medium- or high-throughput format is protein kinase inhibition [10]. After G protein-coupled receptors (GPCRs), protein kinases are the second most important anticancer drug target being pursued today.

Unlike MTS-type assays, which are cell based, these assays are based on detecting the degree to which the potential drug compound can inhibit an enzyme's activity. Most drugs used today demonstrate their bioactivity by acting as either receptor antagonists or as enzyme inhibitors. Enzymes are popular drug targets as they play a significant role in a number of disease processes and are susceptible to inhibition by small drug-like molecules [11].

Protein kinases are an abundant group of enzymes in the human body with approximately 518 different protein kinases encoded in the human genome [12]. Kinases catalyze the chemical transfer of a phosphate group from a high energy molecule such as adenine triphosphate (ATP) to a hydroxyl-containing substrate such as serine, threonine and tyrosine, and are divided into different families based on their selectivity for these amino acids.

As one of the most abundant groups of enzymes in the human body, protein kinases are important in almost every major pathway in eukaryotic cells. They play a central role in the regulation of cellular activities and in signal transduction in signal transmission pathways. Moreover, kinases have important roles in metabolism, cell growth, apoptosis, immune response, gene expression, oncogenesis, cell differentiation and proliferation, metabolism, DNA damage repair, and cell motility [2]. As a result, the deregulation of kinases has been identified to be the main cause in an increasing list of diseases [13]. An estimated one-third of pharma drug discovery programs now focus on targeting cancer-related kinases with the aim of developing potent and selective inhibitors with lower side effects than treatments that traditionally focus on DNA and chromosome regulation [14–17]. An example of a kinase inhibitor that has successfully proceeded onto the pharmaceutical market is Imitinib (Gleevec, Novartis) which is a tyrosine kinase inhibitor that has dramatically improved the prognosis for sufferers of chronic myeloid leukemia after being the first small-molecule kinase inhibitor to be approved for human use [13].

c-AMP Dependent protein kinase A (protein kinase A, PKA) is an enzyme involved in the phosphorylation of a wide range of proteins, ion channels, and transcription factors [18]. It has been demonstrated to regulate a number of physiological processes including cardiovascular function, steroid biosynthesis, reproductive function, myogenesis, adipocyte metabolism, exocytotic processes, and immune function [19, 20]. PKA was also found to play a key role in memory processes [21]. The cAMP-PKA pathway has been linked to the promotion of malignant phenotypes of head and neck squamous cell carcinoma [22] and demonstrated to be activated in a range of tumors [23]. Conversely, PKA inhibitors have been found to display both in vitro and in vivo antitumor activity against various human cancer cell lines and to enhance monocyte function in HIV-infected patients [24]. Thus, it is becoming increasingly apparent that the ability to selectively inhibit PKA provides a new way of potentially modulating cancer, immune function, and memory disorders such as Alzheimer's disease, Parkinson's disease, and schizophrenia [19, 21, 25]. Herein, we describe methods for performing both the MTT/MTS cell proliferation cytotoxicity assay and for screening against the enzyme PKA. Both methods are widely used in our laboratory [6, 26, 27].

2 Materials

Carry out all procedures at room temperature unless otherwise specified.

2.1 Protein Kinase Assay

This protocol is for the analysis of 14 samples per plate—each sample is run twice at 100 μg/mL and in triplicate. This gives six data points per % inhibition reading. Alternatively, 14 samples can be run once at two different concentrations (100 μg/mL, 1 μg/mL) in triplicate. This will give three data points per % inhibition reading (adjust protocol accordingly). The volumes provided in this protocol are intended for a 96-well plate. The actual volumes used can be adjusted as needed.

1. Prepare sample solutions: Prepare a 5 mg/mL stock solution of sample in 100 % dimethyl sulfoxide (DMSO).
2. Sample dilution: Prepare 1 mL of a 1 mg/mL secondary stock solution (in 20 % DMSO) of the sample by taking 200 μL of the above 5 mg/mL stock and adding 800 μL of Milli Q water. During the assay, this will be diluted by 1:10 to give a final concentration of 100 μg/mL in 2 % DMSO (*see* **Notes 1** and **2**).
3. Prepare kinase reaction buffer: Dissolve 48.5 mg of *tris*(hydroxymethyl)aminomethane (Tris), 19.0 mg of magnesium chloride ($MgCl_2$) and 1.0 mg of bovine serum albumin (BSA) in 2 mL of ultrapure H_2O and then make up to 10 mL in a Falcon tube. Adjust the pH to 7.5 using aqueous hydrochloric acid (HCl) if required. This will make 10 mL of reaction buffer that is sufficient for 1 × 96-well plate (i.e., 45 μL × 96 wells = 4.3 mL, plus 5 mL for adenosine triphosphate (ATP) stock) (*see* **Note 3**).
4. Enzyme preparation: The PKA enzyme (Promega) used herein is supplied in buffer as 2,500 units at 114 units/μL. Dissolve whole contents of tube containing enzyme in a total of 2 mL of kinase reaction buffer (*see* **item 3**). Aliquot out 10 × 200 μL of this solution into labeled Eppendorf tubes and store at −70 °C, to give 250 units of activity per tube. The assay described herein utilizes 2.50 units per well, i.e., 2.50 units × 96 wells = 240 units in total. Thus the supplied enzyme will enable testing of 10 × 96-well plates (*see* **Note 4**).

5. Kemptide (PKA) specific substrate (Promega, Australia) (10 mg/mL). The substrate is used as supplied and stored at –20 °C (*see* **Note 5**).
6. Prepare ATP solution: Label three tubes: ATP stock 1, ATP stock 2, and ATP stock 3 (the last one is used in the assay). Weigh 5.51 mg of ATP and make up to 500 μL with kinase reaction buffer to give a 20 mM solution of ATP (stock 1). Take 100 μL of ATP stock 1 and dilute to 500 μL in kinase reaction buffer to give a 4 mM solution. Take 15 μL of the ATP stock 2 and dilute to 3,000 μL in kinase reaction buffer to give a 20 μM solution. This last solution is used in the assay to give a final concentration of 10 μM (*see* **Note 6**).
7. 96-well opaque white (non-sterile) plates (Corning #3912).
8. Micropipettes.
9. Luminometer compatible for 96-well plates.
10. 20 % DMSO solution: Add 800 μL of distilled water to 200 μL of pure DMSO.
11. A known kinase inhibitor standard reference for positive control, e.g., staurosporine, H-9 or H-89 (Sigma-Aldrich) (*see* **Note 7**).
12. Kinase-Glo® Reagents: Kinase-Glo® Buffer and Kinase-Glo® Substrate (Promega Corporation, Australia). Store at –20 °C.

2.2 MTS Assay

Plates containing compound dilutions should be discarded in cytotoxic waste bins. All pipette tips that came into contact with the test compounds should also be disposed of in cytotoxic waste bins

1. Sample preparation: Prepare a 4 mg/mL stock solution of sample in DMSO.
2. MTS reagent: Should be stored at –20 °C for long-term storage and protected from light. Reagent should be warmed up to room temperature before use in the assay.
3. RPMI tissue culture medium containing 5 % fetal bovine serum (FBS) (*see* **Note 8**).
4. Stock cultures of cells should be maintained in RPMI medium containing 5 % FBS (fetal bovine serum).
5. Clear, sterile 96-well microplates.
6. Trypan blue solution (0.4 %, liquid, sterile-filtered, suitable for cell culture).
7. Dimethyl sulfoxide (DMSO).
8. Human, leukemic, monocyte-like, histolytic lymphoma (U937) and human, metastatic breast adenocarcinoma (MDA-MB-231) cancer cell lines, or other cell lines as required (*see* **Note 9**).

Table 1
Kinase assay plate preparation: wells A-F[1-12] are the test compounds; wells G1-6 are positive controls; wells H1-6 are negative controls; and H7-12 and G7-12 are internal reference standards (Std)

	1	2	3	4	5	6	7	8	9	10	11	12
A	1a	1b	2a	2b	3a	3b	4a	4b	5a	5b	6a	6b
B	1a	1b	2a	2b	3a	3b	4a	4b	5a	5b	6a	6b
C	1a	1b	2a	2b	3a	3b	4a	4b	5a	5b	6a	6b
D	7a	7b	8a	8b	9a	9b	10a	10b	11a	11b	12a	12b
E	7a	7b	8a	8b	9a	9b	10a	10b	11a	11b	12a	12b
F	7a	7b	8a	8b	9a	9b	10a	10b	11a	11b	12a	12b
G	Positive control (100 % luminescence)						Std1a	Std1a	Std1a	Std2a	Std2a	Std2a
H	Negative control (0 % luminescence)						Std1b	Std1b	Std1b	Std2b	Std2b	Std2b

The plate above shows the setup for 12 test compounds run twice (samples a and b) in triplicate. This gives six data points per % inhibition reading

9. Hemocytometer.
10. Centrifuge.
11. Sterile incubator.
12. UV spectrophotometer compatible for 96-well plate.

3 Methods

The kinase assay described here utilizes the Kinase-Glo® Luminescent Kinase Assay Platform from Promega and is performed according to the manufacturer's instructions [28] with minor modifications as used in our laboratory. The method below is relevant to screening against PKA; however, a wide range of kinases can be used in this assay including GSK-3β, PI3K, Src, and MAPK [28, 29]. In theory, potentially any kinase could be used, provided the appropriate substrate is also utilized.

3.1 Kinase Assay

1. To wells A/F1-12 of a 96-well plate, add 5 μL of the 1 mg/mL stock solution of the test compounds, which will give a final concentration of 100 μg/mL sample in 2 % DMSO (Table 1). To the positive control wells (G1-6) and the negative control wells (H1-6), add 5 μL of the 20 % DMSO solution.
2. To wells G/H7-12, add 5 μL of the internal standard (e.g., staurosporine).

Table 2
Kinase assay summary of reagent concentrations and volumes added to plate

	Initial stock conc.	Standard wells (μL)	Positive control (μL)	Negative control (μL)	Final conc. in 50 μL
Sample	1 mg/mL	–	–	–	100 μg/mL
DMSO solution	20 %	–	5	5	2 %
Enzyme/no substrate (2.5×)	125 units/mL	–	20	–	50 units/mL
Enzyme/substrate (2.5×)	140 μM	20	–	20	56 μM
ATP solution (2×)	20 μM	25	25	25	10 μM
Total reaction volume		50	50	50	
Kinase-Glo® reagent		50	50	50	

3. Thaw a single tube of the enzyme (250 units per tube) and make up to 2 mL with the kinase reaction buffer to give a concentration of 125 units/mL. This will give a final concentration of 50 units/mL (or 2.50 units/well).
4. To six of the positive control wells (G1-6), add 20 μL of the reaction mixture containing 2.5× the optimal concentration of kinase in kinase reaction buffer. The positive control should provide 100 % luminescence. There will be 1,880 μL of this reaction mixture remaining.
5. Add 20 μL of the kemptide (PKA peptide substrate, 10 mg/mL) to the remaining 1,880 μL of kinase mixture. This will provide 200 μg of the kinase substrate in 1,900 mL to give a 140 μM substrate/enzyme solution in buffer. This amount is sufficient for 1×96-well plate, giving a final concentration of 56 μM in each well per 50 μL reaction.
6. To all remaining wells, add 20 μL of the above reaction mixture (**step 5**) containing 2.5× the optimal concentration of kinase and kinase substrate in 1× kinase reaction buffer (Negative controls, 0 % luminescence).
7. To all wells, add 25 μL of the ATP solution (20 μM, stock 3 ATP solution). This will give a final concentration of 10 μM in each well per 50 μL reaction. *See* Table 2 for a summary of the reagent volumes added to the 96-well plate.
8. Using a plate shaker, gently shake the plate and incubate at room temperature for the optimal amount of time (*see* **Note 10**).

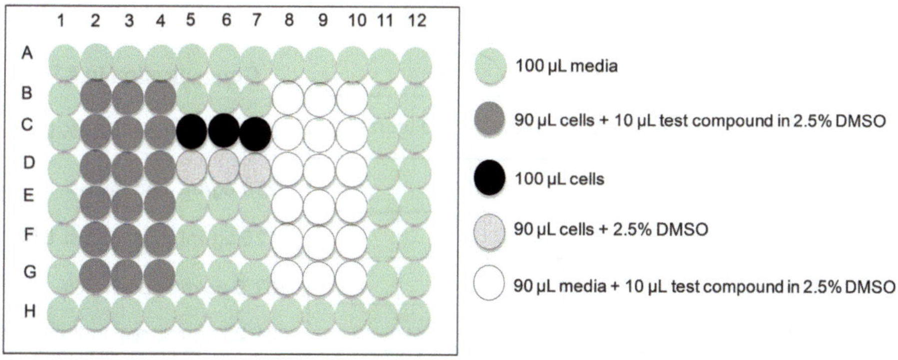

Fig. 2 96-Well microplate setup for the MTS assay

9. Prepare Kinase-Glo® reagent: The kinase buffer should be stored in the freezer and thawed at room temperature. Add the Kinase-Glo® Substrate to the Kinase-Glo® Buffer and add 50 µL of this mixture to each well (*see* **Note 11**).
10. Mix the plate and incubate for 10 min at room temperature. Due to the long half-life of the Kinase-Glo® signal, the plates may be left longer before reading, if desired. Record luminescence, which will be directly proportional to percent inhibition of the controls (*see* **Notes 12–14**).

3.2 MTS Assay

The MTS assay described here utilizes Promega's MTS CellTiter 96® AQueous One Solution Cell Proliferation assay and is performed according to the manufacturer's instructions [30, 31].

1. Determination of cell number and viability: Place 20 µL cells + 20 µL Trypan blue on parafilm and place 20 µL of this mixture under a coverslip on the hemocytometer. Count the number of cells using the equation provided (*see* **Note 15**) to determine the volume required to get a cell concentration of 111,000 cells/mL.
2. Day 1: Setting Up the Plate: Pipette 610 µL of cells (as determined in **step 1** of the procedure) into a new Falcon tube and centrifuge at 1,333×*g* for 5 min (for four plates, pipette 2.44 mL of cells).
3. Drain off supernatant and resuspend the pellet into 2.5 mL of media. This will give you 2.5 mL of cell solution at 111,000 cells/mL (For four plates, you would resuspend in 10 mL of media).
4. Pipette 100 µL of media into wells A1-A12, H1-H12, B1-G1, B11-G11, B12-G12, B5-B7, E5-E7, F5-F7 and G5-G7 as per Fig. 2. These wells are used as blanks to stop any interference that may occur when the spectrophotometer reads the absorbance values for the test wells.

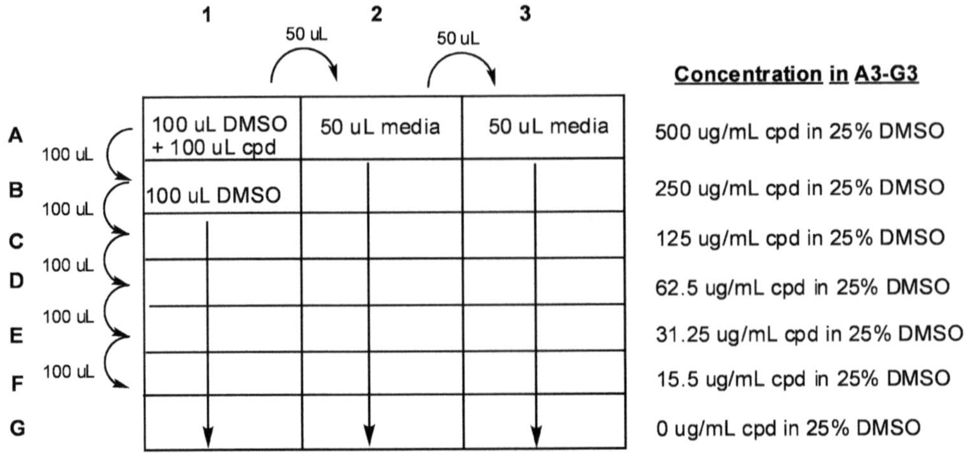

Fig. 3 Preparation of microtiter plate for 1:2 serial dilutions. Add 100 μL of DMSO and 100 μL of the test compound (4 mg/mL in DMSO) to well A1. Next, add 100 μL of DMSO to wells B1-G1. Then, 100 μL is transferred from A1 to B1; B1-C1 and so forth until F1, where 100 μL is removed and discarded. G1 should contain just 100 μL of 100 % DMSO. Next 50 μL of media is added to wells A2-G2 and A3-G3 and 50 μL transferred from A1 to A2 and then from A2 to A3 to give 100 μL of the sample at a concentration of 500 μg/mL in 25 % DMSO in media. This is repeated for wells B2-B3, C2-C3, etc. to give a series of 1:2 dilutions of the sample

5. Pipette 90 μL of media into wells B8-G8, B9-G9, B10-G10. These wells will form the sample background controls.
6. Pipette 90 μL of cell solution into wells B2-G2, B3-G3, B4-G4. These wells will form the sample wells.
7. Pipette 90 μL of cell solution into wells D5-D7. These wells will form the 2.5 % DMSO controls.
8. Pipette 100 μL of cell solution into wells C5-C7. These wells will form the cell controls.
9. Incubate plates for 24 h at 37 °C with 5 % carbon dioxide (CO_2).
10. Day 2: Diluting Test Compounds: Prepare a microtiter plate for serial dilutions of the test compounds as shown in Fig. 3 for each compound to be tested. Wells A3-G3 should all contain a final volume of 100 μL.
11. Adding serial dilutions of samples to cells: Using the plate of cells prepared on Day 1, pipette 10 μL of each sample dilution of the test compound in 25 % DMSO (i.e., A3-F3, **step 11**) in triplicate into the 18 wells containing 90 μL of cell solution (i.e., 10 μL of the sample at 500 μg/mL in triplicate into wells B2-B4, 250 μg/mL sample into C2-C4 and so forth, down to the 15.5 μg/mL sample into G2-4). Repeat this step, by adding 10 μL of the serial dilutions of the sample in triplicate to the 18 wells containing 90 μL of media (e.g., wells B8-G8,

B9-G9, B10-G10). The test compounds will now have a final concentration of 2.5 % DMSO in all wells.

12. Pipette 10 μL of the 0 μg/mL in 25 % DMSO (i.e., G3 from **step 11**) into the three remaining wells containing 90 μL cells (wells D5-D7). This will be the 2.5 % DMSO control.
13. Incubate plate for 24 h at 37 °C and 5 % CO_2 (*see* **Note 16**).
14. Day 3: Adding MTS Reagent and Plate Reading: Thaw MTS reagent before use (~1 mL MTS reagent required per plate).
15. Pipette 20 μL of the MTS reagent into all sample wells (e.g., B2-G2, B3-G3, B4-G4, B5-D5, B6-D6, B7-G7, B8-G8, B9-G9, B10-G10) (*see* **Note 17**).
16. Incubate for 3 h at 37 °C and 5 % CO_2.
17. Wrap the plates in aluminum foil and take to the spectrophotometer. Read the absorbance of the whole plate at 490 nm (*see* **Note 18**).

4 Notes

1. A final concentration of 2 % DMSO was used for the sample and was found not to interfere with results at this concentration. Standards and controls used in the assay were also dissolved in a final concentration of 2 % DMSO. Other solvents such as 2 % ethanol can be used, provided all standards/controls are also prepared in this solvent.
2. We found that a starting stock solution of 1 mg/mL of sample (diluted to 100 μg/mL in the assay) was appropriate to identify kinase inhibitory activity. Samples can be tested at lower concentrations or over a range of different concentrations to determine IC_{50} values.
3. We found that it is best to prepare the buffer solution fresh each time and to store the buffer at room temperature.
4. For kinases other than PKA, the optimum amount of kinase enzyme will need to be determined and the quantity of other reagents adjusted accordingly [28].
5. Kemptide (PKA Peptide Substrate) is a synthetic peptide substrate for PKA derived from the PKA phosphorylation site in liver pyruvate kinase.
6. This assay can be used to identify whether the kinase inhibitor is ATP competitive or noncompetitive, by utilizing various ratios of ATP (e.g., <10 μM ATP for ATP-competitive inhibitors and >100 μM for ATP noncompetitive inhibitors). In general, we found it is best to prepare ATP solutions fresh, while the ATP reagent itself should be stored in the freezer.

7. We include an internal reference of a known kinase inhibitor such as staurosporine in our assays (IC_{50} = 7 nM vs PKA; IC_{50} = 0.7 nM vs PKC) by adding 5 μL of a 1 mg/mL solution of the inhibitor (in 20 % DMSO) to six of the wells in place of a sample set. Staurosporine is a potent inhibitor of PKC and can induce apoptosis in Jurkat cells. Other PKA inhibitors include the isoquinoline sulfonamides H-9 (IC_{50} = 2 μM) and H-89 (IC_{50} = 48 nM).
8. RPMI-1640 was developed by Moore et al. [32, 33] at Roswell Park Memorial Institute.
9. Human, leukemic, monocyte-like, histolytic lymphoma (U937) and human, metastatic breast adenocarcinoma (MDA-MB-231) cancer cells were obtained from American Type Culture Collection (ATCC, VA, USA) distributed by Cryosite, NSW, Australia. Cells were regularly cultured in vitro in culture medium consisting of RPMI-1640 medium, along with 2 mM l-glutamine, 5.6 % (2 g/L) $NaHCO_3$, and 5 % fetal calf serum. The cells were maintained in a Huracell incubator (Kendro Laboratory Products, Langenselbold, Germany) at 37 °C with a humidified atmosphere containing 5 % CO_2. However, a wide range of both normal and cancer cell lines can be used in this assay.
10. We found that thawing the Kinase-Glo® reagents (Kinase-Glo® substrate and Kinase-Glo® buffer) generally took 1 h and we also allowed an hour for the incubation time. We strongly advise optimizing the kinase reaction conditions so that the reaction can be run at room temperature to avoid formation of temperature gradients. Information regarding optimization of kinase reaction conditions can be found in the manufacturer's Technical Bulletin [28].
11. The choice of Kinase-Glo® reagent will depend on the desired ATP concentration to be used in the assay. This information can be found in the manufacturer's protocol, Promega Corporation, Australia [28, 34].
12. We find that it is best to read the plate within 15 min after the addition of the Kinase-Glo® reaction mixture.
13. Instrument settings depend on the manufacturer. An integration time of 0.25–1 s per well should serve as a guideline. We found that the best plate reading is immediately after adding the Kinase-Glo® Buffer and when the incubation is performed at room temperature.
14. The most common assay detection methods are colorimetric or fluorescence-based. However, luminescent-based detection is particularly desirable for colored products such as natural products, as colored extracts and compounds can lead to false

results when using absorbance as the end point. Utilization of luminescence can be more beneficial for biological screening by identifying hits with a lower number of false positives [35].

15. An example calculation of cell concentration of which the values obtained will be used in the method protocol described herein:

$$\text{No. of cells} = \frac{(\#\text{ of cells in 4 grids})}{4} \times 2 \times 10{,}000.$$

For example if we count 91 cells:

$$\text{No. of cells} = \frac{91}{4} \times 2 \times 10{,}000 = 455{,}000 \text{ cells.}$$

Therefore, to get a cell concentration of *111,000 cells/mL*, you will need:

$$\frac{111{,}000 \text{ cells / mL}}{455{,}000 \text{ cells}} = 0.244 \text{ mL} = 244\ \mu\text{L}.$$

For the preparation of one plate the cells will be resuspended in 2.5 mL media therefore the volume of cells required would be:

$$244\ \mu\text{L} \times 2.5 = 610\ \mu\text{L}.$$

Adjust accordingly if more plates are required.

For example, the preparation of four plates will require:

$$244\ \mu\text{L} \times 10 \text{ mL}(2.5 \text{ ml per plate} \times 4 \text{ plates}) = 2.44 \text{ mL}.$$

16. Longer incubation times of 48–72 h are also routinely employed in this assay.
17. The MTS reagent is light sensitive so this step should be performed with the lights off in the cytotoxic cabinet.
18. As this is a colorimetric assay, it should be kept in mind that both colored compounds and natural product extracts may interfere with the absorbance reading, and appropriate background controls should always be performed.

Acknowledgment

This work was supported by the Center of Medicinal Chemistry and the School of Chemistry, University of Wollongong.

References

1. Almeida CA (2010) Cancer: basic science and clinical aspects. Wiley-Blackwell, London
2. Hanahan D, Weinberg RA (2000) The hallmarks of cancer. Cell 100:57–70
3. Altman R, Sarg M (2000) The cancer dictionary. Facts on File, New York, NY
4. Berg K, Zhai L, Chen M et al (1994) The use of watersoluble formazan complex to quantitate the cell number and mitochondrial function of *Leishmania major* promastigotes. Parasitol Res 80:235–239
5. Malich G, Markovic B, Winder C (1997) The sensitivity and specificity of the MTS tetrazolium assay for detecting the *in vitro* cytotoxicity of 20 chemicals using human cell lines. Toxicology 124:179–192
6. Matesic L, Locke JM, Bremner JB et al (2008) N-phenethyl and N-naphthylmethyl isatins and analogues as in vitro cytotoxic agents. Bioorg Med Chem 16:3118–3124
7. Boyd MR, Paull KD (1995) Some practical considerations and applications of the National Cancer Institue *in vitro* anticancer drug discovery screen. Drug Dev Res 34:91–109
8. Shoemaker RH (2006) The NCI60 human tumour cell line anticancer drug screen. Nat Rev Cancer 6:813–823
9. Collins JM (2012) Developmental therapeutics program NCI/NIH. http://dtp.nci.nih.gov. Accessed 29 Jun 2012
10. von Ahsen O, Bomer U (2005) High-throughput screening for kinase inhibitors. Chembiochem 6:481–490
11. Copeland RA (ed) (2005) Evaluation of enzyme inhibitors in drug discovery: a guide for medicinal chemists and pharmacologists (Methods in biochemical analysis). Methods of biochemical analysis. Wiley, Hoboken, NJ
12. Manning G, Whyte DB, Martinez R et al (2002) The protein kinase complement of the human genome. Science 298:1912–1934
13. Goldstein D, Gray N, Zarrinkar P (2008) High-throughput kinase profiling as a platform for drug discovery. Nat Rev Drug Discov 7:391–397
14. Liao JJL (2007) Molecular recognition of protein kinase binding pockets for design of potent and selective kinase inhibitors. J Med Chem 50:409–424
15. Sharma P, Sharma R, Tyagi R (2008) Inhibitors of cyclin dependent kinases: useful targets for cancer treatment. Curr Cancer Drug Targets 8:53–75
16. Skropeta D, Pastro N, Zivanovic A (2011) Kinase inhibitors from marine sponges. Mar Drugs 9:2131–2154
17. Ono-Saito N, Niki I, Hidaka H (1999) H-series protein kinase inhibitors and potential clinical applications. Pharmacol Ther 82:123–131
18. Shabb JB (2001) Physiological substrates of cAMP-dependent protein kinase. Chem Rev 101:2381–2412
19. Chen AE, Ginty DD, Fan CM (2005) Protein kinase A signalling via CREB controls myogenesis induced by Wnt proteins. Nat Rev Drug Discov 433:317–322
20. Tasken K, Aandahl EM (2004) Localized effects of cAMP mediated by distinct routes of protein kinase A. Physiol Rev 84:137–167
21. Arnsten AFT, Ramos BP, Birnbaum SG et al (2005) Protein kinase A as a therapeutic target for memory disorders: rationale and challenges. Trends Mol Med 11:121–128
22. Suzuki M, Shinohara F, Endo M et al (2009) Zebularine suppresses the apoptotic potential of 5-fluorouracil via cAMP/PKA/CREB pathway against human oral squamous cell carcinoma cells. Cancer Chemother Pharmacol 64:223–232
23. Putz T, Culig Z, Eder IE et al (1999) Epidermal growth factor (EGF) receptor blockade inhibits the action of EGF, insulin-like growth factor I, and a protein kinase A activator on the mitogen-activated protein kinase pathway in prostate cancer cell lines. Cancer Res 59:227–233
24. Bryn T, Mahic M, Aandahl EM et al (2008) Inhibition of protein kinase A improves effector function of monocytes from HIV-infected patients. AIDS Res Hum Retroviruses 24:1013–1015
25. Torgersen KM, Vang T, Abrahamsen H et al (2002) Molecular mechanisms for protein kinase A-mediated modulation of immune function. Cell Signal 14:1–9
26. Indira CV, Matesic L, Locke JM et al (2012) Anti-cancer activity of an acid-labile N-alkylisatin conjugate targeting the transferrin receptor. Cancer Lett 316:151–156
27. Zivanovic A, Pastro NJ, Fromont J et al (2011) Kinase inhibitory, haemolytic and cytotoxic activity of three deep-water sponges from North Western Australia and their fatty acid composition. Nat Prod Commun 6:1921–1924
28. Promega (2009) Kinase-Glo® luminescent kinase assay platform
29. Baki A, Bielik A, Molnar L et al (2007) A high through-put luminsescent assay for glycogen synthase kinase-3 beta inhibitors. Assay Drug Dev Technol 5:75–83

30. Promega (2012) CellTiter 96® AQueous one solution cell proliferation assay. pp 1–13
31. PromegaCorporation (2012) Promega protocols. http://www.promega.com/resources/protocols. Accessed 26 Jul 2012
32. Moore GE, Woods LK (1967) Culture media for human cells – RPMI 1603, RPMI 1634, RPMI 1640 and GEM 1717. Tissue Cult Assn 3:503–508
33. Moore GE, Gerner RE, Franklin HA (1967) Culture of normal human leukocytes. J Am Med Assoc 199:519–524
34. Koresawa M, Okabe T (2004) High-throughput screening with quantitation of ATP consumption: a universal non-radioisotope, homogeneous assay for protein kinase. Assay Drug Dev Technol 2: 153–160
35. Kashem MA, Nelson RM, Yingling JD et al (2007) Three mechanistically distinct kinase assays compared: measurement of intrinsic ATPase activity identified the most comprehensive set of ITK inhibitors. J Biomol Screen 12:70–83

Chapter 15

Screening for Antidiabetic Activities

Rima Caccetta and Hani Al Salami

Abstract

Screening extracts and drug entities for antidiabetic bioactivity is essentially limited to animal models as the processes leading to hyperglycemia and the complications of diabetes involve more than one organ. Further, in vitro results seldom translate into meaningful in vivo outcomes especially in a disease such as Diabetes Mellitus. In vivo studies on specialized animal models have allowed great progress in tailoring research questions towards individualized genetic and biochemical contributors and their effect on the pathogenesis of the disease processes. Various disease models have been used either through genetic-manipulation (transgenic models) or through chemical induction (disease-induced models). Although there is a surplus of animal models (spontaneous and induced) to study Type I and Type II diabetes, there is no ideal or standard model for studying the individualized effects of various classes of antidiabetic drugs. Rodents, most commonly rats and mice, have been used by researchers as animal models of the disease and both normoglycemic and diabetic animals are used to assess the antidiabetic activities of drugs or extracts under investigation. Screening for antidiabetic activities can be achieved by measuring a wide range of biomarkers and end points including blood glucose and insulin levels.

Key words Diabetes mellitus, Drug discovery, Plant extracts, Type I, Type II

1 Introduction

Active constituents of many plant species are isolated for direct use as drugs or lead drug compounds. Historically different plant species have been used to treat diabetes mellitus and these medicinal plants are of prime interest as sources of drug molecules. Unfortunately, taken as plant concoctions, there are many drawbacks and its success in different patients relies on many factors including the type of diabetes they have. Continuously, the doses are variable and commonly other interactive ingredients (which can be toxic) are also present.

Success in screening these extracts for potential antidiabetic activity relies on a number of factors including the extraction technique employed, the experimental design, and the biological model utilized. Most importantly, antidiabetic drugs must reduce blood glucose levels in order to be effective. Further, different animal

Ute Roessner and Daniel Anthony Dias (eds.), *Metabolomics Tools for Natural Product Discovery: Methods and Protocols*, Methods in Molecular Biology, vol. 1055, DOI 10.1007/978-1-62703-577-4_15, © Springer Science+Business Media, LLC 2013

models can expand on the mechanism(s) of action and give an indication whether the drug mimics insulin or acts in some other means to enhance insulin release or its actions. Nonetheless, the initial screening needs to be promising enough to signal pursuit of isolating active constituents.

Diabetes mellitus (DM) is a chronic condition characterized by hyperglycemia (which can cause hypertension) with disturbances of carbohydrate, fat, and protein metabolism resulting from defects in insulin secretion, insulin action, or both [1]. There are two major types of DM: Type I and Type II. Diabetics are genetically predisposed to either type of the disease but both types appear to rely on other, including environmental, factors to trigger or expose the disease. DM is on the rise with Type II diabetes increasing dramatically worldwide.

Type I diabetes (T1D) is believed to be an autoimmune disease of sudden onset usually during childhood. It is characterized by the lack of insulin release due to the destruction of pancreatic β-cells. This appears to be triggered (or accelerated) by an early fetal event such as blood group incompatibility or fetal viral infections or early exposure to cow's milk proteins or nitrosamines. Due to the lack of insulin, the body believes it is starved and thus breaks down biomolecules for the production of glucose and energy. The uncontrolled or poorly controlled diabetic exhibits elevated ketonemia.

Type II diabetes (T2D) is a metabolic disorder believed to be due to a genetic predisposition (involving a number of metabolic genes) which is in most cases subsequently triggered by becoming overweight with a sedentary lifestyle. The pre-diabetic state is characterized by insulin resistance leading to compensatory hyperinsulinemia. The disease is then manifested by impaired insulin secretion from the pancreas and insulin action in the peripheral tissue including skeletal muscle, and adipose tissue [2]. The reduced insulin action is attributed to the decline in the number and/or responsiveness of the insulin receptors. At the microscopic level, the pancreatic islets in T2D are characterized by an insufficiency in β-cell mass and increased β-cell apoptosis [3, 4] with the islets commonly containing large deposits of amyloid derived from islet amyloid polypeptide, also known as amylin (*see* **Note 1**) [5]. All these processes culminate in a deficit in glucose transport (internalization from the blood into peripheral tissues) which may in part be attributed to the decreased expression of the major transporters GLUT 4 (*see* Fig. 1); however, little or no reduction in GLUT-4 levels is observed in type II diabetic patients [6, 7]. Ultimately, uncontrolled T2D will develop to exhibit T1D symptoms and thus the patient in many cases eventually requires insulin injections.

There is a range of animal models available for screening for antidiabetic potential. Each model carries some of the described characteristics of the human disease but no one model alone will

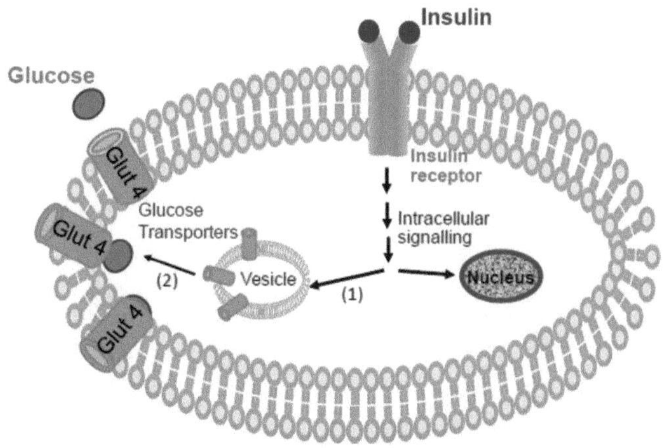

Fig. 1 A representation of how insulin increases glucose uptake by increasing the number of glucose transporters (GLUT-4) in muscle and adipose cell membranes. Insulin binds to the insulin receptor generating a series of internal signals that can continue to generate a response via the nucleus. However, a quicker response is triggered by (*1*) recruitment of insulin sensitive glucose transporters from intracellular pool to the surface of the cell membrane (*2*)

carry all the human manifestations and thus screening needs to incorporate at least a couple of different models to gain a better understanding of the effectiveness of the screened drug or mixture. It is important to use adequate controls (positive and negative controls, e.g., insulin and other drugs on the market) as comparisons but also to minimize the contribution to the noise which, in the case of glucose levels, is adrenaline. Further, it is advisable to visit a statistician in the planning stages of the study to ensure minimum animal use and that the numbers chosen will enable meaningful statistical analyses. This chapter presents a number of common rodent models used (*see* Table 1) for screening from natural sources and validating lead or new therapeutic drugs in vivo for antidiabetic activity. The choice of model is dependent on the assessment of the desired endpoints, including the sample size, but may also be budget driven.

1.1 Normoglycemia

Normoglycemic animals can be used for screening potential hypoglycemic agents. Many studies have used and continue to use normoglycemic controls and these have in some cases mirrored the effects in the diabetic test animals. However, the results have been variable mostly due to the lower effects seen on normoglycemics than diabetics. In an attempt to assess some peptides for insulin-mimetic potential, intravenous administration is necessary but one can use anesthetized normoglycemic rodents, e.g., Wistar rats. Such a model avoids the effects of adrenaline which can otherwise work to mask positive effects of the drug or extract screened.

Table 1
A summary of several mice and rat diabetic models used to assess drugs and/or extracts for antidiabetic activity

Rodent	Model	Strain	Characteristics
Mouse	GLUT-4	C57BL/KsJ-*db/db-Tg* (SLC2a4)	Derived from mutation in leptin receptor from C57BL/KsJ mice; obese, insulin resistant. Develops hyperglycemia and hyperinsulinemia
	T2D		High GLUT-4 expression in cardiac and skeletal muscle, brain, and adipose tissue; not expressed in kidney, spleen, or pancreas
	GSK3B *DM*	FVB/N-Tg(Mic2-tTA-GSK3)1Pfi	Transgenic mouse
	Rip-HAT *DM*	FVB/N-Tg(Ins2-IAPP1)Pfi	Transgenic mouse
	T2D	C57bl6/J (male)	High-fat diet induced
Rat	Goto-Kakizaki (GK) *T2D*	GK/TohiCskCrljCrl	Non-obese Wistar sub-strain
	Spontaneously hypertensive heart failure (SHHF) *T2D*	SHHF/MccGmiCrl-*Lepr* ^{cp}/Crl	Developed from backcrossing the spontaneously hypertensive obese rat to the spontaneously hypertensive rat
	Zucker diabetic fatty (ZDF)	ZDF-*Lepr* ^{fa}/Crl	Genetic mutation in a colony of outbred Zucker rats; obese/lean; hyperlipidemia, glucose intolerance, obesity, hyperinsulinemia
	T2D		*Purina diet to induce programmed and consistent development of T2D in male ZDF rats*
	ZSF1 *T2D*	*Lepr* ^{fa} *Lepr* ^{cp}/Crl	Hybrid rat; cross between ZDF female rat and SHHF male rat; obese /lean
	RIP-HAT *DM*	Crl:CD(SD)-*Tg*(*Ins2-IAPP*)*1Pfi*	Transgenic rat

There are several benefits of using normoglycemic animals which include the ease of caring for these animals and their cost. Further, normoglycemic models offer assessment with an intact pancreas which can provide valuable information on the mechanism of action when compared to a diabetic model (e.g., Streptozotocin induced, *see* Subheading 1.5). Using normoglycemic animals for initial screening is valid; however, one needs to remember that the effectiveness of the compound or extract screened might be diminished.

1.2 Genetically Modified

A good example of an animal model of T2D is Db/db mice. This model mimics T2D and represents an accurate estimation of many disturbed biomarkers including hyperglycemia, insulin resistance, and obesity, and show a marked decrease in skeletal muscle utilization without an accompanying decrease in GLUT 4 expression. This insulin resistant animal model evaluates improved glycemic control through overexpression of GLUT 4. Upregulation of GLUT4 expression has been shown to improve skeletal muscle glucose transport in these insulin-resistant db/db mice [8, 9].

Human transgenic rodents, Rip-HAT Mouse/RIP-HAT Rat, develop DM between 4 and 8 weeks and 5–10 months of age, respectively. These models exhibit selective β-cell apoptosis leading to the onset of type II diabetes in the rodent. Basically, rats and mice have similar islet amyloid polypeptide (IAPP) homology but differ from humans by substitution of proline residues in the amyloidogenic portion of IAPP, and therefore do not form amyloid fibrils nor spontaneously develop diabetes in midlife. Thus, insertion of a human transgene may provide a means to explore the role of the amyloid fibrils in β-cell destruction.

Rodent models that carry a spontaneous single gene mutation in either the leptin or the leptin receptor inbred in either a C57BL/6J or C57BL/KsJ mice (*see* Table 1) are widely used. However, there is concern that these genetic mutations are rare occurrences in humans. In fact leptin administration is ineffective in human T2D patients, except for a rare population with mutations in their leptin and/or leptin receptor genes.

1.3 Selectively Bred

The Goto-Kakizaki (GK) rat is a non-obese Wistar sub-strain which develops T2D early in life. This model was developed in the 1970s by repeated selection of breeders with high blood glucose levels which lead to glucose intolerance after five generations.

1.4 High-Fat Diet Induced

Obesity is a well-established risk factor (~30 %) for T2D and a high-fat diet is a major trigger of insulin resistance and thus the development of T2D in genetically predisposed individuals. Induction of type II DM has been successfully achieved in C57bl6/J male mice and in Golden Syrian hamsters through the administration of highly saturated Purina diet. The mouse develops T2D over 8–12 weeks exhibiting marked obesity, hyperinsulinemic, insulin resistance, glucose intolerance, and peripheral leptin resistance.

1.5 Streptozotocin (STZ) or Alloxan Induced

The induction of T1D through the administration of STZ [10] or alloxan [11] is a common practice. Both are cytotoxic and when administered in appropriate doses (25–50 mg/kg) target the β-pancreatic cells resulting in cell necrosis and eventual cessation of insulin production by the pancreas. Chronic administration of lower doses may result in the development of T2D model, while in higher doses both agents can become toxic and may result in severe tissue damage especially the liver which can result in internal hemorrhage and death of the animal. Some animals may not develop DM and so it is important to assess the animal condition before proceeding with drug experimentation.

2 Materials

2.1 Mice/Rats

Research animal models may be bought from local Animal Resource Centers or from overseas (e.g., Charles River Laboratories International). Animals may need to be housed for a period of time prior to experimentation for them to stabilize in their new environment. Commercial chow and water should be provided ad libitum and the bedding needs to be changed regularly. Frequent monitoring and handling of the animals is necessary and advisable. Animals may gain or lose weight which needs to be taken into consideration as it might affect the experimental procedures.

2.2 Anesthetics

1. Hypnorm–Dormicum or ketamine–xylazine mixture (*see* refs. [12, 13]).
2. Dilute anesthetics in vehicle like 1× Phosphate Buffered Saline (1× PBS). Prepare a stock solution of 10× PBS: weigh 80 g of NaCl, 2 g of KCl, 14.4 g of Na_2HPO_4, 2.4 g of KH_2PO_4 and dissolve in Milli-Q water, make up to 1 L. This 10XPBS solution can be store on the shelf. To achieve 1XPBS, dilute a portion (as needed) of the 10× PBS tenfold and pH check the solution to be ~7.4 (*see* **Note 2**). The 1× PBS requires sterilization either by autoclaving or by filtration (through sterile 0.22 μm pore syringe filters) into sterile containers. The 1× PBS needs to be stored at 4 °C for up to 1 month.
3. Warming box.
4. 1 mL syringes.
5. 26–27 G needles.
6. Thick paper towels.

2.3 Controls

1. Insulin, e.g., Humulin R (100 U) (Diabetes Australia from Eli Lilly and Company).
2. Metformin 1,1-dimethylbiguande hydrochloride (Sigma–Aldrich).
3. PBS (*see* Subheading 2.2, **item 2**).

2.4 Glucose and Insulin Measurements

1. Accu-Chek *Go* Glucometer.
2. Accu-Chek Glucometer strips.
3. HTRF® Insulin assay (Cisbio bioassays) or insulin enzyme-linked immunosorbent assay (ELISA) kits (Cayman Chemicals).
4. Positive controls (*see* **Note 3**).
5. Plate reader.
6. Scalpel.

2.5 Other Materials

1. pH meter.
2. Alloxan monohydrate dissolved in deionized water or STZ (*see* refs. [2, 3]).
3. Analytical balances.
4. Surgical lamp (optional).

3 Methods

Blood glucose and insulin levels are the key indicators for diagnosis of DM. Elevated serum triglycerides and abdominal obesity are commonly associated with insulin resistance and represent valuable clinical markers of the metabolic syndrome [14]. Further, since comorbidity of DM and dyslipidemia is common, cholesterol, triglycerides, inflammatory markers, and hypertension are usually measured in the same experiment. However, for the purposes of this chapter we shall focus on the measurements directly linked to DM, namely glucose and insulin. Blood glucose concentration can be measured using a one-touch meter (e.g., Accu-Chek, *see* Subheading 2.4, **item 1**). Insulin concentration in blood can be measured using by using ELISA kits (*see* Subheading 2.4, **item 3**) following the manufacturer's protocols. Blood samples can be collected from the animal at different intervals before and after treatment and analyzed appropriately.

3.1 Experimental Design

1. Decide whether it is best to assess the acute or repeated use effects of the extract or compound you are screening (*see* **Note 4**).
2. Depending on the properties of the compound or extract to be assessed, you might decide on its mode of delivery to be ad libitum, oral gavage, intravenous, or intraperitoneal. Obviously, oral gavage studies cannot be carried out while animals are

under anesthesia. The route of administration you choose must take into consideration the limitations as well as the advantages in the assessment process. A no effect may be attributed to the mode of administration as the compound may require metabolism to be activated or metabolism may eradicate its activity.

3. Choose satisfactory positive and negative controls (*see* Subheading 2.3). The oral controls may be orally administered drugs like metformin while insulin is ideal for intraperitoneal or intravenous injections. Other marketed antidiabetic medications might also be useful especially to compare the activity and explore the possible mode of action of the drug under investigation.
4. For intravenous or intraperitoneal administration, choose a vehicle with buffering range similar to blood, e.g., PBS. Assess the pH of the samples before injecting as the drug might have modified the pH of the vehicle. If necessary adjust the pH to ~7.3–7.4 before injecting into the animals.
5. Re-solubilized extracts might require filtration or centrifugation (*see* **Note 5**) following dissolution to remove undissolved debris before injecting the clear solutions. It is desirable that the volume to be administered is kept to a minimum (*see* **Note 6**).
6. Measure animal body weight.
7. Measure blood glucose and insulin levels using kits (*see* Subheading 2.4) following the manufacturers' protocols before injecting treatments to establish a baseline (*see* **Note** 7).
8. Measure blood glucose and insulin after injecting treatment samples at specific predetermined times; more frequently (e.g., every 10–30 min) in acute trials but less frequently (e.g., daily or weekly) for longer term trials.

3.2 Animal Housing

Following ethics approval, all animals should be maintained in an experimental animal facility and given standard diet and water ad libitum. Feed the rodents (unless in the case of diet induced diabetics) standard commercial laboratory chow, allow them free access to water and cage either individually or in pairs. House the animals under standard conditions and temperature (20 ± 1 °C), with a regular 12 h dark and 12 h light cycle. Monitor the animals for any signs of agitation, fighting, distress or abnormal drop in weight. Measure their body weights daily allowing frequent handling to reduce nervousness, induce calmness and reduce aggressiveness. Fasting the animals (18 h), food only, before experimentation may be necessary depending on the experimental design [12, 13]. In such cases animals need to be monitored for coprophagia and thus changing bedding when commencing fasting is recommended.

3.3 Animal Numbers and Randomization

The trial size and animal numbers will depend on many factors including the power of the study, the predicted average mean difference and intra and inter individual variation, as well as other factors.

1. If this is a pilot study then decide to do some preliminary analysis of the data after about eight animals to check if there is a significant difference. However, if you have preliminary data then do a power calculation to determine numbers per group. Numbers should be the same in each group including the control group.
2. Randomly allocate animals in their treatment groups.

3.4 Anesthesia

If the study is assessing the acute effects of the drug or extract (e.g., within the first 3–4 h), then the animals can be anesthetized and remain under anesthesia but monitored for normal breathing and for any signs of pain or distress. The choice of anesthesia is important as some can induce hyperglycemia. Also, since deep anesthesia can affect the glycemic levels of the animals a low dose to keep the animals sufficiently anesthetized is advisable. Anesthetics of choice are Hypnorm–Dormicum [12] or a combination of ketamine and xylazine adjusted to suit each animal according to body weight [13]. A successful technique includes the injection of an initial priming dose 30–45 min prior to test-substance administration followed by additional low doses, every 20–30 min, to keep the animal under [12, 13].

3.5 Routes of Administration

When considering the route of administration both the site of action and the solubility of the molecule or concoction under investigation are considered to be major determinants. Further, an effect might be missed depending on the route of administration as drug molecules might be activated or inactivated by metabolism. There are several routes; however for the purposes of this chapter only the most common systemic approaches are briefly outlined below as these are relevant to the end points focused on in this chapter.

1. *Ad libitum*.

   This is an option if the drug or extract is water soluble and does not have an offensive smell or taste which might deter the animals from drinking it. Usually the drug or extract (dry or lyophilized) to be tested is dissolved in the drinking water and the animals would drink it instead of pure water. A positive control such as metformin (*see* Subheading 3.1, **step 3**) can be used via this route of administration.
2. *Gavage*.

   Oral gavage is reserved to conditions where the drug may not be feasibly or accurately administered otherwise.

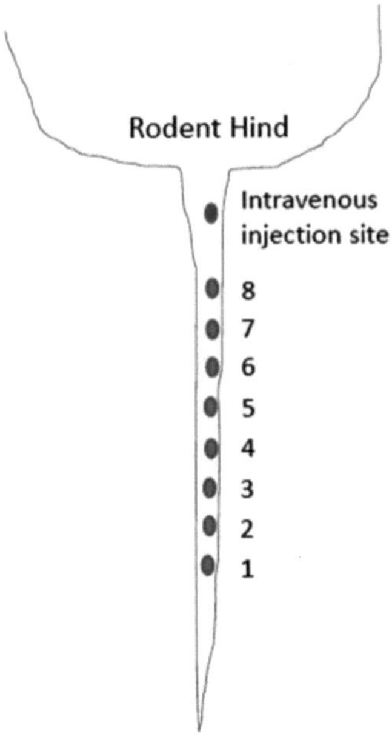

Fig. 2 A representation of a rodent's tail outlining the desired sites and procedure of administering treatments intravenously as well as taking blood samples from the tail vein. *1–8* (*upwards*) is the recommended sampling procedure in short-term trials allowing the avoidance of newly formed clots

Many drugs require testing via the oral route but may not be easily soluble in water or may have a repulsive taste preventing accurate dosing. There are several techniques described in the literature and advised by animal resource centers. Feeding tubes are available commercially, specifically for mice or rats.

3. *Intravenous injection and intraperitoneal injection.*

   Injection in the tail vein should be done at a high point (*see* Fig. 2). Intraperitoneal injection is generally used for fat-soluble drugs and slow release assessments. Both these routes are quite commonly used when assessing extracts which, depending on the extraction procedure, can have compounds that are water and/or lipid soluble [15].

3.6 Blood Sampling

Methods for blood collection include tail-nick, tail snip, saphenous vein, submandibular and retro-orbital bleeding. These methods have been previously outlined [16]. Tail vein sampling is the most effective way and one that will be outlined in this chapter.

1. A sampling schedule needs to be chosen.
2. The rodent can be held in position or if anesthetized, lying down.
3. The tail fur can be shaved slightly.

4. Depending on whether samples are taken daily, weekly or more frequently, incisions/sampling might require a procedure to avoid recently formed clots. The vein is located, and with multiple sampling in an acute study needle pricks (or small incisions with a scalpel) are started from bottom of the tail (*see* Fig. 2). Subsequent blood samples are taken above the initial incision and this continues in that general direction. Less frequent bleeds may not require such precautions.

3.7 Glucose Measurement

Blood samples need to be analyzed immediately. Each sample (~10 μL) can be analyzed by transferring onto Accu-Chek glucose strips (*see* Subheading 2.4). The readings are obtained from the Accu-Chek glucometer within 5 s. Blood glucose values can be obtained in mmol/L or mg/dL.

3.8 Insulin Measurement

There are a number of kits (*see* Subheading 2.4, **item 3**) to measure serum and/or plasma insulin from blood collected via the tail vein. The protocols are provided with the kit. Such methods are quick and easy to perform and require small sample sizes (5–10 μL volumes are required for the HTRF® assay).

4 Notes

1. Amylin (or IAPP) is co-secreted with insulin by β-cells. Certain amylin isoforms can dimerize to form amyloid fibrils or oligomers which can initiate apoptosis [3, 4].
2. The 1× PBS should naturally be pH 7.4 and not require adjustment; however, if a small adjustment is necessary use HCl or NaOH.
3. Compounds that are well known to give significant hypoglycemic levels (commonly insulin) and vehicle to indicate the effects in the absence of a hypoglycemic agent. The route of administration needs to be considered when choosing the controls. The concentrations will need to be adjusted to give hypoglycemia but allow recovery not death of the animals.
4. If acute, anesthesia is recommended especially if you are doing multiple samplings. The animal needs to be kept warm by using a warming box although surgical light might also give heat.
5. Extracts using solvents can be dried down under nitrogen or using a rotary evaporator. These extracts can be stored at −20 °C and reconstituted in suitable buffers (e.g., PBS) for injection. Reconstituted solutions, and also depending on the pH change, may have particulate undissolved residues which will require centrifugation or filtration to sediment and remove particulate matter before injection of the solutions in animal models.

6. The volume should be one that can be accurately measured and also one that easily and fully solubilizes the drug of interest.
7. This measurement might need to be done a few times before the blood glucose levels stabilize especially as the animal adrenaline levels drop with the animal relaxing.

References

1. King H (1999) WHO and the International Diabetes Federation: regional partners. Bull World Health Organ 77:954
2. Wilcox G (2005) Insulin and insulin resistance. Clin Biochem Rev 26(2):19–39
3. Butler AE, Jang J, Gurlo T, Carty MD, Soeller WC, Butler PC (2004) Diabetes due to a progressive defect in beta-cell mass in rats transgenic for human islet amyloid polypeptide (HIP rat): a new model for type 2 diabetes. Diabetes 53:1509–1516
4. Matveyenko AV, Butler PC (2006) Beta-cell deficit due to increased apoptosis in the human islet amyloid polypeptide transgenic (HIP) rat recapitulates the metabolic defects present in type 2 diabetes. Diabetes 55:2106–2114
5. Goldsbury C, Goldie K, Pellaud J, Seelig J, Frey P, Muller SA, Kistler J, Cooper GJ, Aebi U (2000) Amyloid fibril formation from full-length and fragments of amylin. J Struct Biol 130:352–362
6. Garvey WT, Maianu L, Hancock JA, Golichowski AM, Baron A (1992) Gene expression of GLUT4 in skeletal muscle from insulin-resistant patients with obesity, IGT, GDM, and NIDDM. Diabetes 41:465–475
7. Pedersen O, Bak JF, Andersen PH, Lund S, Moller DE, Flier JS, Kahn BB (1990) Evidence against altered expression of GLUT1 or GLUT4 in skeletal muscle of patients with obesity or NIDDM. Diabetes 39:865–870
8. Brozinick JT Jr, McCoid SC, Reynolds TH, Nardone NA, Hargrove DM, Stevenson RW, Cushman SW, Gibbs EM (2001) GLUT4 overexpression in db/db mice dose-dependently ameliorates diabetes but is not a lifelong cure. Diabetes 50:593–600
9. Gibbs EM, Stock JL, McCoid SC, Stukenbrok HA, Pessin JE, Stevenson RW, Milici AJ, McNeish JD (1995) Glycemic improvement in diabetic db/db mice by overexpression of the human insulin-regulatable glucose transporter (GLUT4). J Clin Invest 95:1512–1518
10. Wu KK, Huan Y (2008). Streptozotocin-induced diabetic models in mice and rats. Curr Protoc Pharmacol Chapter 5, Unit 5:47
11. Satyanarayana S, Sekhar JR, Kumar KE, Shannika LB, Rajanna B, Rajanna S (2006) Influence of selenium (antioxidant) on gliclazide induced hypoglycaemia/anti hyperglycaemia in normal/alloxan-induced diabetic rats. Mol Cell Biochem 283:123–127
12. Schaffer L, Brissette RE, Spetzler JC, Pillutla RC, Ostergaard S, Lennick M, Brandt J, Fletcher PW, Danielsen GM, Hsiao KC, Andersen AS, Dedova O, Ribel U, Hoeg-Jensen T, Hansen PH, Blume AJ, Markussen J, Goldstein NI (2003) Assembly of high-affinity insulin receptor agonists and antagonists from peptide building blocks. Proc Natl Acad Sci USA 100:4435–4439
13. Caccetta R, Ireng A, Helmerhorst E, Parsons R (2013) Teucrium polium significantly lowers blood glucose levels acutely in normoglycaemic male Wistar rats; a comparative to insulin. Submitted for publication
14. Grundy SM (1999) Hypertriglyceridemia, insulin resistance, and the metabolic syndrome. Am J Cardiol 83:25F–29F
15. Jung M, Park M, Lee HC, Kang YH, Kang ES, Kim SK (2006) Antidiabetic agents from medicinal plants. Curr Med Chem 13:1203–1218
16. Baribault H (2010) Mouse models of type II diabetes mellitus in drug discovery. Methods Mol Biol 602:135–155

Chapter 16

Screening for Antibacterial, Antifungal, and Anti quorum Sensing Activity

Elisa J. Hayhoe and Enzo A. Palombo

Abstract

The plate-hole diffusion assay is an invaluable screening tool to evaluate the antibacterial potential of natural products. It relies on the diffusion of test material from pre-cut wells through agar seeded with bacteria. Samples that are capable of inhibiting bacterial growth will produce a clear zone surrounding their well. For the evaluation of antifungal activity of natural products, we describe the broth microdilution method. This assay is performed using a 96 well microtiter tray containing fungal inoculum, test medium and natural product material. Samples demonstrating antifungal activity will prevent any discernible growth as detected visually. A disk diffusion assay, utilizing the pigmented indicator strain *Chromobacterium violaceum*, is described here for the screening of natural products for anti quorum sensing activity. Inhibition of quorum sensing results in growth of non-pigmented bacteria.

Key words Antimicrobial, Antibacterial, Antifungal, Anti quorum sensing, Screening, Disk diffusion, Plate-hole diffusion, Broth microdilution, Natural product, Extracts

1 Introduction

There are several methods reported in the literature to evaluate antibacterial activity of natural products. They involve either introducing the test compounds to a finite area of set agar (disk diffusion [1] and plate-hole diffusion assay [2]) or incorporating the test compounds into the molten agar or liquid media (broth [3] and agar dilution [4] assays). The advantage of using the diffusion methods for screening is that they require only a small volume of test material and the color or opaqueness of a sample will not interfere with interpretation of results. The plate-hole diffusion assay involves the inoculation of molten agar with test bacteria which is then set in a petri plate. Wells are cut into the agar which are then filled with the natural product material and suitable controls. Following incubation, zones of inhibited bacterial growth surround wells containing antibacterial compounds. The diameter of this inhibition zone is

Ute Roessner and Daniel Anthony Dias (eds.), *Metabolomics Tools for Natural Product Discovery: Methods and Protocols*, Methods in Molecular Biology, vol. 1055, DOI 10.1007/978-1-62703-577-4_16, © Springer Science+Business Media, LLC 2013

affected by the rate of diffusion of the test material through the agar and therefore serves only as a qualitative interpretive result [5]. However, it is still an effective method for screening processes.

Broth dilution methods, rather than agar-based methodology, are considered the most reliable for the screening of natural products for antifungal activity [3, 6]. Here, we describe the broth microdilution method [3, 7]. This assay uses a 96-well microtiter tray which allows several samples and/or fungal isolates to be screened simultaneously. The procedure first involves the preparation of a conidial suspension from the test fungi containing approximately 10^5 conidia per milliliter. This inoculum is then used to prepare the microtiter tray with the test material. Samples demonstrating antifungal activity will prevent any discernible growth as detected visually.

One approach to bacterial inhibition that is gaining interest in the field of natural products research is inhibition of quorum sensing (QS). Multicellular behaviors controlled by QS include swarming, bioluminescence, and biofilm formation which are often involved in bacterial pathogenesis [8]. Therefore, QS inhibitors (QSI) can potentially reduce the virulence of bacteria without necessarily affecting their growth. This is a significant advantage compared with conventional "killing" chemotherapeutics as it is less likely to result in changes to the host microbiota and the development of resistance to treatments [9, 10]. Anti quorum sensing (AQS) activity has been reported in samples from natural product sources such as plant essential oils and solvent extracts [8, 11–13], and fungal secondary metabolites [9, 14]. The method detailed here is based on the antibiotic susceptibility method, described in 1966 [1], and takes advantage of the indicator strain *Chromobacterium violaceum* which produces a purple pigment (violacein) under *N*-acylhomoserine lactone mediated QS control [12, 15].

2 Materials

2.1 Antibacterial Assay: Plate-Hole Diffusion

1. Mueller Hinton agar (MHA) (Becton, Dickinson and Company, Sparks, MD, USA).
2. Sterile petri dishes, 9 cm diameter.
3. Overnight cultures of test bacteria, adjusted to 1×10^8 CFU/mL (0.5 McFarland standard).
4. Cork borer (*see* **Note 1**).
5. Natural product test material (*see* **Note 2**).

2.2 Antifungal Assay: Broth Microdilution

1. Filamentous fungi for susceptibility testing (*see* **Note 3**).
2. Sterile loop.

3. Sterile saline.
4. Potato Dextrose Broth (PDB).
5. Sterile 96 U-shaped well microtiter trays.
6. Natural product test material.
7. Antifungal agent such as amphotericin B.

2.3 Anti quorum Sensing Assay: Disk Diffusion Assay

1. Petri plates containing 15 mL of Mueller Hinton Agar.
2. Overnight cultures of *Chromobacterium violaceum* (UNSW collection 040100), adjusted to 1×10^8 CFU/mL (0.5 McFarland standard).
3. Natural product test material.
4. Sterile filter paper disks, 6 mm diameter.
5. Glass spreader.
6. Forceps.

3 Methods

3.1 Plate-Hole Diffusion

1. Prepare MHA in purified water as per the manufacturer's directions.
2. Heat the solution with frequent agitation and boil for 1 min to ensure that the powder has completely dissolved.
3. Decant the solution into small glass bottles [15 mL per petri dish (*see* **Note 4**)] and sterilize by autoclave (121 °C, 15 min).
4. Allow agar to cool to approximately 45 °C (*see* **Note 5**).
5. Aseptically inoculate each aliquot with 150 μL of bacterial suspension.
6. Mix the contents by gently inverting the bottle four times and then pour each into a sterile petri dish.
7. Allow the plates to set for 5–10 min and then use the cork borer to stamp wells into the agar (*see* **Note 6**).
8. Remove and dispose of the agar plugs (*see* **Note** 7). Label each well.
9. Pipette 20 μL of test material and controls into each well (*see* **Note 8**). Leave the plates at room temperature for 30 min to allow evaporation of any solvents and pre-diffusion of material into the agar.
10. Incubate plates according to test bacteria requirements (*see* **Note 9**).
11. Record the diameter of zone of inhibition from the underside of the plate with a ruler or sliding calipers (*see* **Note 10**).

Table 1
Example of variation in optical density and correlating inoculum sizes between fungi [3, 16]

Species	OD_{530} range	10^6 CFU/mL range
Aspergillus niger	0.20–0.50	0.5–4.5
A. fumigatus	0.09–0.11	0.6–5
Fusarium oxysporum	0.15–0.17	0.8–5
Scedosporium apiospernum	0.15–0.17	0.4–3.2

3.2 Broth Microdilution

1. Flood the PDA plate with 1 mL of sterile saline. Using the loop, lightly probe the surface of the mycelia to introduce conidia (and hyphal fragments) into the solution.
2. Transfer the solution to a sterile tube and vigorously vortex the suspension for at least 1 min to produce a homogenous solution.
3. Allow it to stand for 3 min and then transfer the upper portion of liquid to a sterile tube.
4. Using a spectrophotometer, adjust to the optical density that equates to a stock solution of $0.4–5 \times 10^6$ viable conidia or sporangiospores per milliliter (CFU/mL) (*see* **Note 11**) (Table 1).
5. Dilute the fungal suspension 1:50 in PDB to create a $0.8–1 \times 10^5$ CFU/mL range.
6. Fill microtiter tray wells with 100 μL of fungal suspension and 100 μL of natural product test material.
7. Growth control (GC) wells contain 100 μL of fungal suspension and 100 μL of distilled water (*see* **Note 12**).
8. Add the antifungal agent to the fungal suspension for a positive control and the test material solvent to the fungal suspension for a negative control.
9. Incubate the tray without agitation at 35 °C for up to 70 h, checking for visible growth in the GC wells after 24 and 48 h.
10. Natural product material demonstrating antifungal activity will prevent any discernible growth of the fungi.

3.3 Disk Diffusion Assay

1. Add 20 μL of each test material to a sterile paper disk (*see* **Note 13**).
2. A tetracycline or gentamycin disk (10 μg per disk) can be used to compare AQS activity with antibiotic effect. Purified halogenated furanone can be used as a positive control for AQS activity.
3. Allow the disks to dry for 10 min.
4. Using a sterile glass spreader evenly spread 150 μL of *C. violaceum* onto the surface of a petri plate.

5. Allow the inoculated plate to dry for 5 min.
6. Using sterile forceps, place each disk on the surface of the agar. Gently press the disk with the forceps to ensure complete contact with the agar (*see* **Note 14**).
7. Leave the plates at room temperature for 30 min to allow pre-diffusion of material into the agar.
8. Incubate the agar plate at 30 °C overnight.
9. Inspect the bacterial growth surrounding each disk (*see* **Note 15**).

4 Notes

1. A cork borer is a hollow metal tube that is often used in chemistry labs to bore holes in cork or rubber stoppers for the insertion of tubing. Choose a cork borer with a diameter of 5–7 mm and use this same tool for all subsequent assays.
2. Test material may be in the form of a crude extract, a separated fraction or a purified compound. If required, concentrate the material (via lyophilization or rotary/centrifugal evaporation) or dilute (in a carrier solvent) prior to testing.
3. The fungal suspension is prepared from cultures of filamentous fungi grown on PDA plates. Incubation conditions will depend on the species chosen. For example, *Aspergillus* spp. will generally produce conidia after 7 days growth at 35 °C.
4. We have found that McCartney bottles (28 mL volume) are an appropriate vessel for these aliquots.
5. If the temperature of the agar is too high when you add the bacterial suspension, you will risk damage to the cells. If the temperature is too low, the agar will begin to solidify before it is poured into the plate, which will result in very uneven growth. Place the bottles into a 45 °C water bath following autoclave and allow them to cool. Bacteria must, therefore, be heat stable at 45 °C.
6. When stamping wells into the agar, make sure that the cork borer comes straight down into the agar, not on an angle. Sterilize the cork borer before each plate assay by swilling the end in 75 % (v/v) ethanol, shaking off the excess liquid and placing over a Bunsen burner flame. Make sure that your finger is not covering the open hole at the top of the tool. Allow the cork borer to cool before using again to ensure a clean well is cut in the agar.
7. With one hand, open the lid of the petri plate from the far side. The lid will serve as protection in case you flick the agar plug (containing bacteria) towards yourself. With the other hand, use a wire loop to carefully lift each agar plug from the plate. Dispose of as contaminated culture media.

8. Use the test material solvent as a negative control. A standard antimicrobial agent such as ampicillin can be used as the positive control.
9. Incubation conditions must be kept consistent for subsequent assays. Do not stack more than five plates together, to ensure that all assays reach the incubation temperature at approximately the same time.
10. A clear zone surrounding the well signifies antibacterial activity by the test material. Keep in mind that this assay is a screening method only and that for a quantitative evaluation of antibacterial activity, concentration dependent methods will need to be carried out.
11. Due to the variation in spore sizes between species, an optical density value will equate to different inoculation densities (CFU/mL) depending on the fungal species chosen.
12. The growth control well/s confirms that the fungal isolate is viable and demonstrates 100 % growth of the isolate for comparison with test wells after incubation.
13. Keep disks in an empty sterile petri plate that is labelled to avoid confusion between different test materials.
14. Avoid placing disks closer than 20 mm together (center to center) as zones may overlap. Do not place disks too close to the edge of the petri plate. Once in contact with the agar, do not move disks (even if they have been placed in the wrong position) as the test material will have already started diffusing into the agar.
15. Magnification of the agar surface with an adequate light source will help to differentiate between colorless growth (indicating interruption of quorum sensing), purple growth (no effect), and no growth (indicating antibacterial activity).

References

1. Bauer AW, Kirby WM (1966) Antibiotic susceptibility testing by a standardized single disk method. Am J Clin Pathol 45:493–496
2. Rahman A, Choudhary M, Thompson W (2001) Bioassay techniques for drug development. Informa Healthcare, EBL EBook Library. Accessed 1 Feb 2012
3. Espinel-Ingroff A, Canton E (2007) Antifungal susceptibility testing of filamentous fungi. In: Schwalbe R, Steele-Moore L, Goodwin A (eds) Antimicrobial susceptibility testing protocols. CRC Press, Boca Raton, p 209
4. Hanlon A, Taylor M, Dick J (2007) Agar dilution susceptibility testing. In: Schwalbe R, Steele-Moore L, Goodwin A (eds) Antimicrobial susceptibility testing protocols. CRC Press, Boca Raton, p 91
5. Reade E (1994) Techniques manual: department of microbiology. University of Melbourne Publishing, Melbourne
6. Institute CaLS (2008) Reference method for broth dilution antifungal susceptibility testing of filamentous fungi. Approved standard M38-A2, vol 28, no. 16. Clinical and Laboratory Standards Institute, Wayne, PA
7. Espinel-Ingroff A, Bartlett M, Bowden R, Chin NX, Cooper C Jr, Fothergill A, McGinnis MR, Menezes P, Messer SA, Nelson PW, Odds FC, Pasarell L, Peter J, Pfaller MA, Rex JH, Rinaldi MG, Shankland GS, Walsh TJ, Weitzman I (1997) Multicenter evaluation of proposed standardized procedure for antifungal susceptibility testing of filamentous fungi. J Clin Microbiol 35:139–143

8. Choo JH, Rukayadi Y, Hwang JK (2006) Inhibition of bacterial quorum sensing by vanilla extract. Lett Appl Microbiol 42:637–641
9. Rasmussen TB, Skindersoe ME, Bjarnsholt T, Phipps RK, Christensen KB, Jensen PO, Andersen JB, Koch B, Larsen TO, Hentzer M, Eberl L, Hoiby N, Givskov M (2005) Identity and effects of quorum-sensing inhibitors produced by *Penicillium* species. Microbiology 151:1325–1340
10. Al-Hussaini R, Mahasneh AM (2009) Microbial growth and quorum sensing antagonist activities of herbal plants extracts. Molecules 14:3425–3435
11. Kociolek MG (2009) Quorum-sensing inhibitors and biofilms. Anti-Infect Agents Med Chem 8:315–326
12. Adonizio AL, Downum K, Bennett BC, Mathee K (2006) Anti quorum sensing activity of medicinal plants in southern Florida. J Ethnopharmacol 105(427):435
13. Szabó MÁ, Varga GZ, Hohmann J, Schelz Z, Szegedi E, Amaral L, Molnár J (2010) Inhibition of quorum-sensing signals by essential oils. Phytother Res 24:782–786
14. Van Rij ET, Wesselink M, Chin-A-Woeng TFC, Bloemberg GV, Lugtenberg BJJ (2004) Influence of environmental conditions on the production of phenazine-1-carboxamide by *Pseudomonas chlororaphis* PCL1391. Mol Plant Microbe Interact 17:557–566
15. McClean KH, Winson MK, Fish L, Taylor A, Chhabra SR, Camara M, Daykin M, Lamb JH, Swift S, Bycroft BW, Stewart GSAB, Williams P (1997) Quorum sensing and *Chromobacterium violaceum*: exploitation of violacein production and inhibition for the detection of *N*-acylhomoserine lactones. Microbiology 143:3703–3711
16. Petrikkou E, Rodríguez-Tudela JL, Cuenca-Estrella M, Gómez A, Molleja A, Mellado E (2001) Inoculum standardization for antifungal susceptibility testing of filamentous fungi pathogenic for humans. J Clin Microbiol 39:1345–1347

Chapter 17

Metabolomics and Dereplication Strategies in Natural Products

Ahmed Fares Tawfike, Christina Viegelmann, and RuAngelie Edrada-Ebel

Abstract

Metabolomic methods can be utilized to screen diverse biological sources of potentially novel and sustainable sources of antibiotics and pharmacologically-active drugs. Dereplication studies by high resolution Fourier transform mass spectrometry coupled to liquid chromatography (LC-HRFTMS) and nuclear magnetic resonance (NMR) spectroscopy can establish the chemical profile of endophytic and/or endozoic microbial extracts and their plant or animal sources. Identifying the compounds of interest at an early stage will aid in the isolation of the bioactive components. Therefore metabolite profiling is important for functional genomics and in the search for new pharmacologically active compounds. Using the tools of metabolomics through the employment of LC-HRFTMS as well as high resolution NMR will be a very efficient approach. Metabolomic profiling has found its application in screening extracts of macroorganisms as well as in the isolation and cultivation of suspected microbial producers of bioactive natural products.

Metabolomics is being applied to identify and biotechnologically optimize the production of pharmacologically active secondary metabolites. The links between metabolome evolution during optimization and processing factors can be identified through metabolomics. Information obtained from a metabolomics dataset can efficiently establish cultivation and production processes at a small scale which will be finally scaled up to a fermenter system, while maintaining or enhancing synthesis of the desired compounds. MZmine (BMC Bioinformatics 11:395–399, 2010; http://mzmine.sourceforge.net/download.shtml) and SIEVE (http://www.vastscientific.com/resources/index.html; Rapid Commun Mass Spectrom 22:1912–1918, 2008) softwares are utilized to perform differential analysis of sample populations to find significant expressed features of complex biomarkers between parameter variables. Metabolomes are identified with the aid of existing high resolution MS and NMR records from online or *in-house* databases like AntiMarin, a merger database of Antibase (Laatsch H. Antibase Version 4.0 – The Natural Compound Identifier. Wiley-VCH Verlag GmbH & Co. KGaA, 2012) for microbial secondary metabolites as well as higher fungi and MarinLit for marine natural products (Blunt J. MarinLit. University of Canterbury, New Zealand, 2012). This is further validated through available reference standards and NMR experiments. Metabolomics has become a powerful tool in systems biology which allows us to gain insights into the potential of natural isolates for synthesis of significant quantities of promising new agents and allows us to manipulate the environment within fermentation systems in a rational manner to select a desired metabolome.

Key words Metabolomics, Endosymbionts, Endophytes, Dereplication, Chemical profiling, MZmine, SIEVE, HRFTMS, NMR, Principal component analysis (PCA), Partial least squares discriminant analysis (PLS-DA)

Ute Roessner and Daniel Anthony Dias (eds.), *Metabolomics Tools for Natural Product Discovery: Methods and Protocols*, Methods in Molecular Biology, vol. 1055, DOI 10.1007/978-1-62703-577-4_17, © Springer Science+Business Media, LLC 2013

1 Introduction

The use of metabolomics is a new means to guide the isolation of compounds, as well as to help improve the productivity of downstream fermentation methods. Metabolomics is relatively a new field of "*omics*", adopting to the system biology approach, with the goal of qualitatively and quantitatively analyzing all metabolites contained in an organism at a specific time and under specific conditions. Metabolomics is considered as the most functional approach in monitoring gene function and identifying the biochemical status of an organism [1] and will confirm the results of the presence of biosynthetic gene clusters involved in the production of the biologically active components. Metabolomics in combination with genomics enhances the production of important secondary metabolites (natural products) which is one of the expressed phenotypes in a living organism. Literature has shown that gene clusters are involved in every step of a biosynthetic pathway, as in the production of biologically active polyketides [2]. With genomics, gene clusters can be manipulated to control a biosynthetic pathway. The procedure of employing metabolomics together with genomics to optimize a biosynthetic pathway to selectively produce biologically active secondary metabolites has yet to be explored.

To identify and quantify metabolites in natural product extracts is a challenging task [3, 4]. This is due to the fact that secondary metabolites have diverse atomic arrangements which results in variations in chemical and physical properties. They can also be found in a wide range of concentrations. Reliable, robust, selective, and high resolution analytical methods are therefore required in identifying and quantifying multiple chemical groups of natural products. Mass spectrometry and NMR spectroscopy are complementary analytical methods and are commonly employed in tandem as metabolomics tools. Mass spectrometry is sensitive even at femtogram levels but may not be reproducible between instrument types and ionization capabilities of the metabolites. On the other hand, NMR data is reproducible; it may not be sensitive enough to detect metabolites at lower concentrations.

Efficient high-throughput gradient flash and/or medium pressure chromatography, where gram materials of a microbial extract can be loaded on a column, are employed to isolate the bioactive natural products from microbial extracts. High-throughput gradient medium pressure chromatography is capable of delivering reproducible isolation schemes with high product yield, which is optimum in the purification of microbial extracts obtained from multiple batches and has great advantage over conventional column chromatography [5, 6]. Structure elucidation are accomplished utilizing pulse-field gradient 2D NMR that would be able to provide high resolution data to determine the

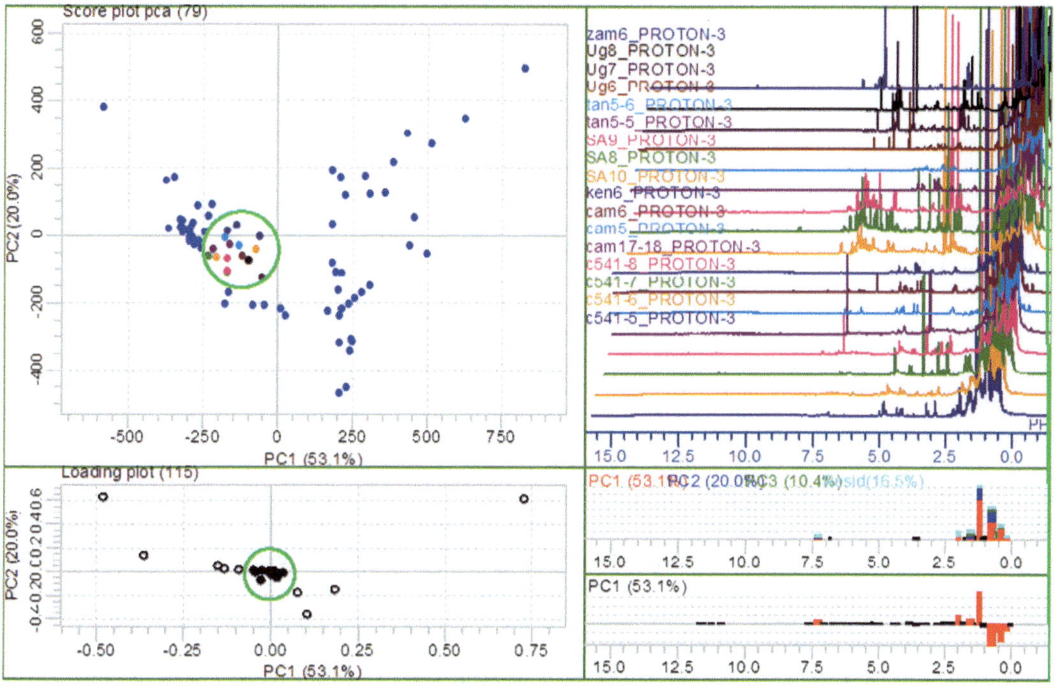

Fig. 1 Principal component analysis of proton NMR data of active vs. inactive fractions of different African propolis samples. The active fractions and the type of functional groups common to these fractions are encircled in green as further indicated by the black bars. Here we could conclude the presence of the active components to be concentrated in the nonpolar fractions to which were later identified as prenylated phloroglucinone derivatives (data presented in the first German-Brazilian Symposium on Neglected Diseases, Edrada-Ebel 2011)

structure of complex molecules with multiple chiral centers as well as higher molecular weight peptides and oligosaccharides [4].

Metabolomics also provides statistical and computational tools to this standard approach of rapid HPLC fractionation which would identify the active entities at an earlier stage. The goal of HPLC fractionation is to get to higher purities of active components which, however, is not achievable in the initial chromatographic isolation work. With metabolomics tools, it will be possible to pin-point the active components at the first fractionation step as well as identify the functional groups involved in the bioactivity which would be present in a series of fractions as implied by the bioactivity screening results. This can be chemometrically achieved by such NMR-metabolomic/PCA (principal component analysis) software like ALICE [7] as shown in an example presented (Fig. 1) where the presence of isoprenyl units and hydroxylated groups among a series of active fractions were found to be necessary to achieve anti-trypanosomal activity. The use of metabolomics will aid in prioritizing the fractions that will proceed to purification, which should save time and resources in isolating the target compounds.

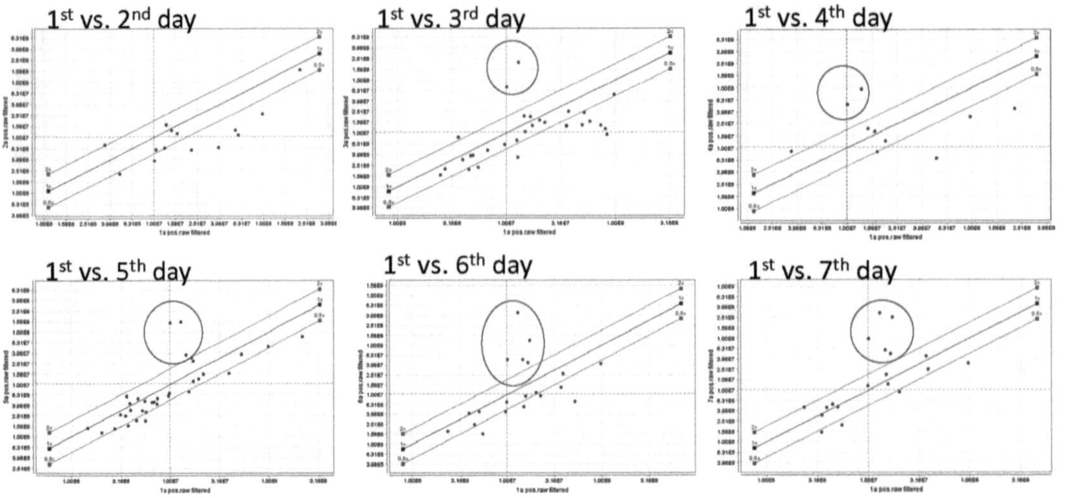

Fig. 2 Production of anti-mycobacterial compounds from a marine actinomycete at different growth phases as illustrated by an S plot. Biological activity was observed on the sixth and seventh day as indicated by the circle

Metabolomics is also used for the quality control of natural products and isolates to monitor the manifestation of a different metabolic profile between individuals, environmental alterations during growth and harvesting, postharvesting treatment, extraction, and method of isolation, all of which can affect the efficacy of natural products. Through this metabolomics approach, the link between the chemical profile and bioactivity pattern of the secondary metabolites is correlated. To date, metabolomics is not yet widely applied in bioactive screening of natural products although it has several advantages over the reductionist approach. Metabolomics enhances identification and dereplication steps, as in bioassay-guided isolation work [8]. Metabolomics can be applied to dereplicate the biosynthesis of the natural product at different development stages of their biological source as well as simultaneously screen for the bioactivity. By using combinations of different analytical methods, the bioassay-guided isolation route is shortened and rapid dereplication of known activities is efficiently delivered [2]. Algorithms have been developed to efficiently detect the production of potential novel secondary metabolites during the cultivation and production processes that would assist in maintaining or enhancing the biosynthesis of the desired compounds (Fig. 2). These algorithms are coupled to differential expression analysis software like SIEVE and MZmine which can then integrate the results to an *in-house* database that includes MarinLit and Antibase to further identify microbial secondary metabolites. Figure 2 illustrates the HRMS data as processed with MZmine, a web-based software developed by VTT (Technical Research Centre of Finland) [9]. The experiment analyzed the production of anti-mycobacterial metabolites from a novel marine actinomycete sponge symbiont collected from the Red Sea.

Metabolomics is applied at two levels: to identify and track active compounds highlighted by screening assays and to optimize the biotechnological production of active compounds in the later stages of the pipeline. Dereplication of secondary metabolites from promising isolates will be achieved by HRFTMS (High Resolution Fourier Transform Mass Spectrometry), for example, using the LTQ-Orbitrap and high resolution NMR. Through multivariate analysis, this will enable Fourier transformation of the FID (Free Induction Decay) data of multiple samples to statistically validate the parameters in the production of pharmacologically novel secondary metabolites. Metabolomes are identified with the aid of existing high resolution MS and NMR records from online and *in-house* databases like the Dictionary of Natural Products (DNP) [10]; MarinLit, a database for marine natural products [8]; AntiBase, a database of microbial secondary metabolites [11]; and KEGG (Kyoto Encyclopedia of Genes and Genomes), a collection of online databases dealing with genomes, enzymatic pathways, and biological chemicals [12]. MZmine [9], MzMatch [13], and SIEVE [14] softwares are utilized to perform quantitative differential expression analysis of sample populations to find significant expressed features of complex biomarkers between parameter variables. These are further validated through available reference standards and two-dimensional (2D) homonuclear NMR, e.g., TOCSY (TOtalSpectroscopY), DOSY (DiffusionOrderedSpectroscopY), and *J*-resolved NMR experiments to classify unknown by-products or degradants which may affect the quality of the desired product. The NMR-metabolomic software ALICE [7] is employed for metabolome recognition as well as to statistically validate the occurrence of metabolic by-products at the different physiological states. The algorithms are utilized to analyze the vast data set generated by the dereplication study as well as metabolomics profiling in monitoring and exploring the relationship between culture methods, diversity, bioactivity, and metabolome evolution in selected isolates. Efficient cultivation and production processes at a small volume scale fermenter are developed through real-time metabolomic-assisted optimization. Samplings are analyzed at *on the fly* for detailed metabolome analysis to fully characterize intermediates, by-products, and degradants. Applying metabolomics for real-time analysis will, in parallel, check the stability of the production of the desired components when altering certain fermentation parameters prior to scale-up. In addition, chemometric studies will be accomplished to support and develop algorithms that will be adapted and optimized to target the secondary metabolites with the desired effects. Metabolomics has become a powerful tool in systems biology and allows insight into the potential of natural isolates for synthesis of significant quantities of promising new agents, and guides the manipulation of the environment within fermentation systems in a rational manner to select a desired metabolome.

1.1 Brief Description of Work

Natural products metabolomics embarks on the dereplication studies of biologically active extracts and interesting isolates from plant, animal, and microbial origin as well as provides the technology to validate the presence and expression of biosynthetic clusters from gene-based screening. Extracts, fractions, and purified secondary metabolites are evaluated for their applicability as sources of potential novel drugs. The chemical profile of a biologically active extract is established by high resolution mass spectrometry and NMR spectroscopy to detect the active compounds of interest at an early stage and prevents unnecessary efforts and resources in working-up extracts. By utilizing metabolomic tools and optimizing the much-needed computational tools of quantitative differential expression as well as multivariate statistics, data sets could be analyzed to identify biomarkers in the production of biologically active extracts. These biomarkers could then be further employed to optimize fermentation and cultivation conditions which would ascertain both primary and secondary biosynthetic pathways in the production of novel secondary metabolites as well as consequently provide sustainability of the interesting metabolites for genetic and metabolic engineering.

1.2 Dereplication Studies

The first step will deal with dereplication studies of secondary metabolites from ca. 100–200 mg of ethyl acetate culture extracts that will be prepared from microorganisms from variable extreme environments. These extracts, 1 mg of each, will be analyzed by HRFTMS (high resolution Fourier transform mass spectrometry) using the Exactive-Orbitrap, which is capable of obtaining high resolution data for both positive and negative modes. Simultaneous experiments done at both modes are advantageous for the nontarget approaches in metabolic profiling. Multivariate analysis using Fourier transformation of FID (free induction decay) data of multiple samples by SIEVE and MZmine along with PCA (principal component analysis) and OPLS-DA (orthogonal partial least square-discriminant analysis) will statistically validate the parameters that govern the occurrence of relevant metabolites from bioactive extracts. Metabolomes will be identified with the aid of existing high resolution MS and NMR records from Antibase or DNP. These results will also validate the presence of biosynthetic gene clusters as previously identified. High resolution NMR will be applied to promising isolates in terms of occurrence of novel biosynthetic gene clusters, expanse of expected novel compounds, and bioactivity (Fig. 3). 1D-^{1}H, *J*-resolved and pulse-field gradient 2D total correlation (TOCSY), diffusion (DOSY), as well as heteronuclear single quantum coherence (HSQC) NMR experiments will be performed to confirm the expected type of chemical groups and determine the method of the isolation work. Multivariate analysis on the NMR data will be done using the software ALICE, which can identify essential functional groups for certain bioactivity [7]. Characterization of the metabolites in the bioactive extract

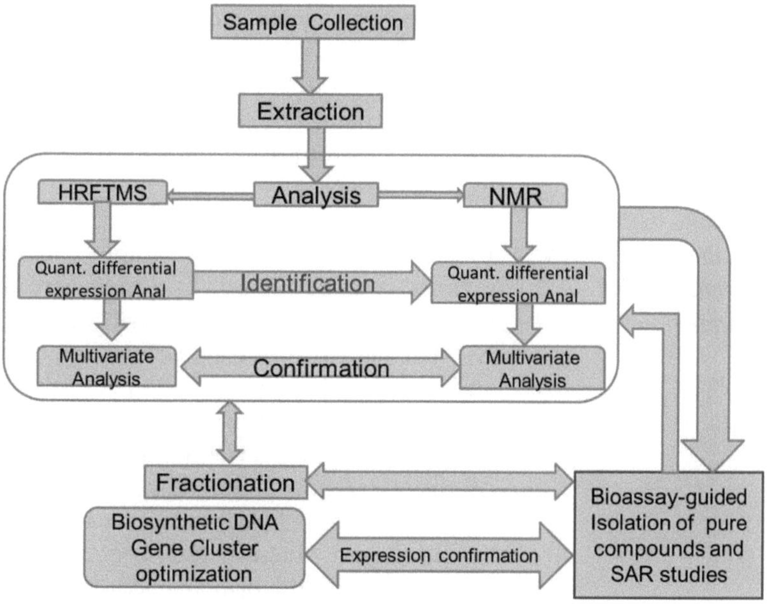

Fig. 3 Flowchart of activities integrating metabolomics to other fields of drug discovery

can be accomplished by MS fragmentation experiments which can also be done on the LTQ-Orbitrap HRMS. From this data, it would be possible to anticipate necessary precautions if expected instability of the compounds could occur. At this stage, the computational tools can be optimized by adapting or creating new algorithms according to the diversity of the samples and complexity of the data set generated from the dereplication study and metabolomics profiling. In order to effectively utilize information from the environment data, genetic variation and metabolic profiles, appropriate statistical methods can be applied and developed aiming at deciphering the associations between the various datasets. Correlations between the bioprocessing parameters and various datasets are first assessed and regularized regression methods are applied among the most relevant variables from each dataset. The aim is to identify which bioprocessing parameters are related to the individual datasets and also assess the similarities and differences between them.

1.3 Metabolomic-Assisted Optimization (Small to Medium Scale)

From the bioactivity results, bioactive microbial isolates will be chosen for further cultivation optimization work. Efficient cultivation and production at a small to medium volume scale fermenter can be developed through real-time metabolomic-assisted optimization. Sampling will be done at real time for detailed metabolome analysis to fully characterize intermediates, by-products, and degradants by rapid off-line LC–MS/GC-MS, high resolution FTMS, and ^{1}H NMR. Sample probes will be collected at different growth phases and subjected to small scale organic extraction and HRFTMS. Metabolomic analysis will be performed to determine changes in metabolite production, which will be further correlated with the bioactivity data. The optimized cultivation method will be chosen to go into medium-scale cultivation (20–40 L) to be used for the isolation work.

1.4 Bioassay-Guided Chromatographic Separation

Bioassay-guided chromatographic separation of active crude extracts is achieved utilizing high-throughput gradient flash and medium pressure liquid chromatography (Buchi Sepacore system). Normal- and reversed-phase columns are employed for nonpolar and polar active extracts, respectively. HP20 and/or Sephadex LH-20 columns are used to separate out salts and other constituents used in the media from the bioactive metabolites. Fractionation of the extracts will be monitored by chromatographic (TLC), LC-HRMS, and NMR spectroscopy (^{1}H NMR) analyses; subsequently each fraction will be subjected to bioassay testing. Chemometric and principal component analysis of active versus inactive fractions provide information of the class of compound responsible for the respective bioactivity. The active fractions are purified by preparative and/or semi-preparative HPLC. All active pure metabolites are subjected to spectroscopic analyses based on NMR, MS, (ultraviolet visible absorption spectroscopy) UV, IR, and circular dichroism (CD) methods. The obtained spectra are compared to available data in the literature in order to identify previously isolated compounds. Structures of new compounds are elucidated mainly by extensive application of 1D and 2D NMR spectroscopy, ^{1}H and ^{13}C assignments can be obtained from COSY, HSQC, HMBC, and ROESY spectral data. Homonuclear ^{1}H connectivity can be determined by the phase-sensitive, double-quantum filtered COSY experiments. One-bond heteronuclear ^{1}H–^{13}C connectivities will be determined by pulse-field gradient-enhanced proton-detected HSQC experiments. Two- and three-bond ^{1}H–^{13}C connectivities will be determined by gradient-enhanced proton-detected HMBC experiments. Experiments are carried out in deuterated solvent ($CDCl_3$, CD_3OD, d_5-pyridine, and d_6-DMSO) solutions of the isolated compounds. Advanced mass spectrometric techniques (EI, CI, ESI, and APCI) for structure assignment experiments are performed on Thermoquest LCQ Deca ESI-Ion Trap, LTQ-Orbitrap, and Exactive mass spectrometers. In cases of high MW peptides, this can be accomplished on a Shimadzu MALDI instrument. Stereochemical information can be provided also by CD studies, computational methods, and chemical derivatizations.

1.5 Metabolomic-Assisted Industrial Bioprocessing

Applying metabolomics for real-time analysis will in parallel check the stability of the production of the desired components when changing certain fermentation parameters prior to industrial scale-up.

Metabolomes of each steady state are identified with the aid of existing high resolution MS and NMR records from Antibase, a database of microbial secondary metabolites together with KEGG, a collection of online databases dealing with genomes, enzymatic pathways, and biological chemicals. SIEVE and MZmine softwares will be used to perform differential analysis of sample populations to find significant expressed features of complex biomarkers between parameter variables. This will be validated through available reference standards and also by 2D NMR (e.g., TOCSY and *J*-resolved) as well as DOSY NMR experiments for unknown by-products or degradants which may affect the quality of the desired product. NMR metabolomics software ALICE is employed for metabolome recognition as well as to statistically validate the occurrence of metabolic by-products at the different physiological states.

1.6 Large-Scale Chromatographic Purification

The desired compounds are obtained by large-scale purification of 100–500 g utilizing high-throughput gradient flash and medium pressure liquid chromatography utilizing either the Buechi Sepacore or Biotage Isolera systems. The purified compounds will be sent for biological activity. The isolation procedure of active natural products will again be validated qualitatively and quantitatively by high resolution NMR spectroscopy and mass spectrometry.

2 Materials

All solutions are prepared using ultrapure water (deionized water is purified through filters of 18 MΩ cm at 25 °C) and HPLC grade reagents as well as solvents. All reagents are prepared and stored at room temperature. Follow all waste disposal regulations when disposing waste materials.

2.1 Biological Materials

1. Collected sponges and plant materials were freeze-dried, sealed under vacuum and subsequently stored at −20 °C until the extraction of its metabolites.
2. Upon receipt, microbial samples were reinoculated to appropriate medium enumerated under Subheading 2.2.
3. Endophytes were isolated from fresh parts of indigenous medicinal plants.

2.2 Reagents, Media, and Chemicals

1. M1 agar: soluble starch (Sigma Chemical Co. USA) 10.0 g, yeast extract (Oxoid, England) 4.0 g, peptone (Fisher Scientific UK Ltd, UK) 2.0 g, agar (Oxoid, England) 18.0 g, add artificial seawater to make a volume of 1.0 L. Autoclave and allow to cool slightly before pouring into petri dishes.
2. Artificial seawater: NaCl (234.70 g), Na_2SO_4(39.20 g), $MgCl_2{\cdot}6H_2O$ (106.40 g), $CaCl_2$(11.0 g), $NaHCO_3$(1.92 g),

KCl (6.64 g), KBr (0.96 g), H_3BO_3(0.26 g), $SrCl_2$(0.24 g), NaF (0.03 g), and add H_2O to make a volume of 10.0 L.

3. Malt agar (MA) medium for short-term storage of fungal cultures or fresh seeding for preparation of liquid cultures: agar 15.0 g, malt extract 15.0 g, distilled water added to make a volume of 1.0 L and pH adjusted to 7.4–7.8 with NaOH or HCl. For the isolation of endophytic fungi from plant tissues, add 0.2 g chloramphenicol or 0.1 g streptomycin to the medium to suppress bacterial growth.
4. Wickerham medium for liquid cultures: yeast extract 3.0 g, malt extract 3.0 g, peptone 5.0 g, glucose10.0 g, add distilled water to make a volume of 1.0 L. Adjust pH to 7.2–7.4 with NaOH or HCl.
5. Rice medium: rice 100 g, distilled water 100 mL. Let it stand overnight before autoclaving.
6. Solvents for extraction: acetone (HPLC grade), acetonitrile (HPLC grade), dichloromethane (HPLC grade), ethyl acetate (HPLC and Analytical Reagent grade), *n*-hexane (HPLC grade), methanol (HPLC and Analytical Reagent grade).
7. Sample adsorbents: Celite 545AW-Reagent Grade and Diaion® HP20 (Supelco, USA).
8. Solvent for NMR: chloroform ($CDCl_3$), methanol (CD_3OD), and dimethylsulfoxide (DMSO-d_6), use 750 μL of appropriate solvent to dissolve sample with minimum weight of 1 mg and maximum weight of 20 mg.
9. Pre-coated TLC plates: Silica Gel 60 F_{254}layer thickness 0.2 mm, RP-18, F_{254} S, layer thickness 0.25 mm.
10. Column chromatography adsorbents: Sephadex LH 20, 0.25–0.1 mm mesh size, VersaPak® Silica and C18 Cartridges 20–45 μm, matrix active group Spherical Silica at various I.D. × L of 23 × 53 mm, 23 × 110 mm, and 40 × 150 mm.

2.3 Instruments and Equipment

1. Analytical miller (model: IKA A11 Basic) was from IKA, Germany.
2. Laminar flow hood (BioMAT2), Medical Air Technology.
3. Homogenizer (IKA T18 Basic Ultra-Turrax), IKA, Germany.
4. Rotary evaporators (model no: R-110 and R-3), BUCHI, Switzerland.
5. Centrifuge used was Force 7 from Fisher Scientific.
6. Sonicator (Ultrawave), Scientific Laboratory Supplies, Ltd.
7. UV lamp (UVGL-55 Handheld UV Lamp), UVP, Cambridge, UK.
8. Flash/Medium Pressure Liquid Chromatography instrument Sepacorepump manager C-615, pump module C-601 from

BUCHI, Switzerland. Column loading anchorages, VersaFlash/ Supelco, Sigma-Aldrich, Germany or Biotage Isolera Spektra One and Four Flash Purification System.

9. Fraction collector (CF2), Spectrum Labs.
10. Freeze dryer (model: Christ Alpha 2-4), Martin Christ Gefriertrocknunsanlagen GmbH, Germany.
11. Handheld dryer HL 2010 E, Type 3482, Steinel.
12. High Resolution Mass Spectrometry (Orbitrap-Exactive) instrument coupled to high pressure liquid chromatography (DionexUltimate-3000), Thermo Scientific, Germany.
13. Nuclear Magnetic Resonance Spectrometer instrument, JNM-LA400 model from JEOL, Japan and the magnet NMR AS400 model EUR0034 is from Oxford Instruments, England. The NMR is equipped with a Pulse-Field Gradient "Autotune" ™ probe 40TH5AT/FG broadband high sensitivity probe to accept 5 mm tubes. Fitted with FG coils, 2H lock channel and capable of variable temperature operation.

3 Methods

3.1 Dereplication of Extracts Using NMR and HRFTLCMS

1. Dissolve respective biologically active plant, sponge, and microbial extracts as well as the extract of the medium as blanks in deuterated DMSO-d_6 and acquire a ^{1}H NMR of 5 mg to 10 mg of sample. Process each of the NMR spectra to correct the baseline. Properly phase the spectra at zero and first order. Apodize the spectrum at Gaussian = 1.0 (*see* Fig. 4). Use the software ALICE to do multivariate analysis of the corrected spectrum. With the software, select peaks for integration from the summed spectra, while solvent peaks or other impurity peaks are deselected prior to analysis.
2. Dry the same extracts and dissolve in 50:50 H_2O:CH_3OH HPLC grade, to make a concentration of 1 mg/mL and submit for high resolution LC-HRFTMS. The samples are eluted through a C18 column (ACE) with a length of 75 mm and internal diameter of 3.0 mm using a mobile phase of 0.1 % formic acid in a gradient of HPLC grade water (solvent A) and acetonitrile (solvent B) and a flow rate of 300 μL/min. The elution begins with 10 % B, which is increased to 100 % B over 30 min which is maintained for 5 min before decreasing to 10 % B in the next min. The column is then equilibrated with 10 % B for 4 min until the next run.
3. Mass spectral data are processed using differential expression analysis software like SIEVE or MZmine 2.2 (*see* **Note 1**).
4. The peaks from the media and solvent blanks were then subtracted from those of the extracts, and the remaining peaks

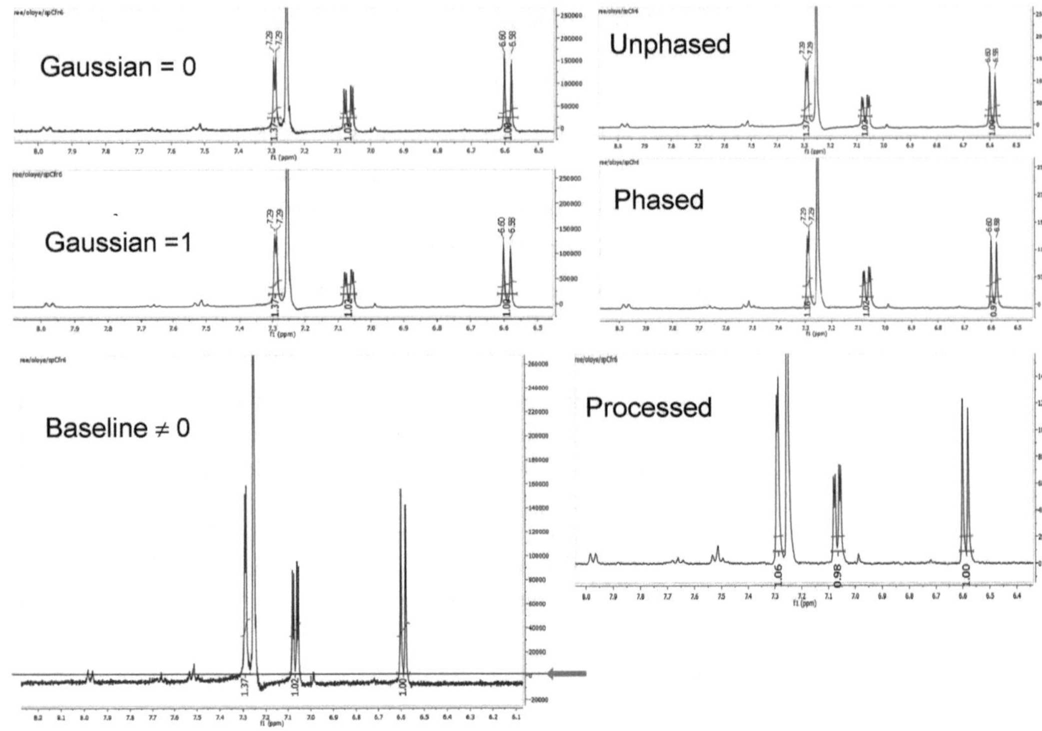

Fig. 4 A properly processed ^{1}H NMR spectrum is a requirement for multivariate analysis of NMR data sets

are compared to the AntiMarin database for identification as demonstrated in Fig. 5a, b.

5. SIEVE and MZmine results can be exported to an EXCEL data sheet which can used for PCA by employing SIMCA P + for pattern recognition as shown in Fig. 6.

3.2 Marine Bacterial Culture on Agar Plates and Broth Media

1. Inoculation and extract preparation. Bacterial isolates are initially cultured at 30 °C on M1 agar containing artificial seawater (*see* **Note 2**). A gram stain is performed and the bacteria plated using the single streak method to obtain a pure colony in order to confirm that the bacteria being grown was identical to that sent.
2. After 7 days of growth, agar plates are transferred into conical flasks where they were covered with 200 mL of EtOAc or CH_3OH.
3. After 24 h, the samples are homogenized and vacuum-filtered and the aqueous layer extracted thrice with EtOAc. The EtOAc fractions are then collected and dried under vacuum.
4. For flasks containing CH_3OH, an equal amount of water is added to saturate the CH_3OH layer before extracting with EtOAc. The EtOAc layers are collected and dried under vacuum.
5. The EtOAc extract is then eluted through an HP-20 column using 100 % water, 50:50 CH_3OH:H_2O, and 100 % CH_3OH in order to remove salt.

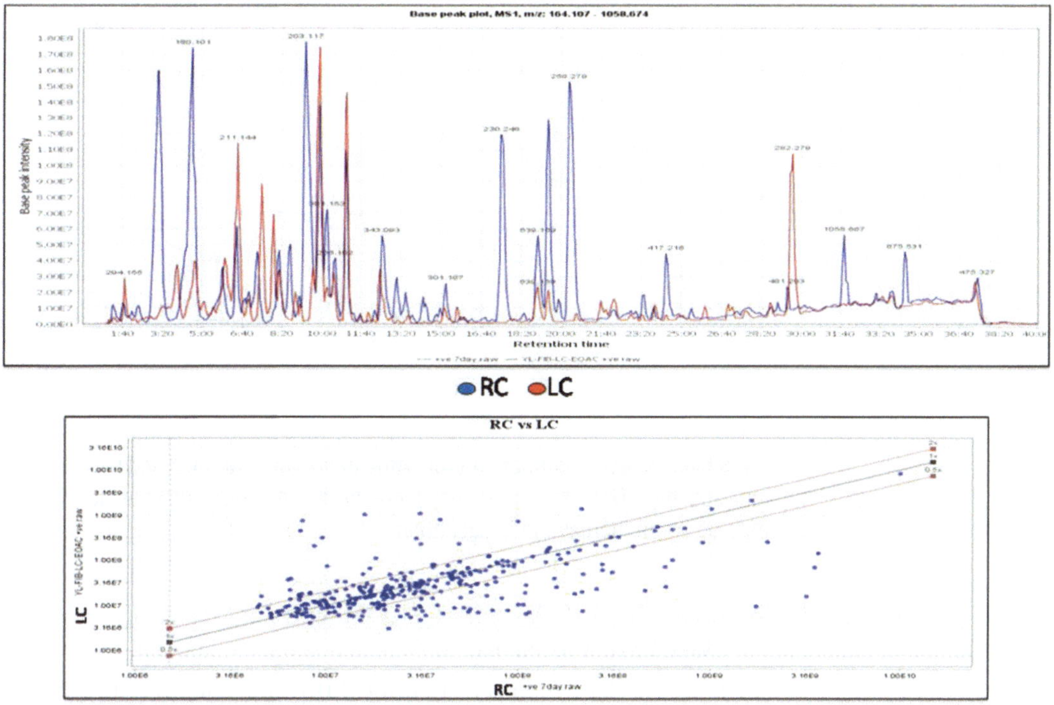

Database search for an Endophytic *Aspergillus*

	MS m/z	row retention time	Name	RC peak m/z	RC peak height	RC peak area	LC peak m/z	LC peak height	LC peak area
Identified	203.118	9.36935	Nb-Acetyltryptamine	203.118	1.78E+08	3.22E+09	203.118	1292067	1.59E+07
Identified	180.1019	4.5859	N-Acetyl-tyramine	180.1019	1.74E+08	3.71E+09	180.1019	5466609	1.43E+08
Identified	235.1188	3.19349	cyclo-(L-prolyl-L-histidine)	235.1188	1.60E+08	3.56E+09	235.1188	3918087	6.92E+07
	258.2791	20.3209		258.2791	1.53E+08	2.68E+09	N/A	N/A	N/A
Identified	436.198	9.907385	Neoxaline	436.1979	1.38E+08	1.66E+09	436.1982	1.74E+08	2.14E+09
Identified	639.1711	19.4446	Secalonic acid	639.1711	1.29E+08	3.26E+09	N/A	N/A	N/A
	230.2478	17.5803		230.2478	1.19E+08	1.75E+09	230.2478	895430.5	9434694
	478.2707	11.0389	'Paraherquamide E; VM 54159; Paraherquamide VM-54159	478.2712	1.10E+08	1.02E+09	478.2702	1.46E+08	1.36E+09
Identified	381.1543	10.2111	Dinaphtho[2,1-b:1',2'-d] furan-5,9-dione	381.1543	7.23E+07	1.04E+09	N/A	N/A	N/A
Identified	360.2281	6.413705	Asperparaline A; Aspergillimide; VM-55598	360.228	6.17E+07	6.41E+08	360.2281	9320451	1.01E+08
Identified	1058.673	31.8234	Niphithricin B	1058.673	5.85E+07	8.16E+08	N/A	N/A	N/A
Identified	343.0926	12.54905	Griseolutein A	343.0926	5.58E+07	9.18E+08	343.0925	1.76E+07	2.43E+08
	164.1073	8.677255		164.1075	5.03E+07	5.93E+08	164.1071	3076721	2.59E+07
	875.5355	34.3874		875.5355	4.58E+07	6.49E+08	N/A	N/A	N/A
	367.1387	7.27488		367.1387	4.56E+07	7.99E+08	N/A	N/A	N/A

Fig. 5 (**a**) An S plot differentiated the production of secondary metabolites by an *Aspergillus* species in a liquid and solid media. (**b**) Dereplication of known metabolites in both media was identified from the AntiMarin database

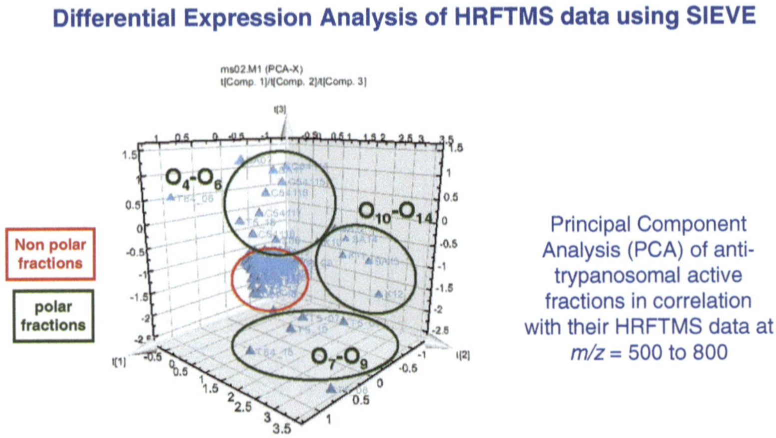

Fig. 6 Mass spectral data of different African propolis samples and fractions was analyzed by SIMCA to reveal that increase in anti-trypanosomal activity was affected by an increase in oxygenation

6. Both the EtOAc and CH_3OH extracts are analyzed using LC-FTMS and NMR.
7. Four flasks, each containing 500 mL of M1 broth, are inoculated with a quarter of an M1 agar plate previously incubated for 5 days at 30 °C with the bacterial isolate. Another four flasks are inoculated in the same manner with a similar bacterial isolate incubated for 10 days.
8. Two flasks each from the 5- and 10-day cultures are shaken at (orbital shaker) 150 rpm at 30 °C for 7 days. The other four flasks, two of which contained 5-day incubated bacteria and the other two containing 10-day incubated bacteria, were allowed to stand at 30 °C for 7 days.
9. After 7 days of incubation, the broths were then covered with EtOAc and the metabolites are extracted using the procedure outlined above, dereplicated and submitted for bioassay.

3.3 Scale-up and Extract Preparation from Marine Microbial Isolates

1. Bioactive extracts are cultured in 1 L conical flasks containing 500 mL of M1 broth media for 5 days (30 °C, shaken at 150 rpm).
2. 20 mL of the inoculum is transferred into 16 out of the 17–500 mL conical flask containing 200 mL of M1 broth media.
3. The flasks spanned duplicates from day 0 to 7 with one blank media-only flask. The cultures are then grown for 1–7 days (30 °C, shaken at 150 rpm) and taken out on their respective days.
4. An equal volume of EtOAc is added to the broth and left overnight for cell lysis. Culture extraction is performed using EtOAc as the organic phase.

5. The extract is collected in tared vials through solvent evaporation. A similar process was repeated to obtain extracts for days 8, 15 and the accompanying blank media extract.
6. The extracts are dereplicated and submitted for bioassay.

3.4 Isolation and Purification of Fungal Endophyte Cultures

1. Isolation of fungal endophytes from plant source (*see* **Note 3**). With a sterile scalpel, the outer tissue was removed from the plant samples and the inner tissue was carefully dissected under sterile conditions and cultivated onto malt agar (MA) plates containing chloramphenicol to prevent the bacterial growth.
2. After 3–4 weeks of incubation at room temperature, hyphal tips of the fungi are removed and transferred to fresh MA medium. Plates are prepared in duplicates to eliminate the possibility of contamination.
3. Pure strains are isolated by repeated inoculation.
4. Pure fungal strains are grown on MA medium at room temperature for 14 days.
5. Pure cultures are then stored at 4 ° C for a maximum period of 6 months, and then reinoculated onto fresh media.

3.5 Small Scale Extraction of Endophyte Cultures for Screening and Metabolomic Profiling

1. One petri-dish of each fungal species is transferred into a 250 mL flask, then left overnight with EtOAc and extracted (3×150 mL EtOAc) followed by filtration. The filtrate is then dried under vacuum, suspended in 200 mL H_2O and partitioned between (EtOAc 3×200 mL) in a separating funnel.
2. The H_2O-soluble phase is passed on a Sephadex LH-20 column using methanol as eluant. CH_3OH and EtOAc soluble portions are concentrated and subjected to MS and NMR analysis for metabolic profiling and dereplication studies using AntiMarine databases.

3.6 Cultivation of Fungal Endophyte for Large-Scale Extraction and Isolation of Secondary Metabolites

1. Fungal cultures are transferred into 1 L Erlenmeyer flasks containing 500 mL of Wickerham medium for liquid cultures or 100 g rice for solid cultures.
2. The cultures are then incubated at room temperature (no shaking) for 15 and 30 days, respectively.
3. Large-scale cultivation is then carried out using 30 and 10 1 L Erlenmeyer flasks for liquid and solid rice cultures, respectively.

3.7 Extraction of Fungal Liquid Culture

1. 250 mL of EtOAc and *n*-BuOH are added alternately to the culture after overnight maceration with EtOAc.
2. Culture media and mycelia were then homogenized in the Ultraturrax for 10 min for cell destruction, followed by vacuum filtration using Buchner funnel.

3. The mycelium residue is then discarded while EtOAc and *n*-BuOH culture filtrates are collected and further extracted with EtOAc (200 mL × 3) through a separating funnel.
4. The pooled EtOAc extracts are then subjected to evaporation.
5. The water phase is then chromatographed over Sephadex LH-20 using CH_3OH as eluent.

3.8 Observation of Metabolite Progression by High Performance Liquid Chromatography–Mass Spectrometry (HPLC–MS) and Nuclear Magnetic Resonance (NMR) Spectroscopy

All analyses requiring HPLC–MS are carried out using the same Dionex UltiMate 3000-Thermo Scientific Exactive system and method as mentioned above. All 1D and 2D spectra are measured on the aforementioned Jeol-LA400 FT-NMR spectrometer.

4 Notes

1. Differential expression analysis software can be summarized by the following steps: import raw data, chromatogram building (peak picking), chromatogram deconvolution, deisotope, alignment, and identification which can be accomplished through adduct search, complex search, fragment search, and formula prediction. MZmine is a free ware and can be downloaded through http://mzmine.sourceforge.net/. Detect the peaks in the solvent blank and media using the chromatogram builder. Mass detection is performed using an exact mass detector threshold greater than the noise level set at 1.00E2. Filter the peaks using Fourier transform mass spectrometry shoulder peaks filter at a mass resolution of 50,000 with the Gaussian peak model. The chromatogram builder uses the highest data point function. The minimum time span is set at 2 s, and the minimum height and m/z tolerance at 1.00E2 and 0.001 m/z, respectively. Chromatogram deconvolution is then performed to detect the individual peaks. The baseline cut-off method (minimum peak height: 1.00E2, minimum peak duration: 0:02, and baseline level 1.00E1) is applied. Crop filter the spectra of the extracts between 2:00 and 35:00 min. Chromatogram building and deconvolution then followed, while the noise and baseline levels are set higher (noise level: 1.00E5, minimum time span: 0:10, minimum height: 1.00E5, and m/z tolerance: 0.001 m/z, baseline level: 1.00E3). Isotopes are also identified in the extracts as well as the blanks using the isotopic peaks grouper (m/z tolerance: 0.050 m/z, RT tolerance: 0.05, monotonic shape, maximum charge: 1, and representative isotope: most intense). Duplicate peaks are filtered

out using the duplicate peak filter (m/z tolerance: 0.050 m/z, RT tolerance: 0:05 s, requiring the same identification). An adduct search is performed for Na–H, K–H, and NH_3 (RT tolerance: 0:10, m/z tolerance: 0.100 m/z, max adduct peak height: 20 %) and a complex search is performed for both the positive and negative modes of ionization (RT tolerance: 0:10, m/z tolerance: 0.100 m/z, max complex peak height: 20 %).

Finally, the peak lists of the solvent blank, media, and the extracts are aligned using the join aligner parameters set at m/z tolerance: 0.100 m/z, weight for m/z: 70, RT tolerance type: Absolute, RT tolerance: 0:30, Relative RT tolerance: 15 %, Weight for RT: 10, require same charge state, require same ID state, compare isotope pattern, and isotope pattern score threshold level: 65 %.

2. The presence of salt can result in broad NMR peaks and it is not advisable for further mass spectrometric analysis as it can block heated capillaries. Salt from samples should be removed using HP20 Chromatography. The dried extract is reconstituted in 20 mL methanol and 20 mL acetone. The flasks are then sonicated to aid in the dissolution of the sample. The sample is then mixed thoroughly with enough Celite and placed in the fume hood to dry.

Approximately 400 cm^3 of HP20 is covered in analytical grade CH_3OH and poured into the column, which is plugged with cotton that had been previously soaked in CH_3OH. Until the column is half-full, more CH_3OH is added and the HP20 is activated for 3 h. The CH_3OH is then drained from the column. A cotton plunger is placed above the HP20 bed. Fresh CH_3OH is run through the column to wash it. The washing is collected and run through the column for a second time. The column is then washed with pure-grade water twice. The column is then equilibrated in pure-grade water overnight.

The Celite containing the adsorbed extract is loaded at the top of the HP20 column. Purified sand is added to reduce the caking of the Celite. Cotton is placed above the Celite. Fractions are eluted using a gradient of 100 % pure-grade water to 100 % CH_3OH (HPLC grade) in 10 % increments. The volume for each eluant is 250 mL. This was allowed to pass through the column completely before the next eluant is added. 250 mL of 50:50 acetone:methanol (HPLC grade), followed by two additional runs of 100 % methanol are used to wash the column. Two milliliter aliquots of each fraction are collected for LC-HRFTMS analysis. The fractions are dried and placed in the vacuum desiccators prior to weighing and NMR analysis.

3. Plant materials are cut into small pieces, washed with sterile water, then thoroughly surface sterilized with 70 % isopropanol for 1–2 min and ultimately air dried under a laminar flow hood. This is done in order to eliminate surface contaminating epiphytic microbes.

Acknowledgments

This work was supported by the University of Strathclyde Starter Grant, Scottish Funding Council Studentship, and Royal Society Research Grants–2011 R2.

References

1. Yuliana ND, Khatib A, Choi YH, Verpoorte R (2011) Metabolomics for bioactivity assessment of natural products. Phytother Res 25:157–169
2. Moldenhauer J, Chen XH, Borriss R, Piel J (2007) Biosynthesis of the antibiotic bacillaene, the product of a giant polyketide synthase of the trans-AT type. Angew Chem Int Ed Engl 46:8195–8197
3. Ebada SS, Edrada-Ebel RA, Lin WH, Proksch P (2008) Methods for isolation, purification and structural elucidation of bioactive secondary metabolites from marine invertebrates. Nat Protoc 3:1820–1831
4. Kjer J, Debbab A, Aly AH, Proksch P (2010) Methods for isolation of marine-derived endophytic fungi and their bioactive secondary products. Nat Protoc 5:479–490
5. Murata M, Oishi T, Yoshida M (2006) State-of-art methodology of marine natural products chemistry: structure determination with extremely small sample. In: Fusetani N, Clare AS (eds) Antifouling compounds (Marine Molecular Biotechnology), vol 42. Springer, Heidelberg, pp 203–220
6. Gray AI, Igoli JO, Edrada-Ebel RA (2012) Natural products isolation in modern drug discovery programs. Methods Mol Biol 864:515–534
7. (a) JEOL (2005) Latest information and future for ALICE2 software - application for metabonomics. JEOL News 40:43–36. (b) Kawaguchi H, Hirakawa K, Miyauchi K, Koike K, Ohno Y, Sakamoto A (2010) Pattern recognition analysis of proton nuclear magnetic resonance spectra of brain tissue extracts from rats anesthetized with propofol or isoflurane. PLoS One 5: e11172. doi:10.1371/journal.pone.0011172
8. Blunt J (2012) MarinLit. University of Canterbury, New Zealand
9. (a) Pluskal T, Castillo S, Villar-Briones A, Orešič M (2010) MZmine 2: Modular framework for processing, visualizing, and analyzing mass spectrometry-based molecular profile data. BMC Bioinformatics 11:395–399. (b) Sourceforge. (2013) "MZmine 2 - framework for mass spectrometry data processing". http://mzmine.sourceforge.net/download.shtml
10. Buckingham J (consultant ed) (2012) Dictionary of natural products on DVD. Version 20:2. London, UK, Chapman and Hall/CRC
11. Laatsch H (2012) Antibase Version 4.0 – The natural compound identifier. Wiley-VCH Verlag GmbH & Co. KGaA
12. Kanehisa Laboratories (2012) "KEGG: Kyoto Encyclopedia of Genes and Genomes." http://www.genome.jp/kegg/
13. (a) Sourceforge (2013) "mzMatch/PeakML: Metabolomics Data Analysis." http://mzmatch.sourceforge.net/. (b) Jankevics A, Merlo ME, de Vries M, Vonk RJ, Takano E, Breitling R (2012) Separating the wheat from the chaff: aprioritization pipeline for the analysis of metabolomics datasets. Metabolomics 8: 29–36. (c) Creek DJ, Jankevics A, Breitling R, Watson DG, Barrett MP, Burgess KEV (2011) Towards global metabolomics analysis with liquid chromatography–mass spectrometry: improved metabolite identification by retention time prediction. Anal Chem 83: 8703–8710
14. (a) Vast Scientific (2010) http://www.vastscientific.com/resources/index.html. (b) Kamleh A, Barrett MP, Wildridge D, Burchmore RJ, Scheltema RA, Watson DG (2008) Metabolomic profiling using Orbitrap Fourier transform mass spectrometry with hydrophilic interaction chromatography: a method with wide applicability to analysis of biomolecules. Rapid Commun Mass Spectrom 22:1912–1918

Chapter 18

Bridging the Gap: Basic Metabolomics Methods for Natural Product Chemistry

Oliver A.H. Jones and Helmut M. Hügel

Abstract

Natural products and their derivatives often have potent physiological activities and therefore play important roles as both frontline treatments for many diseases and as the inspiration for chemically synthesized therapeutics. However, the detection and synthesis of new therapeutic compounds derived from, or inspired by natural compounds has declined in recent years due to the increased difficulty of identifying and isolating novel active compounds. A new strategy is therefore necessary to jumpstart this field of research. Metabolomics, including both targeted and global metabolite profiling strategies, has the potential to be instrumental in this effort since it allows a systematic study of complex mixtures (such as plant extracts) without the need for prior isolation of active ingredients (or mixtures thereof). Here we describe the basic steps for conducting metabolomics experiments and analyzing the results using some of the more commonly used analytical and statistical methodologies.

Key words Chromatography, Nuclear magnetic resonance, Multivariate statistics, Secondary metabolism, Chemical synthesis, Metabolomics

1 Introduction

Metabolomics may be simply defined as the comprehensive quantitative and qualitative analysis of all metabolites present in a specific biofluid, cell, tissue, or organism [1]. Since its foundation in the late 1990s [2, 3] it has proven to be a valuable tool in the analysis of biological systems where it is has been used in an ever increasing number of diverse applications such as disease diagnosis, environmental science, toxicology and health science [4–8]. In the majority of these cases metabolomics/metabonomic analysis is used to determine how a particular treatment, toxicant or disease state has altered a distinct set of primary metabolites (compounds essential for or directly involved in normal growth, development, and reproduction). Increasingly however, the technique is also being utilized in the study of secondary metabolites, particularly in plants and fungi. Secondary metabolites are chemicals such as

Ute Roessner and Daniel Anthony Dias (eds.), *Metabolomics Tools for Natural Product Discovery: Methods and Protocols*, Methods in Molecular Biology, vol. 1055, DOI 10.1007/978-1-62703-577-4_18, © Springer Science+Business Media, LLC 2013

pigments produced by plants for which no role has yet been found in growth, photosynthesis, reproduction, or other "primary" functions. Each plant family, genus, and species produces a characteristic mix of these chemicals. Many thousands have been identified and plant and fungal secondary metabolites have been a fertile area of chemical investigation driving the development of both analytical chemistry and of new synthetic reactions and methodologies for many years [9]. Indeed the use of whole plants or extracts as medicines via the isolation of active compounds began in the early nineteenth century with the isolation of morphine (and later codeine) from opium. The first synthesis of morphine (by Gates in 1952) is considered a classic in the field [10]; although there is still a need for developing morphine substitutes without the addictive properties of the parent compound. Other examples of plant metabolites used in human health include digitalis and quinine, while penicillin, cephalosporin, ergotrate and many statins are equally well used fungal based compounds [11]. Despite their wide use however, understanding how plants and fungi produce such intricate compounds from relatively simple starting points is a surprisingly complex challenge. The vast number of metabolites typically present in natural products and their large variation in molecular weight, polarity and concentration makes their analysis particularly challenging and there is high potential for many potentially useful compounds to go unnoticed [12].

Interestingly, both natural products chemistry and metabolomics have as their goal the identification of active compounds, either as a purified active component (natural products chemistry) or as a group of compounds used as an indicator of a particular biological state (metabolomics). Natural products chemistry has a long tradition of sophisticated techniques that allow identification of complex molecular structures but it often fails when dealing with complex mixtures. In contrast, metabolomics deals well with mixtures and uses the power of multivariate analysis to isolate the "statistical needle" from the stack of other needles but it is often limited in the identification of all the compounds involved [13]. Chemical synthesis together with metabolomics has the potential to significantly simplify the research pathway to new product discovery. Here we outline protocols for basic metabolomics studies using standard operating procedures for the more the commonly used analytical and statistical methodologies.

2 Materials

2.1 Metabolite Extraction (Methanol–Chloroform– Water)

The first step in any metabolomics study is to extract the metabolites from the sample matrix which may be almost any form of biological material; from fluids, such as blood, milk, and urine, to cell cultures, tissue biopsies and even whole organisms such as *Daphnia* [14]. A major challenge in metabolomics is to address the extremely diverse and complex nature of the organism. Metabolites may have a concentration range of the order of ~10^9, a mass range of the order of ~1,500 amu and polarity ranges of ~10^{20} [15]. This means that no one analytical approach is capable of detecting and analyzing all the metabolites that may be present in an organism or tissue. Therefore, multiple analytical techniques and sample preparation strategies are necessary to enable complete coverage of the metabolome. This problem is further complicated by the fact that high throughput metabolic profiling technologies are still under active development and frequently require the user to undergo specialist training. Whichever method is used, it is essential that both metabolite extraction and sample preparation are as efficient and reproducible as possible (*NB*: it is worth spending time and resources on this issue at the start of a study since it is likely to save significantly more than it costs of both in the long run).

Relatively simple liquid–liquid extraction procedures are commonly utilized for metabolic extractions/separations. The method chosen should be able to liberate a broad range of metabolites from the sample as possible. Multiple solvent combinations can be applied depending on the target compounds. However, as recently confirmed by Lin et al. [16] the most effective and commonly applied is the methanol/chloroform/water based method [17]. This creates a biphasic solvent partitioning system, where the miscibility of methanol/water partitions polar metabolites into an upper aqueous phase, whilst retaining nonpolar molecules in the chloroform–methanol partition (nonaqueous phase). Proteins and other cellular material tend to sit conveniently in pellet form between the aqueous and organic layers, creating a useful visible boundary. Factors of interest include the nature, volume, and ratio of the organic and aqueous solvents used and for sensitive material (e.g. plant pigments) one may also want to consider the pH, light levels and temperature at which the extraction is carried out [18].

Some workers advocate the use of deuterated solvents in place of the standard, non-deuterated versions for NMR analysis, since this negates the requirement for a drying step (e.g., lyophilization/speed vacuum drying) and thus retains volatile metabolites in the polar phase that are susceptible to evaporation. However, such losses are exceedingly minor and the extra expense involved in using large amount of deuterated solvents means that their use is not common practice. The methods assume (including drying and solvent exchange) non-deuterated solvents for the majority of steps outlined below. Actual samples should be stored at around −80 °C prior to analysis and processed as quickly as possible once thawed to minimize metabolite degradation. Some sample types (e.g. urine) can be stored at −20 °C for a number of years with no negative effects. However, one needs to confirm if this is suitable for the sample type under study, thus −80 °C is the default safety value. All solutions outlined below should be prepared using ultrapure water (18 MΩ) and analytical grade reagents. All reagents should be stored at room temperature (unless otherwise indicated). All national and internationally relevant regulations should be followed when disposing of waste materials.

1. Sample.
2. Steel or ceramic mortar and pestle (*NB*: ceramic equipment does not stand up well to liquid N_2).
3. Dry ice (solid CO_2) or liquid N_2.
4. 2 mL Eppendorf (or equivalent) snap top tubes.
5. Gilson (or equivalent) pipettes, properly calibrated.
6. Analytical grade methanol, chloroform and water.
7. Centrifuge.
8. Speed vacuum concentrator.
9. Sonicator and foam floats for Eppendorf tubes.

2.2 Analysis of Metabolite Extracts Via ^1H NMR Spectroscopy

Biofluid and organ NMR spectroscopy is a multivariate method in as much as all small molecules containing the nucleus under study—usually but not exclusively the proton—produce a composite response over the range of detected frequencies. Following necessary pre-processing, this frequency/intensity response can be converted into a histogram which itself is then put into tabular form suitable for chemometric analysis. NMR spectroscopy of fluids normally gives sharp, well-resolved signals in the spectra. This is not the case for solids or tissues and a specialized technique, Magic Angle Spinning, is required to give informative spectra from such samples [19].

Proton Nuclear Magnetic Resonance (^1H NMR) spectroscopy is a versatile and widely used tool in metabolomics. It has the advantage of being quick and easy to perform as well as being cheap on a per sample basis. The concept of metabolic profiling is not new. For example, ^{31}P, ^1H and ^{13}C NMR spectroscopy have been used extensively to profile metabolites in a range of organisms including humans [20]. NMR is however, limited in sensitivity and cannot easily be utilized to look at organic phase metabolites. As such, while NMR is good for a first look at a sample it may miss metabolites that are present in low concentrations. However, since it is a nondestructive technique, samples may be recovered and run via GC or LC-MS to provide further information.

Below we have described a simple generic method for 1D ^1H NMR analysis. Such data may be further augmented by the use of 2D NMR spectroscopy (e.g. COSY, HSQC, HMBC, JRES, and TOCSY pulse sequences) for the purposes of metabolite identification and structural information.

1. Dried extract in 2 mL Eppendorf (or equivalent) snap top tubes.
2. NMR tubes and labels.
3. Gilson (or equivalent) pipettes, properly calibrated.
4. Buffer solution (formulated as outlined below).
5. Analytical grade D_2O.
6. NMR spectrometer.

2.3 Analysis of Metabolite Extracts Via Gas Chromatography–Mass Spectrometry (GC-MS)

Gas Chromatography–Mass Spectrometry (GC-MS) has also been widely used as a metabolic profiling tool since its first use to study steroids, amino acids and drug metabolites in the early 1970s [21]. GC-MS is a popular technique in metabolomics analysis because of its superior sensitivity and increased separation and identification capabilities when compared to NMR. In GC-MS compounds are first separated by GC with eluting compounds then being detected via MS, traditionally by using electron-impact ionization. Newer techniques such as GC-TOF-(Time Of Flight)-MS and two dimensional GC×GC analysis can theoretically identify even greater numbers of metabolites [22].

One downside to this technique is that many samples must be in the gas phase for analysis. Since many (but by no means all) biological metabolites are not thermally stable or volatile, a derivatization step is usually required in order to employ the method successfully. Although costly and time consuming this step improves the chromatographic separation [23]. This step can of course be avoided if one is conducting metabolomics on volatile metabolites such as those in breath [24].

Derivatization methods can be classified into four main groups according to the reagents used and the reaction achieved, namely silylation, acylation, esterification and alkylation [25]. Silylation is the most widely used technique, it involves the replacement of an acidic hydrogen with an alkylsilyl group, for example—$SiMe_3$, to form tri-methyl-silyl (TMS) derivatives [23]. This reaction is however, not without its problems. Trimethylsilylation of primary amino groups can fail to produce one single product and is very sensitive to even minor changes in conditions. In addition, the resulting *N*-trimethylsilyl (TMS) derivatives are very susceptible to hydrolysis and, as a consequence, removal of excess silylation agent and extraction of the derivatives is often impossible. Direct injection of excess silylation agent is therefore often required. The reaction for TMS derivatives usually occurs cleanly without artifact or by-product formation and there are a number of reagents available for this purpose [26]. Brief summaries of some of the more popular are given below. However, most metabolomics studies make use of MSTFA or (more rarely) BSTFA.

2.3.1 BSA [N,O-Bis(Trimethylsilyl) Acetamide]

BSA is a strong silylation reagent, which can be used to form very stable TMS derivatives of a wide variety of compounds such as alcohols, amines, carboxylic acids, phenols, steroids, biogenic amines and alkaloids. However, it is not recommended for use with hydrocarbons or very low molecular weight compounds. BSA is a good solvent for polar compounds but frequently it is used in combination with a solvent (e.g. pyridine) or together with other silylation reagents.

2.3.2 BSTFA [N,O-Bis(Trimethylsilyl) Trifluoroacetamide]

BSTFA is a powerful trimethylsilyl donor with approximately the same donor strength as its unfluorinated analogue, BSA. Reactions of BSTFA are similar to those of BSA. The major advantage of BSTFA over BSA is the greater volatility of its reaction products. This property is particularly useful for the GC analysis of some lower boiling point TMS amino acids. BSTFA is non-polar (less polar than MSTFA), it can be mixed with acetonitrile for improved solubility. BSTFA with 1 % trimethylchlorosilane is also widely used as a derivatizing agent. Zhang and Zuo [27] found the best ratio for the formation of a single completely derivatized product, to be 1:1 BSTFA + 1 % TMCS and pyridine (v/v), although that study was looking at estrogens as opposed to metabolites.

2.3.3 DMCS [Dimethyldichlorosilane]

DMCS is used to form dimethylsilyl (DMS) derivatives. DMS derivatives are much more susceptible to hydrolysis than TMS derivatives; therefore, strictly anhydrous conditions are very important during the reaction. As a result this reagent is not commonly used.

2.3.4 HMDS [Hexamethyldisilazane]

HMDS is a weak TMS donor. Used alone its action is slow and not very effective. However, after addition of a TMS catalyst such as TMCS (Trimethylchlorosilane), e.g. 1 %, it becomes a fast and quantitative reagent for trimethylsilylation of organic compounds. Aprotic solvents like acetonitrile, dimethylformamide, carbon disulfide and dimethylacetamide are recommended for use with HMDS.

2.3.5 MTBSTFA [N-Methyl-N-Tert-Butyldimethylsilyl-Trifluor-Acetamide]

MTBSTFA is a silylation reagent which donates a *tert-butyldimethylsilyl* group (TBDMS) for derivatizing active hydrogen atoms in hydroxyl, carboxyl and thiol groups, as well as primary and secondary amines. Reactions proceed fast (typical reaction times of 5–20 min) with high yields (>96 %). The by-products are neutral and volatile. TBDMS ethers are 10^4 times more stable than the corresponding TMS ethers. Due to the large protecting group retention times are longer, which may improve some separations. In addition MTBSTFA derivatives are more resistant to hydrolysis than their TMS counterparts. Some workers use MTBSTFA in preference to MSTFA because of the greater thermal and hydrolytic stability of the *tert.*-butyldimethylsilyl derivatives [28]. This reagent has also been successfully used in an alternative procedure to the BF_3-methanol lipid derivatization technique [29].

2.3.6 MSTFA [N-Methyl-N-Trimethylsilyl-Trifluoroacetamide]

MSTFA is the most volatile trimethylsilyl amide available. BSA and BSTFA, which, in the past, have been used most frequently in GC silylation, can often be replaced by MSTFA. The already excellent solution characteristics of this reagent can be improved by the addition of sub-molar quantities of protic solvents (e.g. trifluoroacetic acid for extremely polar compounds, such as hydrochlorides, or pyridine for carbohydrates). MSTFA can be used for trimethylsilylation of carboxylic acids, hydroxy and ketocarboxylic acids, amino acids, amines, alcohols, polyalcohols, sugars, mercaptans, and similar compounds with active hydrogen atoms. Even amine hydrochlorides can be silylated directly. Most studies use MSTFA on its own for derivatization. However, some workers have used a two-step silylation. In such cases the sometimes unstable *N*-silyl groups formed by MSTFA derivatization are converted to the more stable *N*-trifluoroacetyl form by the addition of *N-methyl-N-*(trifluoroacetamide) (MBTFA) [30].

2.3.7 TMCS [Trimethylchlorosilane]

TMCS is often used as a catalyst with other trimethylsilyl reagents. Without additives it can be used for preparing TMS derivatives of organic acids.

2.3.8 TMSDEA [N-Trimethylsilyl-Diethylamine]

TMSDEA is a strongly basic silylating reagent which is particularly useful for derivatizing amino acids, antibiotics, urea-formaldehyde condensates and steroids. The reaction by-product diethylamine is very volatile. TMSDEA is ideal for the synthesis of standards, as the reaction can be easily driven to completion by evaporating the diethylamine.

2.3.9 TSIM [N-Trimethylsilyl-Imidazole]

TSIM is considered to be the strongest hydroxyl silylator and is the reagent of choice for carbohydrates and most steroids. The reagent is unique in that it reacts quickly and smoothly with hydroxyl (even *tert.* OH) and carboxyl groups but not with amines. This characteristic makes TSIM particularly useful in multi-derivatization schemes for compounds with different functional groups, which are to be derivatized differently (e.g., -O-TMS/-N-HFB derivatives of catecholamines). TSIM is used in the trimethyl-silylation of alcohols, phenols, organic acids, steroids, hormones, glycols, nucleotides and narcotics.

2.3.10 MSHFBA [N-Methyl-N-Trimethylsilyl-Heptafluorbutyramide]

MSHFBA is similar to MSTFA in reactivity and chromatography. It may be used for the general purpose trimethylsilylation of carboxylic acids, alcohols, phenols, primary and secondary amines and amino acids. MSHFBA is used either alone or in combination with a catalyst (TMCS, TSIM) or another silylation reagent with or

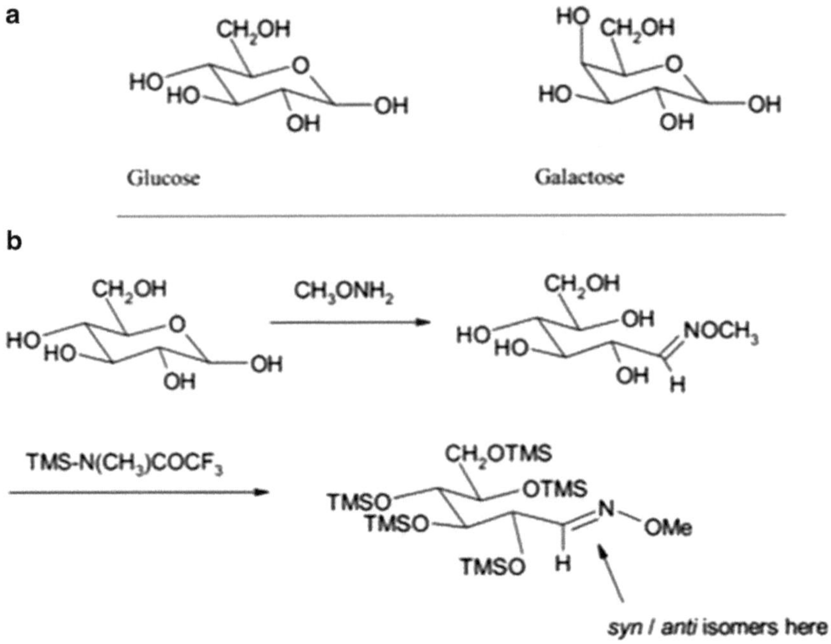

Fig. 1 *Panel A* shows diagrams of glucose illustrating their structural similarity and *panel B* illustrates the derivatization reaction of glucose

without a solvent. The by-product *N*-methyl-heptafluorobutyric amide has a lower retention time than the silylating reagent itself. The large ratio of fluorine to silicon of 7:1 in MSHFBA is especially useful for flame ionization detection, since degradation of the excess MSHFBA does not produce SiO_2, which can clog up this type of detector.

2.4 Structural Isomers

The GC analysis of many biological compounds, such as sugars, is also complicated by the fact that many are isomeric. For example for instance glucose, galactose and mannose are isomeric aldohexoses, arabinose and xylose are aldopentoses and sorbitol is the aldehyde-reduced derivative of glucose. With the exception of sorbitol, these compounds exist in solution mainly in ring-closed form, with various ratios of pyranose (6-membered) to furanose (5-membered) rings. The derivatization process described here initially locks the molecules into the acyclic form by converting the latent aldehyde into an *O*-methyl oxime derivative, using methoxyamine hydrochloride in pyridine. The hydroxyl groups are then trimethyl silylated to increase analyte volatility and thermal stability. The *O*-methoximation reaction normally gives a mixture of *syn*- and *anti*-isomers (Fig. 1) which appear as separate components on the GC profile, hence two retention times are often seen for most of the *aldo*-sugars and these isomers are normally present in unequal amounts.

1. Aqueous and organic fractions from the extraction procedure.
2. 2 mL screw-cap GC vials with bonded PTFE/silicon septa and glass inserts for small sample volumes.
3. Methoxyamine hydrochloride (20 mg/mL in pyridine) (should be stored in fridge at 4 °C when not in use).
4. *N*-methyl-*N*-trimethylsilyltrifluoroacetamide (MSTFA). *NB*: should be stored in fridge at 4 °C when not in use.
5. Boron trifluoride (BF_3). *NB*: should be stored in fridge at 4 °C when not in use.
6. Vortex mixer.
7. Analytical grade *n*-hexane, chloroform, H_2SO_4 and water.
8. Heating block or oven.
9. Glass syringe (for Methoxyamine hydrochloride, MSTFA and BF_3).

2.5 Analysis of Tissue Extracts Using Liquid Chromatography–Mass Spectrometry (LC-MS)

Liquid Chromatography–Mass Spectrometry (LC-MS) operates on a similar principal to GC-MS except that the chromatography stage utilizes a liquid mobile phase rather than a gas. This eliminates the need for metabolite volatility so there is no requirement for sample derivitization (although sample modification can still be useful for improving chromatographic resolution). The lack of derivitization means that a much wider range of analytes can potentially be measured and the overall analysis time per sample is often much shorter than for GC-MS. However, standard operating procedures are complicated by the fact that there are a large amount of instruments on the market each with differing set up requirements. In addition a multitude of ionization sources are in use, often differing from those used in GC-MS. As a result, the fragmentation libraries used for GC-MS based metabolite identification cannot be used for LC-MS whilst spectral libraries specific for LC-MS data are still under development; making compound identification very difficult. Additional problems include difficulties with chromatographic reproducibility, as well as matrix effects [31].

1. Aqueous and organic fractions from the extraction procedure.
2. 2 mL screw-cap LC vials with bonded PTFE/silicon septa and glass inserts for small sample volumes.
3. Analytical grade mobile phases: acetone, acetonitrile, methanol, ultrapure water etc.
4. Glass syringe (for direct injection).
5. Internal standard such as leucine encephalin.

2.6 Metabolomic Data Processing

Data handling tasks in metabolomics can be roughly divided into two steps: data processing and data analysis. The data processing step consists of low-level processing of raw data with signal processing methods and combining data between measurements. These tasks transform the raw data into format that is easy to use in the subsequent data analysis steps. The data analysis stage includes tasks for analysis and interpretation of processed data. This typically includes multivariate analyses such as clustering of metabolic profiles or discovering important differences between groups of samples [18].

1. Computer.
2. Appropriate software.
3. Time.

3 Methods

3.1 Metabolite Extraction

1. Take approximately 100 mg or 100 mL if using liquid samples (for NMR) (up to ten times) less can be used for GC/LC-MS of frozen samples but the exact amount will be part of optimized method development.
2. Pulverize sample (if solid) with plenty of dry ice using a pestle and mortar to break up the sample. Crushing under liquid N_2 is preferred if at all possible (*see* **Note 1**).
3. Recent metabolomics studies have ventured into the study of soil communities. Here the best method is to take 500 mg of soil and grind it to as fine a powder as possible with liquid N_2. It is also important to sieve the sample and remove bits of clay/twigs and other detritus. **Step 2** can be ignored if sample is liquid (such as blood). In many cases using a liquid sample only requires drying down and reconstituting in buffer solution (*go to* **step 9**).
4. When the sample has been crushed (and is still frozen) transfer to a pre-labelled Eppendorf tube.
5. Quickly add a methanol–chloroform mixture in the ratio of 2:1 using a Gilson Pipette. This should be in the region of 6 mL/g of tissue (so 600 μL if using 0.1 g of sample) (*see* **Note 2**).
6. Shake (or invert) each tube by hand and then sonicate the sample-solvent mixture for 15 min.
7. Add a chloroform–water mixture in the ratio of 1:1 to form an emulsion. This should be in the region of 2 mL of each solvent per gram of sample (again more may be needed depending on the sample type) (*see* **Note 3**).
8. Centrifuge samples for 20 min at around 13,200 rpm (704 × *g*). This will generate distinct organic and aqueous fractions and a cell pellet.

9. Using a Gilson pipette (with a clean tip for each sample) transfer the organic and aqueous layers to separate Eppendorf tubes. The cell/protein pellet may be left in the original tube for proteomic analysis if desired.
10. Aqueous extracts should be dried overnight in an evacuated centrifuge. The cell/protein pellet and organic layers may be left to dry overnight in a fume hood (since they contain chloroform).
11. The aqueous layer should be analyzed using nuclear magnetic resonance (NMR) spectroscopy, and/or Gas/Liquid Chromatography–Mass Spectrometry (GC/LC-MS) for a full coverage of the metabolome.
12. The organic layer should be analyzed by GC/LC-MS since lipids give broad resonances and are not well analyzed via NMR.

3.2 ^1H NMR of Metabolite Extracts Analysis

Rehydrate the dried aqueous layer extracts from the tissue extraction step in 500 μL of D_2O and 100 μL sodium phosphate buffer, pH 7.0. This may be a stock solution and consists of 40 mM NaH_2PO_4, 160 mM Na_2HPO_4, 0.1 % Sodium Azide (to minimize bacterial degradation), and 1.2 mM Trimethylsilyl propionate (TSP), Made up in D_2O (*see* **Notes 4** and **5**).

1. Transfer the sample to pre-labelled 5 mm NMR tubes. Care should be taken not to spill the sample and to ensure that the tubes are correctly labelled.
2. The samples should now be analyzed via an NMR spectrometer (e.g. a Bruker) 500 MHz spectrometer using a 5 mm Broadband TXI Inverse ATMA probe for the ^1H frequency.
3. A solvent suppression pulse sequence based on a 1D NOESY pulse sequence should be used to saturate the residual ^1H water signal (e.g. relaxation delay = 2 s, t_1 = 3 μs, mixing time = 150 ms).
4. The number of transients needed may vary depending on sample type. It may be best to start with 128 transients collected into 16 K data points over a spectral width of 12 ppm at around 27 °C [20] and adjust as appropriate.

3.3 Aqueous Layer

Samples should be derivatized via silyation using the procedure reported by Gullberg et al. [32]. Samples are stable for, at most, 48 h at room temperature but should ideally be analyzed as soon as possible.

1. Take 150 μL of the aqueous metabolites previously suspended in D_2O for ^1H NMR spectroscopy and evaporate to dryness in the evacuated centrifuge (Speedvac).

Table 1
Suggested GC-MS parameters for aqueous metabolites

Injection volume	1 μL (splitless)
GC column	30 m × 0.25 mm i.d. 5 % phenyl polysilphenylene-siloxane column with a chemically bonded 25 μm TR-5 MS stationary phase. Film thickness 0.25 μm; internal diameter. 0.22 mm
Carrier gas	Helium (1.2 mL/min)
GC temperature	70 °C isothermal for 1 min, 10 °C/min to 310 °C, held for 10 min
MS parameters	EI + mode, solvent delay—4 min 340 °C (transfer line); 220 °C (source)
Scans per second	3 (mass range—50–650 m/z)
Electron energy	70 eV

2. Once the samples are dry add 30 μL of methoxyamine hydrochloride (20 mg/mL in pyridine).
3. Vortex mix the samples and then leave for 17 h (making sure the lids of the sample tubes are tightly sealed) to ensure complete methoximation.
4. Silylate the samples by adding 30 μL of *N*-methyl-*N-trimethyl silyltrifluoroacetamide* (MSTFA) and leave the sample to react for 1 h (*see* **Note 6**).
5. Dilute the derivatized sample with hexane in a ratio of 1:10 prior to GC-MS analysis (so add 600 μL of hexane). Add less hexane for a more concentrated sample if desired (*see* **Note 7**).
6. Run samples on GC-MS using the settings outlined in Table 1 (*see* **Note 8**).
7. Identify compounds via retention time matching and or comparison of mass spectral data with known standards (*see* **Note 9**).

3.4 Organic Layer

Organic-phase metabolites can be derivatized by acid-catalyzed esterification [33]. It is possible to silylate these compounds using a reagent such as *N*-Methyl-*N-tert*-butyldimethylsilyl-trifluoracetamide (MTBSTFA) [29]. However, the derivatives from this reaction are usually not in the GC-MS NIST library and so are difficult to identify. For completeness two options for acid-catalyzed esterification are listed here but time wise the second version is much faster so if BF_3 is available go straight to **step 9**. Samples are stable for approximately 72 h at room temperature but should ideally be analyzed as soon as possible.

1. Redissolve extracts in 175 μL of hexane.
2. Add 225 μL of $CHCl_3$ (chloroform).

3. Add 250 μL of methanol containing 3 % v/v H_2SO_4.
4. Leave to react at room temperature for 4 h.
5. Extract the solution by adding 1 mL of water, vortex mixing and then centrifuging for 120 min at 13,000 rpm and discard the water.
6. Repeat **step 5**.
7. Dry the sample in the fume hood overnight.
8. Add 80 μL of hexane, then 10 μL of pyridine and 10 μL of MSTFA and leave for 30 min before running on the GC-MS.

Alternatively, replace **steps 1–8** with the following:

9. Dissolve the organic phase in 750 μL of chloroform/methanol (1:1 v/v).
10. Add 125 μL BF_3 (10 % in methanol) and incubate the vials at 80 °C for 90 min in the oven or the heating block (*see* **Notes 10** and **11**).
11. Leave vials to cool for 5–10 min and then add 300 μL H_2O (milliQ) and 600 μL hexane and vortex-mix each vial.
12. Discard the aqueous (lower) layer.
13. Evaporate the remaining organic layer to dryness in a fume hood overnight (*see* **Note 12**).
14. Reconstitute in 200 μL of hexane prior to GC-MS analysis (*see* **Note 7**).
15. Run samples on GC-MS using the settings outlined in Table 2.
16. Identify compounds via retention time matching and or comparison of mass spectral data with known standards (*see* **Note 13**). Some suggested retention times and diagnostic ions for commonly found lipids are given in Table 3.
17. If running a new sample type run the first sample with a high split value on the GC to ensure you do not overload the detector (*see* **Note 14**).

3.5 Analysis of Tissue Extracts Using Liquid Chromatography–Mass Spectrometry (LC-MS)

Interpretation of LC-MS based metabolomics data is often quite demanding and requires experience and training (although there is considerable potential for targeted analysis of specific biochemical pathways, such as those involving larger compounds such as hormones, which are not amenable to analysis by GC-MS). It is therefore beyond the scope of this chapter to provide protocols of all possible LC-MS set ups which may be encountered. The interested reader is referred instead to a recent excellent publication by Roberts et al. [18].

1. Reconstitute the tissue extracts in 100 μL of a suitable solvent.

Table 2
Suggested GC-MS parameters for organic phase metabolites

Injection volume	1 μL (splitless)
GC column	TR-FAME stationary phase column (Thermo Fisher Scientific—30 m ×0.25 mm ID× 0.25 μm, cyanopropyl polysilphenyl-siloxane)
Carrier gas	Helium (1.2 mL/min)
GC temperature	60 °C isothermal for 2 min, 15 °C/min to 150 °C, 4 °C per minute to 230 °C, hold for 7 min. Injector Temp 230 °C
MS parameters	240 °C (transfer line); 250 °C (source) EI+mode, 4 min solvent delay
Scans per second	3 (mass range 50–650 *m/z*)
MS electron energy	70 eV

2. If possible use a "purge-wash-purge" cycle to minimize sample carry-over.
3. Use LC-MS settings as in Table 4 (*see* **Note 8**).

3.6 Alternative Analytical Techniques

Although the methods outlined above are the principal tools currently in use for metabolomics, any technique capable of generating comprehensive metabolite measurements can potentially be applied. Other analytical methods that have been utilized include but are not limited to; Capillary Electrophoresis-Mass Spectrometry (CE-MS); Fourier Transform-Infrared Spectroscopy (FT-IR). Fourier Transform-Mass Spectrometry (FT-MS); Raman Spectroscopy (RS), and Direct-Injection Mass Spectrometry (DIMS) [1]. Some techniques, such as FT-IR, are non-destructive allowing the same sample to be analyzed using multiple instruments allowing for a more complete coverage of the metabolome.

3.7 Data Processing

3.7.1 NMR Spectra

NMR spectra should be processed using appropriate software, for instance ACD Labs 1D NMR processor (ACD, Toronto, Canada) or NMR Suite Professional, version 5.1 (Chenomx, Alberta, Canada).

1. Individual spectra should be Fourier transformed following multiplication by an exponential line broadening of around 1 Hz (you may wish to also try lower values) and referenced to the TSP singlet at 0.0 ppm (*see* **Note 15**).
2. Phasing and baseline correction may be done automatically or manually (although the latter is usually more suitable) using appropriate software. This may be the manufacturer's proprietary program or something provided by an external company (e.g., Chenomx) which will read the format of various NMR manufactures (*see* **Note 16**).

Table 3
List of commonly identified organic phase fatty acids using GC-MS

Potential retention time (min)	Name	Suggested diagnostic ions
12.42	Decanoic acid (10:0)	101, 143, 155
14.42	Undecanoic acid (11:0)	87, 157, 169
16.25	Lauric acid (12:0)	143, 155
18.03	Tridecanoic acid	143, 185, 197
19.77	Myristic acid (14:0)	157, 169
21.41	Pentadenanoic acid (15:0)	143, 171, 213, 256
21.80	Nonanedioic acid	143, 152, 185
23.00	Palmitic acid	199, 227, 270
23.39	Palmiteleic acid (9c-16:1)	152, 194, 236
24.52	Heptadecanoic acid (17:0)	143, 241, 284
24.95	Heptadecenoic acid (10c-17:1)	152, 208, 250
25.96	Stearic acid (18:0)	149, 255, 298
26.26	Oleic acid (9c-18:1)	180, 222, 264, 296
26.36	Octadecenoic acid (18:1)	180, 222, 264, 296
26.90	Linoleic acid (9c, 12c-18:2)	150, 220, 263, 294
27.32	γ-Linolenic acid (6c, 9c, 12c-18:3)	150, 194, 265
27.37	Nonadecanoic acid (19:0)	143, 269, 312
27.81	α-Linolenic acid (9c, 12c, 15c-18:3)	236, 278, 310
28.73	Eicosanoic acid (20:0)	143, 283, 326
29.99	Di-homo-γ-linolenic acid (8c, 11c, 14c-20:3)	107, 150, 204, 320
30.30	Arachidonic acid (5c, 8c, 11c, 14c-20:4)	150, 207, 263
31.24	Eicosapentanoic acid (5c, 8c, 11c, 14c, 17c-20:5)	131, 147, 215
31.40	Docosanoic acid (22:0)	143, 199, 255, 311
31.73	Docosaenoic acid (13c-22:1)	133, 207, 236, 320

3. Once processed integrate each spectra across 0.04 ppm regions (buckets) between 0.5 and 4.5, and 5.1 and 10.0 ppm (i.e., excluding the water area). Intelligent bucketing may also be used and bucket sizes may be varied if desired (*see* **Note 17**).
4. To account for any difference in concentration between samples, normalize each spectral region either to a specific peak (e.g., the TSP) or to a set value, e.g. 10,000.

Table 4
Suggested liquid chromatography and mass spectrometry parameters

Injection volume	10 μL
LC column	2.1 mm × 10 cm Symmetry C_{18} column at 40 °C
Desolvation gas temperature	250 °C, maintained at a flow rate of 500 L/h
Elution settings	Flow rate of 600 μL/min and a linear gradient of 0–20 % B (0.5–4 min), 20–95 % B (4–8 min). Solvent composition held at 95 % B for 1 min then returned to 100 % A, where A = 0.1 % formic acid (aq.) and B = 0.1 % formic acid in acetonitrile. The column eluent is then split so that approximately 120 μL a minute enters the mass spectrometer
MS parameters	Lock mass of leucine encephalin used via LockSpray interface at a concentration of 0.5 ng/μL in 50:50 acetonitrile–water + 1 % formic acid for positive-ion mode, and 1 ng/μL in 50:50 acetonitrile–water for negative-ion mode, at 30 μL/min
Number of scans per second	Two in both positive and negative ion modes, made over 10 min
Mass range	50–850 *m/z*
Cone gas flow rate	50 L/h
Cone voltage	30 V
Capillary voltage	3.2 kV (positive-ion mode) and 2.6 kV (negative-ion mode)
Solvents	90 % methanol (wash solvent) 0.1 % formic acid (purge solvent)
Lock spray frequency	5 s (lock mass data averaged over ten scans)

3.7.2 Chromatograms

1. GC-MS and LC-MS chromatograms should again be processed using suitable software, for instance QuanBrowser (Thermo Electron Corporation) or MarkerLynx Applications Manager (Waters).
2. Processing GC-MS data will vary slightly depending on the instrument; however, the first step will be to go through several chromatograms to deconvolute, identify and integrate each peak individually, using suitable software, e.g., MarkerLynx Applications Manager Version. An automated macro can then be created using retention time data to speed up the process (*see* **Note 18**).
3. Aligned peaks with the same mass/retention times across the datasets along with their normalized intensities and combine into a single data matrix for further analysis perhaps though the use of free software such as XCMS (http://metlin.scripps.edu/XCMS/).

4. Normalize to a set value such as total peak area, a specific peak area or integer (e.g. 10,000).
5. Export peak area/height/ID data to suitable statistical software such as the SIMCA-P package (Umetrics, Umea, Sweden) for processing (*see* **Note 19**).

3.7.3 Multivariate Data Analysis and Pattern Recognition

The analysis of a large number of biological samples by any technique will usually produce an equally large, and frequently, complex data set. Such "multivariate data" consists of measurements of a range of metabolites (variables) from a number of individuals or samples (observations) under a series of conditions (classes). As a large number of changes in metabolite concentration may correlate with class separation, some form of data reduction technique is usually necessary in order to simplify the data set and make the trends and patterns within it obvious. Thus, in metabolomics, chemical analysis is usually carried out in conjunction with pattern recognition algorithms to decipher the most significant changes that accompany a given modification, pathology or biological response.

Typical metabolomic experiments produce large amounts of data and handling such complex datasets is an important step that has impact on the extent and quality at which the metabolite identification and quantification can be made, and thus on the ultimate biological interpretation of results. Various chemometric data reduction techniques for managing, visualizing and interrogating large multivariate datasets exist but the foremost in metabolomic analysis are Principal Components Analysis (PCA) and the related regression method, Partial Least Squares Discriminate Analysis (PLS-DA).

For GC and LC data individual peak areas may be used as variables. For NMR, each spectrum is data reduced into separate "buckets" (by convention set at a width of 0.04 ppm) spanning the spectral width (but excluding the water peak). The spectra are then reduced into the principal components of greatest variation and then mapped into multidimensional space. PCA creates new variables, or "components", from weighted sums of original variables in such a way that relatively few of the components contain most of the variance in the original data. Graphical data display of the resulting "model" in two or three dimensions using the components as axes may then reveal similarities and differences between samples that would not be obvious from direct perusal of the original variables. An advantage of PCA over some other chemometric methods, e.g. non-linear mapping or neural network analysis, is that the weightings applied to variables ("loadings") in construction of the components can provide information as to which variables are responsible for differences between biological sample groups and hence possibly lead to the identification of specific biomarkers for the biological perturbation under study.

Both univariate and multivariate statistics are utilized in the field of metabolomics. It is an often overlooked fact that the same statistical standards apply to metabolomics experiments as to any other type of quantitative analysis. To obtain meaningful results, one must have a suitable number of biological replicates and appropriate statistical tests for the type of experiment conducted should be used. There are several ways to calculate the "appropriate" number of replicates. Roberts et al. [18] for example recommend the use of a power calculation that integrates technical reproducibility, baseline biological variability, and expected differences.

1. Ensure each variable is (mean) centered (usually automatic) and appropriate scaling is applied. For GC-MS univariate scaling may be used for NMR data *Pareto scaling* is more appropriate but these are not fixed.
2. Analyze datasets using both unsupervised methods such as Principle Component Analysis (PCA) and supervised approaches such Partial Least Squares Discriminant analysis (PLD-DA).
3. Ensure that all models are suitable validated if necessary for instance by using permutation tests or leave one out analysis [34].
4. Identification of major metabolic perturbations within the pattern recognition models can be achieved via analysis of corresponding loadings plots and Variable Importance in the Projection (VIP) scores for the variables.

Many popular software packages exist to carry out such multivariate analyses. These include SIMCA-P, (Umetrics, Umeå, Sweden), Pirouette (InfoMetrix, Woodinville, USA), and Matlab (The MathWorks Inc., Nantick, MA, USA). R (www.r-project.org) is an interesting alternative as both SIMCA and Matlab are rather expensive. The R user page has a very good user manual, references and documentation but programming is still time intensive. Alternatively, there are numerous R packages which are freely available; some even come with graphical user interface (e.g. Bio conductor).

For the interested reader the package developed in the first link includes PCA/PLS-DA, and the program at the second link may be used to carry out PCA in Excel.

1. http://www.biomedcentral.com/1471-2105/10/363.
2. http://sourceforge.net/apps/mediawiki/imdev/index.php?title=Main_Page.

Increasing interest in metabolomics has led to great interest in related data processing. A wide variety of methods and software tools have been developed for metabolomics during recent years, and this trend is likely to continue. New analytical and bioinformatics technologies and techniques are continually being created or optimized, significantly increasing the *cross*-disciplinary capabilities of metabolomics. This wealth of metabolic information can be used to mine the multitude of molecules which exist in nature with the focus of chemical synthesis, providing a platform for *discovery-driven* research to uncover lead compounds for further chemical optimization and development leading to new products. Therefore, once metabolites or groups of metabolites of interest have been identified, synthetic organic, natural product, material, and medicinal chemists are tasked with both discovering new compounds and efficient routes of synthesis.

4 Notes

1. All equipment must be thoroughly cleaned between each sample to avoid cross contamination.
2. The samples must be well covered and thoroughly mixed with the solvents in order for the extraction to work.
3. Unlike methanol, chloroform is immiscible with water, so although they can be stored in the same vial you will need to make sure to add each separately.
4. One of the most impressive uses of ^{1}H NMR spectroscopy was the detection of ~50 metabolites in lymphocytes [35]. A comprehensive review of the use of NMR spectroscopy in metabolic profiling is Fan's review of the subject [36].
5. The TSP serves as an internal chemical shift reference and the spectrometer field frequency lock is provided by the D_2O. The amount of buffer will depend on the sample, the exact amount will vary depending on sample type and may be determined via trial and error during method development. NMR data (peak shifts) are susceptible to changes in pH so it is an important factor to keep in mind.
6. Ensure the lids of the sample tubes are tightly sealed as the TMS derivatives are very susceptible to hydrolysis (e.g. by moisture in the air).
7. Low final sample volumes may require the use for vial inserts· to ensure the autosampler (if used) can reach the sample.
8. All instrument settings given here are designed as starting points. They may be refined by the individual over time as appropriate.
9. While fragmentation patterns may possibly indicate which sub-family of sugars a component may belong to, they are not definitive for specific isomers. Mass spectra without retention times are thus not reliable for identifying sugars in mixtures and care should be taken when using the library search routine.
10. Make sure to only use MilliQ water and to keep it topped-up if you are using a heating block. Tap water will leave mineral deposits as it evaporates which will, at the very least, cause problems with the heating block. Not keeping the water level

topped up will result in unequal distribution of heat and possibly difficulty retrieving the vials from the block if using standard sized equipment.

11. It is advisable to use glass vials rather than plastic Eppendorf tubes for this derivatization step (which involves high temperatures). Plastic tubes often give noisier chromatograms and extra peaks (most likely plasticizers) compared to glass vials.
12. As the density of organic layer will increase the more chloroform is used it is often worthwhile keeping both layers if one is not sure which one is which. As a rule of thumb the organic layer will dry much faster than the aqueous layer.
13. Many lipids exist as structural isomers and thus mass spectra without retention times are not reliable for identifying compounds by themselves and should be used in conjunction with suitable standards for retention time matching (Table 3).
14. Higher injection volumes and/or lower split ratios will give larger peaks in the resulting chromatograms. However, they are also more likely to generate overlapping peaks as well as increase the risk of sample carry over. As a rough guide the correct injection volume/split ratio should give peaks that are well separated from baseline and which no do not overlap (*NB*: This is not always possible).
15. A suitable alternative reference peak may be utilized if required (for example chloroform in organic phase spectra).
16. In most NMR software it is also possible to export spectra or FIDs as text based files which can then be processed with home written software using even quite simple programs such as Tcl/Tk (http://tcl.sourceforge.net/).
17. Some workers recommend buckets of 0.001 ppm. This often works well but note that increasing the number of buckets also increases the number of datapoints/variable which in turn increases the risk of a statistical error.
18. If computer automation is used be sure to check all peaks in all chromatograms are aligned correctly.
19. Any method used must be applied consistently to all chromatograms/samples.

Acknowledgments

OAHJ thanks all members of the Griffin lab at the Biochemistry Department at the University of Cambridge, UK. The help and support of Dr. Julian Griffin, Mr Steven Murfitt, Dr. Helen Atherton, Dr. Jeff Troke, Dr. Lee Roberts, Dr. Mahon Maguire and Dr. Cheng Kian Kai in particular is gratefully acknowledged.

References

1. Jones OAH, Cheung VL (2007) An introduction to metabolomics and its potential application in veterinary science. Comp Med 57:436–442
2. Nicholson JK, Lindon JC, Holmes E (1999) "Metabonomics": understanding the metabolic responses of living systems to pathophysiological stimuli via multivariate statistical analysis of biological NMR spectroscopic data. Xenobiotica 29:1181–1189
3. Oliver SG, Winson MK, Kell DB, Baganz F (1998) Systematic functional analysis of the yeast genome. Trends Biotechnol 16: 373–378
4. Whitfield PD, German AJ, Noble PJ (2004) Metabolomics: an emerging post-genomic tool for nutrition. Br J Nutr 92:549–555
5. Keun HC (2006) Metabonomic modeling of drug toxicity. Pharmacol Ther 109:92–106
6. Spurgeon DJ, Jones OAH, Dorne J-LCM, Svendsen C, Swain S, Stürzenbaum SR (2010) Systems toxicology approaches for understanding the joint effects of environmental chemical mixtures. Sci Total Environ 408: 3725–3734
7. Griffin JL, Shockcor JP (2004) Metabolic profiles of cancer cells. Nat Rev Cancer 4: 551–561
8. Griffin JL, Atherton H, Shockcor J, Atzori L (2011) Metabolomics as a tool for cardiac research. Nat Rev Cardiol 8:630–643
9. Hall RD (2006) Plant metabolomics: from holistic hope, to hype, to hot topic. New Phytol 169:453–468
10. Gates M, Tschudi G (1952) The synthesis of morphine. J Am Chem Soc 74:1109–1110
11. Keller NP, Turner G, Bennett JW (2005) Fungal secondary metabolism: from biochemistry to genomics. Nat Rev Microbiol 3:937–947
12. Yuliana ND, Khatib A, Choi YH, Verpoorte R (2011) Metabolomics for bioactivity assessment of natural products. Phytother Res 25: 157–169
13. Robinette SL, Brüschweiler R, Schroeder FC, Edison AS (2012) NMR in metabolomics and natural products research: two sides of the same coin. Acc Chem Res 45:288–297
14. Vandenbrouck T, Jones OAH, Dom N, Griffin JL, De Coen W (2010) Mixtures of similarly acting compounds in daphnia magna: from gene to metabolite and beyond. Environ Int 36:254–268
15. Atherton H, Bailey N, Zhang W, Taylor J, Major H, Shockcor J, Clarke K, Griffin J (2006) A combined 1H-NMR spectroscopy- and mass spectrometry-based metabolomic study of the PPAR-alpha null mutant mouse defines profound systemic changes in metabolism linked to the metabolic syndrome. Physiol Genomics 27:178–186
16. Lin C-Y, Wu H, Tjeerdema RS, Viant MR (2007) Evaluation of metabolite extraction strategies from tissue samples using NMR metabolomics. Metabolomics 3:55–67
17. Le Belle J, Harris N, Williams S, Bhakoo K (2002) A comparison of cell and tissue extraction techniques using high-resolution 1H-NMR spectroscopy. NMR Biomed 15:37–44
18. Roberts LD, Souza AL, Gerszten RE, Clish CB (2012) Targeted metabolomics. Curr Protoc Mol Biol 98:30.32.31–30.32.24
19. Griffin JL (2003) Metabonomics: NMR spectroscopy and pattern recognition analysis of body fluids and tissues for characterisation of xenobiotic toxicity and disease diagnosis. Curr Opin Chem Biol 7:648–654
20. Griffin JL, Scott J, Nicholson JK (2007) The influence of pharmacogenetics on fatty liver disease in the wistar and kyoto rats: a combined transcriptomic and metabonomic study. J Proteome Res 6:54–61
21. Horning EC, Horning MG (1971) Metabolic profiles: gas-phase methods for analysis of metabolites. Clin Chem 17:802–809
22. Bölling C, Fiehn O (2005) Metabolite profiling of *Chlamydomonas reinhardtii* under nutrient deprivation. Plant Physiol 139: 1995–2005
23. Blau K, Halket JM (1993) Handbook of derivatives for chromatography, 2nd edn. John Wiley & Sons, Chichester, UK
24. Deng C, Zhang X, Li N (2004) Investigation of volatile biomarkers in lung cancer blood using solid-phase microextraction and capillary gas chromatography–mass spectrometry. J Chromatogr B 808:269–277
25. Thurman EM, Mills MS (1998) Solid phase extraction: principles and practice, 1st edn. Wiley Europe, Hoboken, USA
26. Jones OAH, Roberts LD, Maguire ML (2011) Mass spectrometry-based metabolomics Sample preparation, data analysis, and related analytical approaches. In: Ivanov A, Lazarev A (eds) Sample preparation in biological mass spectrometry. Springer, New York
27. Zhang K, Zuo Y (2005) Pitfalls and solution for simultaneous determination of estrone and 17[alpha]-ethinylestradiol by gas chromatography–mass spectrometry after derivatization

with N,O-bis(trimethylsilyl)trifluoroacetamide. Anal Chim Acta 554:190–196
28. Rodriguez I, Quintana JB, Carpinteiro J, Carro AM, Lorenzo RA, Cela R (2003) Determination of acidic drugs in sewage water by gas chromatography–mass spectrometry as tert.-butyldimethylsilyl derivatives. J Chromatogr A 985:265–274
29. Woo K-L, Kim J-I (1999) New hydrolysis method for extremely small amount of lipids and capillary gas chromatographic analysis as *N*(*O*)-tert.-butyldimethylsilyl fatty acid derivatives compared with methyl ester derivatives. J Chromatogr A 862:199–208
30. Ternes TA, Hirsch R, Mueller J, Haberer K (1998) Methods for the determination of neutral drugs as well as betablockers and beta(2)-sympathomimetics in aqueous matrices using GC/MS and LC/MS/MS. Fresenius J Anal Chem 362:329–340
31. McCombie G, Knochenmuss R (2004) Small-molecule MALDI using the matrix suppression effect to reduce or eliminate matrix background interferences. Anal Chem 76:4990–4997
32. Gullberg J, Jonsson P, Nordstrom A, Sjostrom M, Moritz T (2004) Design of experiments: an efficient strategy to identify factors influencing extraction and derivatization of *Arabidopsis thaliana* samples in metabolomic studies with gas chromatography/mass spectrometry. Anal Biochem 331:283–295
33. Morrison WR, Smith LM (1964) Preparation of fatty acid methyl esters and dimethylacetals from lipids with boron fluoride-methanol. J Lipid Res 5:600–608
34. Westerhuis J, Hoefsloot H, Smit S, Vis D, Smilde A, van Velzen E, van Duijnhoven J, van Dorsten F (2008) Assessment of PLSDA cross validation. Metabolomics 4:81–89
35. Sze DY, Jardetzky O (1990) Determination of metabolite and nucleotide concentrations in proliferating lymphocytes by ^{1}H-NMR of acid extracts. Biochim Biophys Acta 1054:181–197
36. Fan TW-M (1996) Metabolite profiling by one- and two-dimensional NMR analysis of complex mixtures. Prog Nucl Magn Reson Spec 28:161–219

Chapter 19

Strategies in Biomarker Discovery. Peak Annotation by MS and Targeted LC-MS Micro-Fractionation for *De Novo* Structure Identification by Micro-NMR

Philippe J. Eugster, Gaëtan Glauser, and Jean-Luc Wolfender

Abstract

In metabolomic studies the identification of biomarkers is a key step but represents a serious bottleneck since the *de novo* identification of natural products is a lengthy process. A strategy for the dereplication and peak annotation of plant biomarkers is presented based on high resolution mass spectra acquired on a quadrupole-time-of-flight mass spectrometry coupled to ultra-high pressure liquid chromatography and chemotaxonomy information. A rational approach for the targeted LC-MS micro-isolation of biomarkers followed by *de novo* identification by NMR at the microgram scale is described, based on gradient transfer from the analytical scale and chromatographic modelling. The methodology is illustrated by the identification of various stress biomarkers of the plant wound response using *Arabidopsis thaliana* as a model.

Key words UHPLC-Q-TOFMS, Biomarker identification, Peak annotation, Dereplication, *Arabidopsis*, Wounding

1 Introduction

Metabolomics plays an increasingly important role in natural product research [1]. Within systems biology, this holistic approach actually provides the most "functional" information of the "omics" technologies [2]. Metabolomics represents a new way of interrogating biological systems since it is an unbiased data-driven approach that may ultimately lead to hypotheses and new biological knowledge.

In typical metabolomic studies, complex crude natural extracts are compared by using various analytical methods that generate metabolic fingerprints [3]. For this, the main approaches are either based on nuclear magnetic resonance (NMR) [4] or mass spectrometry (MS) [5]. MS is often used in hyphenation with chromatographic techniques such as gas chromatography (GC-MS) or high performance liquid chromatography (LC-MS).

Ute Roessner and Daniel Anthony Dias (eds.), *Metabolomics Tools for Natural Product Discovery: Methods and Protocols*, Methods in Molecular Biology, vol. 1055, DOI 10.1007/978-1-62703-577-4_19, © Springer Science+Business Media, LLC 2013

Since primary and secondary metabolites have a wide chemical diversity, no single analytical method that provides a complete survey of all metabolites in a given organism exists at present [6], and combinations of methods are more and more used for metabolomics. Once data are recorded and preprocessed, the comparison of a sufficient number of fingerprints is performed with either unsupervised or supervised multivariate data analysis (MVDA) methods that may reveal features in the dataset that can be linked to biomarkers [7]. These features correspond to peak intensities of NMR chemical shifts, *m/z* ions and/or chromatography retention times according to the approach used.

One of the main difficulties in metabolomics resides in the correct identification of the biomarkers highlighted in the peak lists generated by MVDA. This usually represents a bottleneck in such an approach since, unlike for proteomics, the structure determination of low molecular weight compounds does not follow generic rules, and no freely accessible comprehensive MS/MS database provides unambiguous dereplication for natural products. Despite many efforts, the identification of biomarkers still relies on the interpretation of the MS or NMR data obtained. This is especially true for secondary metabolites that are often species specific [4].

In this chapter, a practical approach for biomarker peak annotation by LC-MS is presented. The strategy proposed relies on a high resolution LC-MS profiling allowing the unambiguous determination of molecular species, the calculation of the corresponding molecular formulae and filtering for validation. Dereplication is then based on database cross searching taking into account chemotaxonomic considerations. Additional information from UV photodiode array (UV-PDA) and/or MS/MS spectra is also integrated to support online identification. For the *de novo* identification of unknown biomarkers, a rational and rapid LC-MS micro-isolation approach is described that provides microgram amounts compatible with further at-line micro-NMR structure determination and/or bioactivity assessment. Due to the complexity of natural extracts, the purification of metabolites present in low concentrations is especially critical [8]. The micro-isolation strategy relies on the optimization of the chromatographic analysis using ultra-high pressure liquid chromatography (UHPLC) thanks to modelling software [9] and further transfer to semi-preparative LC conditions with MS detection [10].

The approach described is based on UHPLC coupled to quadrupole-time-of-flight mass spectrometry (Q-TOFMS), but is generic and may be adapted to other types of high resolution LC-MS instruments.

2 Materials

2.1 Solvents and Reagents

For extractions, sample preparations and dilutions, methanol (MeOH), acetonitrile (ACN), and isopropanol (IPA) of analytical or HPLC grade and water of milli-Q quality (18.2 MΩ cm at 25 °C) are recommended. Solvents and additives of LC-MS quality should be used for UHPLC-Q-TOFMS analyses (*see* **Note 1**). Solvents for semi-preparative LC-MS micro-isolation should be of HPLC grade (*see* **Note 2**).

1. Solvents for sample preparation prior to UHPLC-Q-TOFMS analysis: IPA, MeOH:H_2O 85:15 (v/v) and MeOH 100 %.
2. Final dissolution solvent for sample injection: MeOH:H_2O (70:30).
3. UHPLC and semi-preparative LC-MS mobile phases: A = water + 0.1 % formic acid (FA), B = ACN + 0.1 % FA.
4. Solvents for sample preparation prior to biomarker isolation: IPA, MeOH, MeOH:H_2O (5:95), MeOH:H_2O (70:30).
5. NMR solvent: methanol-d_4 (CD_3OD) (*see* **Note 3**).

2.2 Equipment

2.2.1 Sample Preparation and Extraction for UHPLC-Q-TOFMS Analysis

1. Ball mill with 2 cm diameter stainless steel balls (e.g., Retsch MM200, from Schieritz & Hauenstein AG, Arlesheim, Switzerland).
2. Centrifuge.
3. Centrifugal or nitrogen evaporator.
4. Ultrasonic bath.
5. SPE cartridge 100 mg (Sep-Pak C18, Vac 1 cc, 100 mg, Waters, Milford, USA) with vacuum manifold.
6. Weighed 2 mL microcentrifuge tubes.
7. Glass vials (5 mL).
8. Pipettes and pipette tips.
9. HPLC vials and caps.

2.2.2 UHPLC-Q-TOFMS Analysis

1. UHPLC system able to withstand a maximal pressure of 1,000 bar and equipped with a binary or quaternary pump and a column oven maintaining constant temperature (e.g., Waters Acquity UPLC).
2. Q-TOFMS instrument hyphenated with the UHPLC system through an electrospray (ESI) interface (e.g., Synapt G2 from Waters) (*see* **Note 4**).
3. Acquity UPLC BEH C18 (150 × 2.1 mm I.D., 1.7 μm particle size) column (Waters) with an Acquity UPLC BEH C18 Van guard (5 × 2.1 mm I.D., 1.7 μm particle size) pre-column (Waters).

2.2.3 Data Processing

1. MassLynx 4.1 for LC-MS raw data processing (Waters) or any software adapted to the MS analyzer used.
2. Seven Golden Rules Excel-based software, freely available [11] (http://fiehnlab.ucdavis.edu/projects/Seven_Golden_Rules/Software/).
3. Databases for dereplication based on molecular formulae (The Dictionary of Natural Products (DNP) (Chapman & Hall/CRC Press)) (*see* **Note 5**).
4. Osiris 4.2 for HPLC modelling (Datalys, Saint-Martin-d'Hères, France) (*see* **Note 6**).

2.2.4 Sample Preparation and Extraction for LC-MS Isolation

1. Important amount of fresh plant tissue for separation scale-up (e.g., 500 g in the case of *Arabidopsis* leaves) (*see* **Note 7**).
2. Big mortar and pestle.
3. Erlenmeyer (2 L).
4. Agitation plate or magnetic agitator.
5. Large size filtration paper and filter funnel.
6. Rotative evaporator.
7. Ultrasonic bath.
8. LiChroprep RP-18, 40–63 μm (50 g) (Merck, Darmstadt, Germany).
9. Glass column for open chromatography with frit (porosity 4) and vacuum assembly.

2.2.5 Optional Pre-fractionation and Targeted LC-MS Isolation

1. For pre-fractionation: XBridge BEH C18 (150 × 19.0 mm I.D., 5.0 μm particle size) column, or another column with the same phase chemistry than the analytical column previously used (*see* **Note 8**).
2. Fraction collector that can hold 120 × 10 mL tubes.
3. For final purification: Two XBridge BEH C18 (250 × 10.0 mm I.D., 5.0 μm particle size) columns, or other columns with the same phase chemistry than the analytical column previously used (*see* **Notes 8** and **9**).
4. Semi-preparative HPLC system, such as the Varian 9012 (Varian, Palo Alto, CA, USA), or another semi-preparative LC system able to deliver a 10 mL/min flow.
5. Column oven adapted to semi-preparative columns (40 °C) (*see* **Note 10**).
6. 96-well deep plates (1 mL per well).
7. Splitter able to divide the flow and maintain a 50 μL/min flow rate in the MS instrument (*see* **Note 11**).
8. For LC-MS isolation monitoring: MS detector, such as TSQ 7000 (Thermo Fisher Scientific, Inc., Waltham, MA).

3 Methods

The following protocol has been devised for biomarker identification either based on UHPLC-Q-TOFMS data only (peak annotation) or complete *de novo* structure identification based on LC-MS targeted micro-isolation and subsequent micro-NMR analysis. A prerequisite for this protocol is the localization of a biomarker of interest in the metabolite profiling chromatogram of a representative crude extract. Localization is performed by extracting the ion trace corresponding to a specific feature [$m/z \times$ retention time (RT)] found in the loadings after the MVDA of a classical LC-MS-based metabolomics study (*see*, for example, ref. 12 and **Note 12**). We will refer to it as "biomarker of interest" in the following steps. The process is illustrated by the identification of biomarkers in the aerial parts of the model plant *Arabidopsis thaliana* (Biomarkers **A** and **B** for the dereplication process and Biomarker **C** for the isolation procedure) but may be applied to other plants and metabolite types provided that slight adaptations are made.

3.1 Sample Preparation for Dereplication Based on UHPLC-Q-TOFMS

3.1.1 Extraction Procedure

1. Harvest and weigh approximately 500 mg of the fresh plant tissue of interest (*see* **Note 13**).
2. Put the fresh plant tissue, 5 mL of IPA and a 2 cm diameter ball in the jar of the ball mill (*see* **Note 14**).
3. Extract for 2 min at 30 Hz using the ball mill.
4. Collect the content of the jar in centrifuge tubes, without the balls, centrifuge for 4 min.
5. Collect the supernatant; dry it under nitrogen at 40 °C or using the centrifugal evaporator to obtain the crude extract.

3.1.2 Sample Preparation

1. Dissolve the crude extracts (2–10 mg) in 0.7 mL MeOH:H_2O (85:15), with ultrasonic bath if needed.
2. Place a C18 SPE cartridge on the manifold chamber. Prepare a glass vial under the cartridge to collect the conditioning and equilibrating solvents.
3. Condition and equilibrate the cartridge by washing with 1 mL MeOH 100 % and then 1 mL MeOH:H_2O (85:15), under vacuum. Adjust the pressure to obtain an elution rate of about one drop per second. Discard the eluate.
4. Place a weighed 2 mL microcentrifuge tube in the 5 mL glass vial, under the cartridge.
5. Load the dissolved extract on the cartridge (*see* **Note 15**) and elute it. Wash with 0.8 mL MeOH:H_2O (85:15).
6. Evaporate the collected eluate to dryness using a centrifugal evaporator or under a gentle nitrogen flow.

7. Weigh the 2 mL microcentrifuge tube to assess the yield of the SPE extraction.
8. Dissolve the extract in MeOH:H_2O (70:30) (*see* **Note 16**).

3.2 UHPLC-Q-TOFMS Analysis

3.2.1 UHPLC Gradient Conditions

For metabolite localization and dereplication, a generic high resolution UHPLC-Q-TOFMS (3 %/min slope) 30 min gradient is used (*see* second column of Table 1 and **Note 17**). Other parameters are as follows:

1. Flow rate: 460 μL/min.
2. Column and autosampler temperatures: 40 °C and 10 °C, respectively.
3. Injection volume: 2.0 μL.
4. PDA: 10 Hz at least, over 210–600 nm.

3.2.2 Q-TOFMS Conditions

Perform calibration using for example a sodium formate solution in the 100–1,000 *m/z* range, in both positive (PI) and negative ionization (NI) modes (*see* **Note 18**). Check the mass accuracy by subsequent injection of any selected molecule.

Run two separated analyses using both PI and NI ESI modes (*see* **Note 19**), with alternating scans at low and high collision energies (e.g., MS^E) (*see* **Note 20**). For the Synapt G2 Q-TOFMS (Waters), generic source parameters are as follows:

1. Voltages: capillary 2,500 V, cone 25 V, extraction cone −4.5 V and 3.0 V in NI and PI modes, respectively.
2. Temperatures: source 120 °C, desolvation gas 350 °C.
3. Gas flows: desolvation gas 800 L/h, cone gas flow 20 L/h.
4. Mass range: 85–1,200 Da.
5. Scan time: 0.2 s.
6. Collision energy: 4 eV and ramp of 10–30 eV, both applied on the transfer region of the collision cell.
7. Collision gas: argon, at a flow rate of 2.1 mL/min (pressure inside the collision cell 7.0×10^{-3} mbar).
8. Internal calibration (Lockspray™): infusing a 500 ng/mL solution of leucin-enkephalin at a flow rate of 7.5 μL/min.
9. Lockspray™ scan time and frequency: 0.5 s and 15 s, respectively, with data averaged over five scans for mass correction.

A typical UHPLC-NI-ESI-TOFMS metabolite profiling chromatogram (BPI trace) of the leaf extract of *A. thaliana* is presented in Fig. 1. All biomarkers that are identified by the procedures described below (Biomarkers **A**–**C**) have been labelled by their corresponding *m/z* ions.

3.3 Data Processing

Figure 2a–h illustrates some steps of the dereplication of the Biomarker **A** (*m/z* 545.2596), which was detected in NI mode in the metabolite profile of *A. thaliana*, following leaf wounding (Fig. 1).

Table 1
Gradients used for UHPLC-TOFMS analyses and pre-fractionation on semi-preparative scale

%B	Time (min)			Time (min)
	90 min UHPLC gradient	30 min UHPLC gradient		Pre-fractionation
5.0	0.0	0.0	transfer	0.0
5.0	0.0	0.0		1.0
95.0	90.0	30.0		113.9
95.0	100.0	40.0		152.0
5.0	100.5	40.5		155.0
5.0	110.0	50.0		190.0
	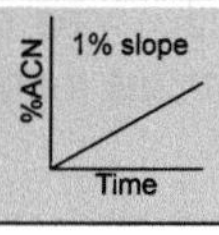	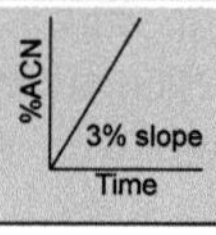		

3.3.1 *Determination of the Molecular Weight*

In a given ESI spectrum, molecular species ions may be present either as protonated or deprotonated molecules ($[M+H]^+$ or $[M-H]^-$) or may form dimers or higher oligomers, or adducts with solvents, molecules present in the solvents, or LC-MS additives. Moreover, the most intense peak does not always correspond to $[M+H]^+$ or $[M-H]^-$. In addition, ions that are not related to the biomarker of interest may be present. The procedure below presents ways to find the MW of an unknown peak based on both PI and NI MS spectra in typical situations.

1. Highlight the peak of interest from the chromatogram by extracting its trace with an adapted mass range window (e.g., 0.02 Da) (*see* Fig. 2a–d and **Note 21**).
2. Combine the spectra containing the mass of interest (*see* **Note 22**).
3. Determine the presence of adducts and/or dimers, and ensure the molecular weight (MW) based on the following rules:
   (a) Look for adducts, detected by the presence of both $[M+H]^+$ and $[M+\text{adduct}]^+$ or $[M-H]^-$ and $[M+\text{adduct}]^-$, depending on the ionization mode (*see* Fig. 2e, f and **Notes 22** and **23**). Table 2 provides a list of the most frequently encountered adducts using ACN + FA mobiles phases and an ESI source.
   (b) Check for the presence of dimers, characterized by $[2M+H]^+$ or $[2M-H]^-$ depending on the ionization mode, or higher oligomers.

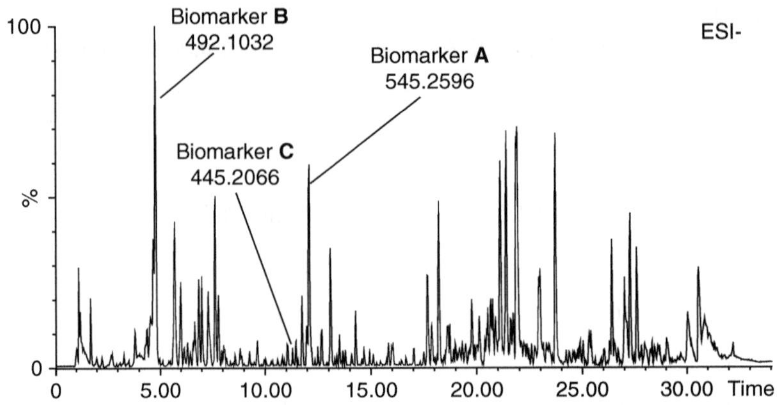

Fig. 1 High resolution UHPLC-TOFMS metabolite profiling (BPI trace) of *Arabidopsis thaliana* (crude leaf isopropanol extract). Separation was carried out on an Acquity BEH C18 (150 × 2.1 mm; 1.7 μm particle size) column with a 5–95 % ACN gradient in 30 min at 40 °C, and detection was performed by TOFMS in the NI mode. Biomarkers **A–C**, used to illustrate the metabolite identification process, are labelled on this chromatogram

(c) Compare results of PI and NI modes, if both are available: the resulting molecular weight (MW) should be the same.

(d) If only one single ion species is present, check the corresponding MS^E spectrum to verify the possible loss of adducts.

3.3.2 Extraction of Molecular Formulae

1. Use the elemental composition tool provided in MassLynx (or another MS software) to obtain putative molecular formulae from the molecular ion species recorded. Work on combined spectra only (*see* **Note 22**).

   (a) Set the minimum number of elements to 0 and the maximum to 200 for C, H, O, N, and S (*see* **Note 24**).

   (b) Set the mass tolerance as three times the usual mass accuracy of the instrument (e.g., 3 × 1–2 = 5 ppm for the Synapt G2). Table 3 illustrates the number of molecular formulae corresponding to the $[M–H]^-$ and $[M+H]^+$ ions (Fig. 2e, f) (*m/z* 499.2549 and *m/z* 501.2708) of Biomarker **A** used as example in Subheading 3.3.1, as well as for Biomarker **B**, containing CHONS. As shown, the number of calculated molecular formulae for a given exact mass around 500 Da may exceed 100 if a large tolerance window (15 ppm) is used and if CHONS elements are considered (*see* **Note 25**).

2. Export the calculated molecular formulae and correct them according to the ionization mode used (add or remove an atom of adducts in NI or PI ESI, respectively).

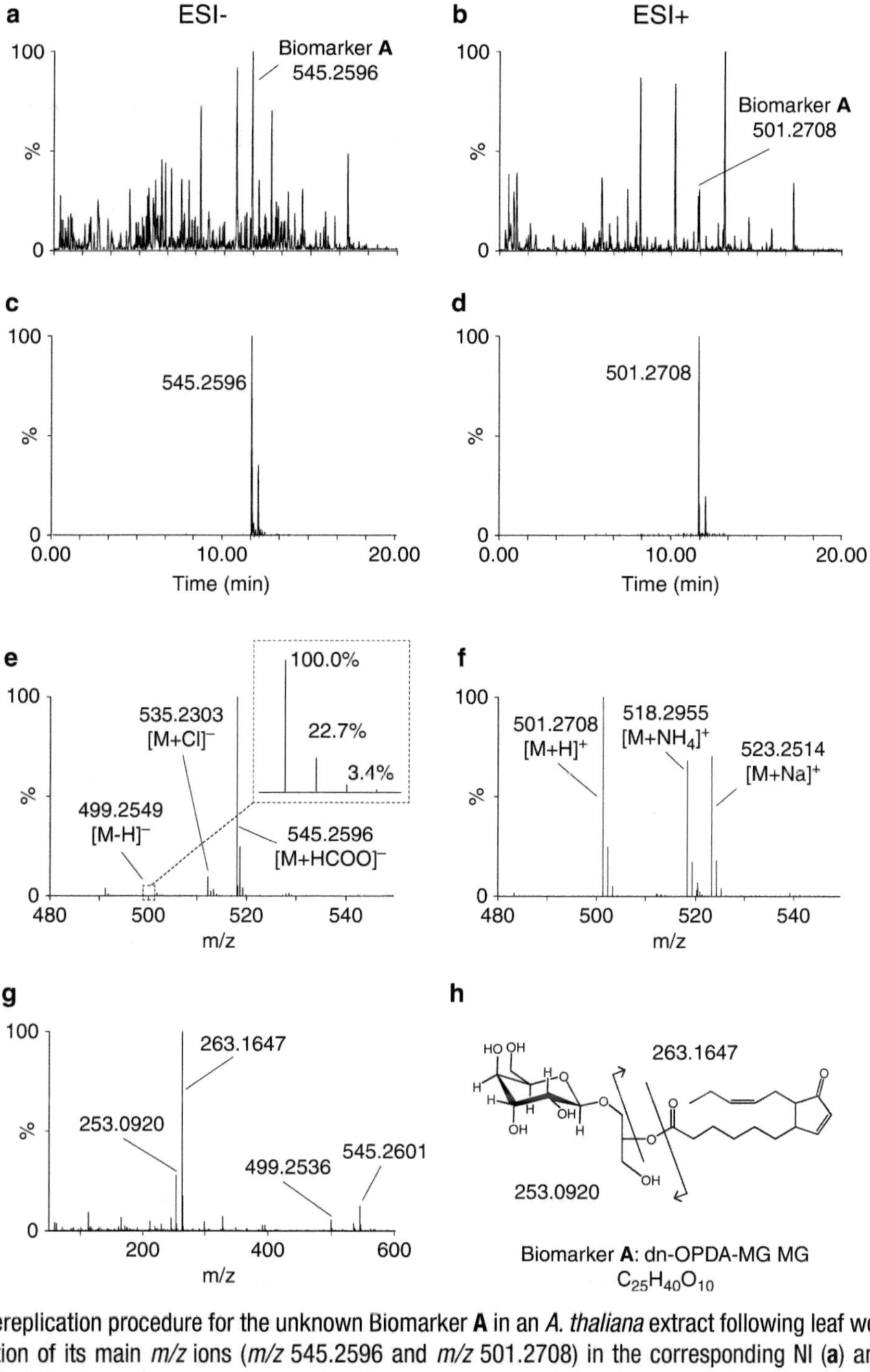

Fig. 2 Dereplication procedure for the unknown Biomarker **A** in an *A. thaliana* extract following leaf wounding. Localization of its main *m/z* ions (*m/z* 545.2596 and *m/z* 501.2708) in the corresponding NI (**a**) and PI (**b**) UHPLC-TOFMS chromatograms and corresponding extracted ion chromatograms (*m/z* ±0.01 Da) (**c** and **d**). The isomer also detected is not discussed in more details. According to the spectra in NI mode (**e**), *m/z* 545.2596 represents the formate adduct $[M+HCOO]^-$, while *m/z* 535.2303 corresponds to $[M+Cl]^-$, and *m/z* 499.2549 to $[M–H]^-$. Comparison with the PI mode (**f**) confirms this, and $[M+H]^+$, $[M+NH_4]^+$, and $[M+Na]^+$ were at *m/z* 501.2708, *m/z* 518.2955, and *m/z* 523.2514, respectively. This combined information provides the unambiguous MW determination of Biomarker **A** (average monoisotopic MW: 500.2620). The inset in (**e**) represents an expansion of the isotopic pattern recorded for the $[M–H]^-$, the peak height ratios were used during heuristic filtering for molecular formulae determination. The hit obtained for Biomarker **A** (**h**) by the database search is confirmed by the fragments recorded in the high energy CID NI-TOFMS^E spectrum (**g**)

Table 2
List of commonly found adducts in ESI PI and NI modes, using acetonitrile as organic modifier and formic acid as additive

Type of ions	Ion mass
Positive ionization mode	
$[M+H]^+$	M + 1.007276
$[M+NH_4]^+$	M + 18.033823
$[M+Na]^+$	M + 22.989218
$[M+K]^+$	M + 38.963158
$[M+ACN+H]^+$	M + 42.033823
$[M+ACN+Na]^+$	M + 64.015765
$[2\,M+H]^+$	2 M + 1.007276
Negative ionization mode	
$[M–H]^-$	M − 1.007276
$[M+Cl]^-$	M + 34.969402
$[M+HCOO–H]^-$	M + 44.998201
$[M+HCOO+Na–2H]^-$	M + 67.987419
$[2\,M–H]^-$	2 M − 1.007276
$[2\,M+HCOO–H]^-$	2 M + 44.998201

All ions are singly charged
Note: comprehensive list of adducts may be found in literature [11, 13, 14]

3.3.3 Heuristic Filtering

As shown above, high mass accuracy measurements alone do not provide unambiguous molecular formula assignment for a given biomarker. Thus additional filtering is needed to reduce the number of possible hits and validate the molecular formula assignment. In this respect the application of heuristic filtering with the Seven Golden Rules [11] represents a rational approach.

1. Import all molecular formulae into the Seven Golden Rules software Excel sheet [11] (*see* **Note 26**).
2. On the spectrum, measure the isotopic pattern, *i.e.*, the height of the $^{13}C_1$, $^{13}C_2$, and $^{13}C_3$ peaks of the biomarker, expressed as a percentage of its main ^{12}C peak. Report it in the dedicated field of the Excel sheet.
3. Set the isotopic pattern error as the ratio of the background to the intensity of the main peak of the marker, as a percentage but at least 3 %.
4. Click the "1) Autofill," "2) Calc" and "3) Check" buttons (*see* **Note 27**). Copy the remaining molecular formulae that are highlighted in blue in the "Pubchem"—"Found" column.

Table 3
Number of potential molecular formulae corresponding to the dereplicated ion and using different mass windows. Biomarker *B* was detected in NI mode only. Spectral accuracy is discussed later

Biomarker A (dn-OPDA-MG MG, MW 500, $C_{25}H_{40}O_{10}$)				
	ESI + (*m/z* 501.2700)		ESI – (*m/z* 499.2543)	
Mass window (ppm)	With CHO	With CHONS	With CHO	With CHONS
15	2	74	2	79
10	1	49	1	53
5	1	26	1	27
3	1	15	1	16
1	1	6	1	4
5 + heuristic filtering	1	2[a]	1	2[a]

Biomarker B (glucohirsutin, MW 493, $C_{16}H_{31}NO_{10}S_3$)			
	ESI – (*m/z* 492.1032)		
Mass window (ppm)	With CHO	With CHON	With CHONS
15	0	36	140
10	0	27	95
5	0	14	47
3	0	7	27
1	0	3	10
5 + heuristic filtering	0	0	1[b]

[a]The two remaining molecular formulae were $C_{25}H_{40}O_{10}$ and $C_{22}H_{32}O_4N_{10}$
[b]The remaining molecular formula was $C_{16}H_{31}NO_{10}S_3$

In the examples discussed in Table 3, application of heuristic filtering reduces the number of possible formulae for Biomarker **A**(MW 500) from 26 to 2 and 27 to 2 in PI and NI modes, respectively ($C_{25}H_{40}O_{10}$ and $C_{22}H_{32}N_{10}O_4$) and from 47 to 1 for Biomarker **B** (MW 493, $C_{16}H_{31}NO_{10}S_3$) (*see* **Note 28**).

3.3.4 Database Search

1. Perform a search of the remaining molecular formulae in the Dictionary of Natural Products:

   This step may be used to discard molecular formulae not corresponding to previously reported natural products. For Biomarker **A**, this discards $C_{22}H_{32}N_{10}O_4$ and the only remaining formula is $C_{25}H_{40}O_{10}$ (*see* **Note 29**). For Biomarker **B**, the unique proposed formula was found to match with existing hits in the Dictionary of Natural Products.

Based on such validated molecular formulae assignment, peak annotation can be made by cross search with chemotaxonomic information of the plant studied.

2. Perform a cross search with validated formulae and chemotaxonomic keywords (e.g., species, genus, family) in the Dictionary of Natural Products or other databases.

For Biomarker **A**, a cross search based on $C_{25}H_{40}O_{10}$ and "*Arabidopsis*" and "*thaliana*" provided one single structure only, the galactolipid dn-OPDA-MG MG (*see* Fig. 2h and **Note 30**). For Biomarker **B**, the same cross search based on $C_{16}H_{31}NO_{10}S_3$ provided one structure only, the glucosinolate glucohirsutin.

3.3.5 Additional Information from MS/MS

The use of collision induced dissociation (CID) spectra may provide additional structural information on the biomarker of interest. With the MS^E acquisition described above, this information can be retrieved as follows:

1. Verify which fragments generated in the high energy MS^E acquisition have identical retention times to the $[M+H]^+$ or $[M-H]^-$ ions and discard those which are not perfectly aligned (*see* **Note 31**). Compare CID MS spectra from the high energy MS^E with those compiled in existing databases such as Massbank or ReSpect for Phytochemicals (*see* **Note 32**).
2. Alternatively, determine whether the fragments obtained may be described by means of rules of possible fragmentation mechanisms.

The CID spectrum of Biomarker **A** displays two main diagnostic fragments (*see* Fig. 2g, h) that confirm the peak annotation made based on molecular formula and chemotaxonomy cross search.

3.3.6 Additional Information from UV-PDA

The use of UV-PDA spectra as an additional filter for dereplication is advantageous provided that the biomarker is present in sufficient amount. Many UV-active natural products such as polyphenols exhibit characteristic chromophores [15] that can be exploited to strengthen the peak annotation.

1. Compare both UV-PDA spectra and wavelength(s) of maximum absorption (λ_{max}) of the biomarker of interest with values reported for the hits previously obtained (Subheading 3.3.4). Discard candidates whose values do not match with the experimental spectrum or λ_{max} (*see* **Note 33**).

No UV information could be obtained for Biomarker **A**, while Biomarker **B** displayed a UV-PDA spectrum comparable to the one of glucohirsutin.

3.3.7 Confirmatory Analysis

At this stage, the number of putative structures for a given biomarker is usually relatively limited. Confirmation of the identity of the analyte may be thus obtained by injection of the standard, if it is commercially available or if synthesis if applicable. The following protocol provides a way to ascertain peak identification based on standard.

1. Prepare a 5 μg/mL solution of the standard in an adapted solvent (preferably in the same solvent as previously used for the analysis of the extract, i.e., MeOH:H_2O (70:30)).
2. Inject 2 μL of this solution in the previously used UHPLC-Q-TOFMS conditions (Subheadings 3.2.1 and 3.2.2). If necessary, adjust its concentration (*see* **Note 34**).
3. Verify that the retention time of the biomarker of interest and of the standard does not vary by more than 1 %, and that adducts and dimers are similar in both cases; provided that the intensity of the main peak is similar (*see* **Note 35**).
4. Further confirm or infirm the identity by spiking the extract with the standard. Only one peak should be visible in the chromatogram when extracting the specific trace.

3.4 Targeted LC-MS Isolation

The targeted LC-MS isolation procedure presented here is used for NMR *de novo* identification of biomarkers that cannot be annotated by the dereplication approach described above. The procedure is only based on HPLC methods and is adapted for isolation of microgram amounts of the biomarker of interest, a scale compatible with full characterization by state of the art micro-NMR methods [16, 17]. This procedure is adapted from ref. [8] and is illustrated by the isolation of Biomarker **C**, a jasmonate synthesized in *A. thaliana* in response to wounding (*see* Fig. 3).

3.4.1 Extraction and Sample Preparation for Biomarker Micro-Isolation

The sample preparation for the micro-isolation of the biomarkers of interest is based on an upscale of the procedure used for the UHPLC-Q-TOFMS metabolite profiling described under Subheading 3.1.1.

1. Harvest (*see* **Note 13**), weigh and grind using a mortar and pestle approximately 500 g of fresh plant tissue (*see* **Note 7**).
2. Extract with 1 L of IPA during 3 h at room temperature, under agitation (*see* **Note 36**).
3. Filter on paper, keep the eluate.
4. Extract the residue once again in the same conditions, filter.
5. Combine and evaporate both eluates using the rotative evaporator to obtain the crude extract.

3.4.2 Extraction and Sample Preparation for Biomarker Micro-Isolation

1. Dissolve the crude extracts (about 5 g) in 40 mL MeOH, with ultrasonic bath if needed.

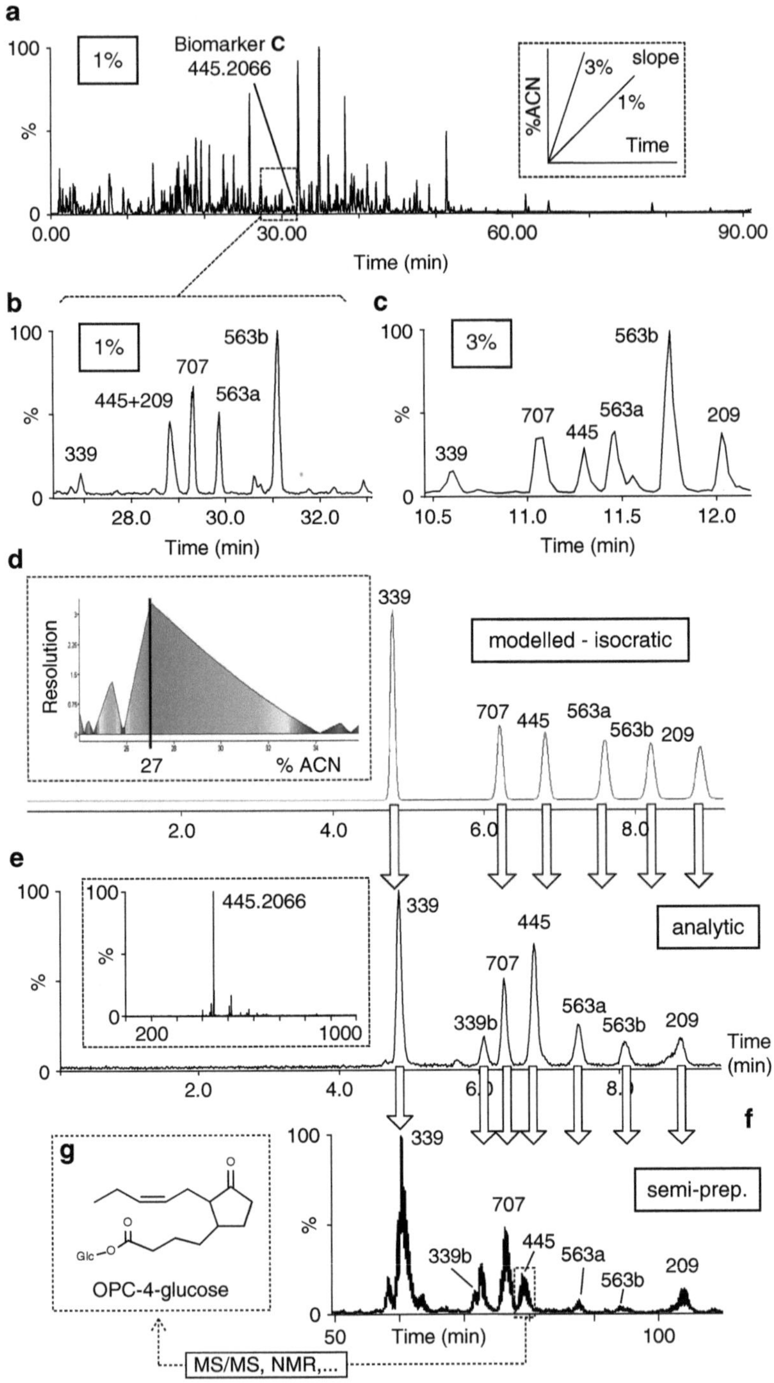

Fig. 3 LC-MS isolation procedure of the Biomarker **C** (*m/z* 445.2066 in NI mode). Two UHPLC-NI-TOFMS gradient runs of 1 %/min slope (**a** and **b**) and 3 %/min slope (**c**) were performed using an Acquity UPLC BEH C18 (150 × 2.1 mm; 1.7 μm particle size) column with a 5–95 % ACN gradient at 40 °C. (**d**) Simulated chromatogram calculated based on the modelled isocratic condition 27 % ACN (inset in **d**) by the Osiris software. (**e**) Isocratic UHPLC-TOF-MS analysis of the enriched **Fig. 3** (con-

2. Add 10 g of LiChroprep RP-18, mix, evaporate to dryness using the rotary evaporator.
3. On the bottom of the open column, place 40 g of LiChroprep RP-18, condition and equilibrate with 250 mL of MeOH and MeOH:H_2O (5:95) respectively, and place above the mixture of extract and LiChroprep RP-18 previously prepared (*see* **Note 37**).
4. Add 250 mL MeOH:H_2O (5:95) and elute by applying an adapted vacuum to obtain an elution rate of about 10 mL/min (*see* **Note 38**).
5. Elute the compounds with 250 mL MeOH:H_2O (70:30) at the same flow rate (*see* **Note 39**).
6. Evaporate to dryness to obtain the enriched extract, weigh.
7. Dissolve the enriched extract in 500 μL MeOH:H_2O (70:30).

3.4.3 Optional Pre-fractionation

A pre-fractionation step in gradient mode is recommended to ensure the optimal purification of minor compounds in the final isolation step. For this purpose, the original UHPLC gradient (3 %/min slope gradient) is transferred to the semi-preparative scale with the following conditions (*see* **Note 40**).

1. Stationary phase: XBridge BEH C18 (150×19.0 mm I.D., 5.0 μm particle size).
2. Mobile phase: water and ACN, both with 0.1 % FA, using the gradient described in the third column of Table 1 (*see* **Note 41**).
3. Flow rate: 10 mL/min.
4. Column temperature: 40 °C.
5. Injection volume: 150 μL (*see* **Note 42**).
6. Divert 0.1–1 % of the flow to the MS detector using a flow splitter (*see* **Note 43**).
7. Collect the fractions every minute (10 mL per fraction) in tubes, for 120 min.
8. Determine which fraction(s) contains the biomarker of interest based on extracted ion chromatogram obtained by MS detection (*see* **Note 44**).
9. Verify the presence of the biomarker of interest in the collected enriched fractions using the UHPLC-TOFMS method described in Subheading 3.2.
10. Combine and evaporate fractions containing the biomarker of interest and dissolve it in MeOH:H_2O (70:30).

---

tinued) fraction containing *m/z* 445. Geometrical transfer to the semi-preparative scale on two XBridge BEH C18 (250×10 mm; 5 μm particle size) columns (**f**) ensured the same selectivity (*see* **d–f**) and provided pure microfractions containing Biomarker **C** finally identified as OPC-4-glucose by further NMR experiments (**g**)

3.4.4 Gradient Modelling and Optimization

To achieve the semi-preparative purification of biomarkers, optimal separation conditions were predicted and tested at the analytical scale using HPLC modelling software (*see* **Note 6**). Figure 3 illustrates the method.

1. Analyze the enriched fraction by UHPLC-TOFMS using the 3 %/min slope gradient (30 min) used previously (*see* Subheading 3.2.1).
2. Run a second gradient in the same conditions but with 1 %/min slope (*see* Fig. 3a, b), according to the gradient described in the first column of Table 1 (*see* **Note 45**). MS conditions are identical to those used in Subheading 3.2.2 (*see* **Note 46**).
3. Localize and note the retention time (RT) of the biomarker of interest, and of all other compounds that elute in the same retention time window (±1 min for the 30 min gradient and ±3 min for the 90 min gradient) in both chromatograms (*see* Fig. 3b, c).
4. In Osiris software, provide the following information:
   (a) The LC conditions as described in Subheading 3.2.1.
   (b) The dwell volume (approximately 120 μL for the Acquity UPLC system).
   (c) The column's dead volume (approximately 375 μL for the Acquity UPLC 150 × 2.1 mm, 1.7 μm column) (*see* **Note 47**).
   (d) A random value for the area of the peak (for example, 1).
   (e) RT of the biomarker of interest and the neighbor peaks in both conditions (*see* Fig. 3b, c).
5. Determine the optimized separation in isocratic mode and using the same column and mobile phase as described in Subheading 3.4.1 (*see* Fig. 3d). If an error occurs, modify the maximal retention factor value (k) up to 99.
6. Inject the pre-fractionated sample in the modelled conditions in UHPLC-TOFMS, and verify the separation of the biomarker of interest (*see* Fig. 3e). If the separation is not satisfactory, model a new separation with slightly different conditions using the Osiris software.

3.4.5 LC-MS Purification

The same setup as used in Subheading 3.4.2 is employed except that a longer column and 96-well plates (2 mL/well) are used micro-collection.

1. Connect two XBridge BEH C18 (250 × 10.0 mm I.D., 5.0 μm particle size) columns together with very short (30 mm) PEEK tubing (*see* **Note 48**).

2. Inject 150 μL of the pre-purified sample using the following conditions:
   (a) Flow rate 3.5 mL/min.
   (b) Isocratic separation as modelled by the Osiris software.
   (c) Oven temperature: 40 °C.
3. Collect fractions every 30 s in 96-well plates.
4. Identify the fractions containing the biomarker of interest, based on extracted ion chromatograms from the MS monitoring of the semi-preparative separation. Figure 3f illustrates the separation obtained after injection of the enriched fraction containing the biomarker of interest.
5. Check the presence of the biomarker of interest in the fractions predicted by the model, using the UHPLC-Q-TOFMS method described in Subheading 3.2.
6. Combine the fractions containing the pure biomarker of interest, dry them using the centrifugal evaporator or under gentle nitrogen flow.

3.5 Micro-Flow NMR Analysis

The micro-isolation procedure described above typically yields a few tenths or hundreds of μg of biomarkers. With such amounts, weighing is often not possible, but the samples are compatible with further micro-NMR analysis. For micro-NMR measurements, samples may be dissolved in 5 μL of deuterated methanol or another appropriate solvent. Such solution can be measured on a microflow probe (Protasis, Marlboro, MA, USA) [18] or in 1 mm capillary tubes. Typical hydroxyjasmonates spectra yielded by micro-flow CapNMR using such a procedure are illustrated in [8]. In the illustrated example, the structure of Biomarker **C** was elucidated as 3-oxo-2-(2Z-pentenyl)cyclopentane-1-butyric acid-1-O-β-glucose (OPC-4-Glucose, Fig. 3g), a new wound biomarker, based on the complementary MS/MS and ^{1}H NMR data obtained [19].

4 Notes

1. Ultra-high purity solvents are required to ensure low background noise in the LC-MS chromatograms and to maintain good instrument performances.
2. For micro-isolation it is important to verify that HPLC solvent purity is good enough to prevent signal interferences in the micro-NMR spectra of semi-preparative LC-MS blank samples.
3. CD_3OD is adapted for the dissolution of most biomarkers eluting in reversed phase C18 separations. Other alternative solvents such as $CDCl_3$ or DMSO may be used to solve solubility issues.

4. A Q-TOFMS system is necessary for selective MS/MS fragmentation experiments. Metabolite profiling may also be recorded on a TOFMS (e.g., LCT Premier from Waters).
5. Other databases containing natural product molecular formulae and information related to their origin (*family*, *genus*, and *species*) can be used (e.g., SciFinder, https://scifinder.cas.org).
6. Other commercially available HPLC modelling software can be used.
7. It is difficult to estimate the amount of plant that is required to finally isolate a few tenth of μg of biomarkers, since MS detection is largely compound dependent. As a rule of thumb, approximately 1,000 times more plant material is needed for the micro-NMR detection of given peaks of the LC-MS metabolite profiles.
8. The same phase chemistry in both analytical (UHPLC) and semi-preparative scales is mandatory to maintain the same selectivity [10, 20].
9. The use of two columns coupled in series provides a good compromise at the semi-preparative scale between high chromatographic resolution and reasonable elution times [21]. If high resolution is not mandatory, shorter columns may be used.
10. To ensure predictable chromatographic transfer, temperature has to be controlled at both the analytical and semi-preparative levels.
11. A splitter with adjustable split ratio or a T-piece with tubing of adapted length may be used after careful measurement of the split ratio. For 10 mL/min flow rate (pre-fractionation step) a 1:200 ratio is advisable, while for 3.5 mL/min (high resolution isolation step) a 1:70 ratio can be used.
12. This metabolite dereplication process is illustrated for specific biomarkers but can be applied to any LC peak in a metabolite profiling chromatogram.
13. If the metabolites that need to be identified are sensitive to enzymatic reactions, the fresh plant tissues have to be frozen in liquid nitrogen immediately after harvesting.
14. Isopropanol was selected since it is miscible with water contained in the fresh leaves and since it extracts both relative apolar and polar metabolites. Other solvents such as MeOH or MeOH–water mixtures may be considered according to the physicochemical properties of the biomarkers of interest.
15. A SPE procedure is advisable for the removal of chlorophyll and other interfering apolar compounds for reversed phase LC-MS metabolite profiling. Such compounds are strongly retained in standard reversed phase conditions and may alter the chromatographic performances and reduce column's lifetimes after multiple injections.

16. Such injection solvent usually dissolves the extract. Its elution strength will not affect UHPLC separation since only a small volume is typically injected (2 μL). In case of solubility issues, the injection solvents should be adapted.
17. This gradient is generic for many separations of complex mixtures, since it represents a good compromise between maximal peak capacity and reasonable gradient time [21].
18. This calibration has to be performed once a week for instruments of new generation such as the Synapt G2 from Waters, or daily for older platforms such as the LCT Premier from Waters. Refer to manufacturer's recommendations.
19. On the Synapt G2 Q-TOFMS, PI and NI modes cannot be acquired in alternating scans during the same analysis. On MS instruments that can alternatively switch to PI and NI modes, MS accuracy, and acquisition frequency are affected and, in our opinion, high quality data can only be obtained by performing separated PI and NI analyses with the current technology.
20. This acquisition mode provides MS fragmentation of all metabolites in an untargeted manner when high collision induced dissociation (CID) energies are used. This information is useful for the dereplication process.
21. Extraction of ion traces at ±0.01 Da (0.02 Da window) reduces sufficiently the noise without loss of data using modern (Q-) TOFMS analyzers. For MS instruments providing a 5 ppm accuracy of higher, this window can be set to 0.05 Da.
22. Combining spectra increases mass accuracy and provides better ion statistics. For high mass accuracy measurement, depending on the instrument used, working with very intense ions may cause saturation and *m/z* shifts. In such a case, combine less intense spectra on the edges of the chromatographic peak only.
23. If usually $[M+H]^+$ or $[M-H]^-$ ions are selected for the determination of elemental composition, any other ion species may be used, but the molecular formula should be corrected according to the type of adducts generated.
24. Most natural products contain only those five elements. According to the adducts observed, one Na^+ or K^+ can be added to the list. Maximum number of CHONS can be overestimated (200) since molecular formulae will be filtered later. If based on chemotaxonomic information the search may be restricted to CHO only, the number of hits is considerably reduced. A link between MW and maximal number of given element for natural products is provided in ref. [11].
25. 15 ppm is selected on purpose here as an extreme case since a 5 ppm tolerance windows may be practically applied in routine for well-calibrated instruments of the last generation.

26. This procedure can alternatively be performed online on freely accessible website: http://maltese.dbs.aber.ac.uk:8888/hrmet/search/gr.html.
27. Chemically nonviable formulae are discarded based on various filters, such as hydrogen/carbon ratio or Lewis rule [11].
28. As shown, molecular formulae assignment may still be ambiguous even after heuristic filtering and require search in natural product databases.
29. This step is valid only if the biomarker of interest was previously reported in databases. One should not completely exclude the presence of an unknown biomarker. In this case, a *de novo* identification procedure is necessary (*see* Subheading 3.4).
30. The structure assignments made are only putative. However, if data previously reported in the database correspond to the same plant species the chances for a correct assignment are important. In the frame of a previous chemotaxomic study, verification of the peak annotation by further isolation of standard revealed a very good prediction of such approach [22]. In the case of Biomarker **A**, another isomer is also present (Fig. 2c, d) with a similar MS/MS spectrum. The dereplication made on MS data alone does not provide information on the stereochemistry of such compounds that are likely diastereoisomers.
31. On the Synapt G2, linkage between precursor and product ions may be automatically made by use of the "MS^E data viewer" software. Selective additional MS/MS experiments may be performed if precursor and fragment ions cannot easily be linked.
32. It should be noted that these databases contain relatively few spectra and the risk of false-positive matches is therefore high. Moreover, CID mass spectra acquired on different mass spectrometers may differ [23]. Unfortunately, no generic MS/MS library with free access exists for natural products and this represents a serious bottleneck for biomarker identification. *In-house* libraries may, however, provide very good match but are limited by the number of standards at hand [24].
33. Absorption maxima may shift according to the solvent used. The UV-PDA spectra obtained in gradient mode may thus be slightly different from those reported in pure solvent in databases.
34. The intensity in ESI strongly depends on the nature of the analyte. The mentioned 5 μg/mL concentration is an average value providing appropriate intensity for the majority of the compounds. On the Synapt G2, appropriate intensity is comprised between 10^3 and 10^6 cps.

35. The spectrum pattern varies with the concentration of the biomarker of interest. For example, a high concentration may result in higher probability of dimerization.
36. For large-scale extraction, a maceration step instead of ball mill extraction is advisable and was shown to provide similar extract composition.
37. This corresponds to a solid introduction of the extract on a large C18 adapted SPE column (3–4 cm diameter).
38. This step aims at eliminating highly polar compounds in order to concentrate the sample.
39. This procedure is similar to that made at the analytical level with SPE (Subheading 3.1.2) and is aimed at the removal chlorophyll and other interfering apolar compounds.
40. Different software packages are available to calculate gradient transfers. Here, a freely available Excel-based program was used [10].
41. An initial hold was introduced to take into account the differences in dwell volume between the systems.
42. Injection volume is adapted in proportion to the volume of the column.
43. A make-up flow may be added to dilute the mobile phase directed to the MS and avoid saturation effects. Delay between MS detection and fraction collection have to be minimized first.
44. MS detection is important at the semi-preparative scale for the monitoring of the separation, to ensure a specific collection of the fraction containing the biomarker of interest based on its characteristic *m/z* ion. It is necessary to verify that the flow reaches the collection tubes and the MS detector at the same time, to ensure proper fraction collection.
45. Osiris software requires at least two chromatograms with elution retention factors (k_e) of 3 and 10 approximately. This corresponds to gradient slopes of 1 %/min and 3 %/min in the present case.
46. This comparison of the 1 %/min and 3 %/min gradient can be made on a specific enriched fraction, but may also be performed on the crude extract directly, to calculate optimized separation conditions for several biomarkers at the same time.
47. To obtain this value, a porosity of 0.7 was assumed.
48. The selected column has the same phase chemistry as the analytical UHPLC-Q-TOFMS used at the analytical level. This is important to ensure similar selectivity as shown by comparing Fig. 3e, f. A very long column (500 mm) was selected to ensure a high chromatographic efficiency (*see* **Note 9**).

Acknowledgments

This work was supported by Swiss National Science Foundation grants no. 205320_135190 and CRSII3_127187. G.G. and J.L.W. are also grateful to the National Center of Competence in Research (NCCR) Plant Survival and to the Swiss Plant Science Web (SPSW).

References

1. Guy C, Kopka J, Moritz T (2008) Plant metabolomics coming of age. Physiol Plant 132:113–116
2. Fiehn O, Kopka J, Dörmann P et al (2000) Metabolite profiling for plant functional genomics. Nat Biotechnol 18:1157–1161
3. Zhou B, Xiao JF, Tuli L et al (2012) LC-MS-based metabolomics. Mol Biosystems 8:470–481
4. Leiss KA, Choi YH, Verpoorte R et al (2011) An overview of NMR-based metabolomics to identify secondary plant compounds involved in host plant resistance. Phytochem Rev 10:205–216
5. Wolfender JL, Glauser G, Boccard J et al (2009) MS-based plant metabolomic approaches for biomarker discovery. Nat Prod Commun 4:1417–1430
6. Dunn WB (2008) Current trends and future requirements for the mass spectrometric investigation of microbial, mammalian and plant metabolomes. Phys Biol 5:011001
7. Boccard J, Veuthey JL, Rudaz S (2010) Knowledge discovery in metabolomics: an overview of MS data handling. J Sep Sci 33:290–304
8. Glauser G, Guillarme D, Grata E et al (2008) Optimized liquid chromatography-mass spectrometry approach for the isolation of minor stress biomarkers in plant extracts and their identification by capillary nuclear magnetic resonance. J Chromatogr A 1180:90–98
9. Heinisch S, Lesellier E, Podevin C et al (1997) Computerized optimization of RP-HPLC separation with nonaqueous or partially aqueous mobile phases. Chromatographia 44:529–537
10. Guillarme D, Nguyen DTT, Rudaz S et al (2008) Method transfer for fast liquid chromatography in pharmaceutical analysis: application to short columns packed with small particle. Part II: gradient experiments. Eur J Pharm Biopharm 68:430–440
11. Kind T, Fiehn O (2007) Seven Golden Rules for heuristic filtering of molecular formulas obtained by accurate mass spectrometry. BMC Bioinformatics 8:105–124
12. Grata E, Boccard J, Guillarme D et al (2008) UPLC-TOF-MS for plant metabolomics: a sequential approach for wound marker analysis in *Arabidopsis thaliana*. J Chromatogr B 871:261–270
13. Huang N, Siegel MM, Kruppa GH et al (1999) Automation of a Fourier transform ion cyclotron resonance mass spectrometer for acquisition, analysis, and E-mailing of high-resolution exact-mass electrospray ionization mass spectral data. J Am Soc Mass Spectrom 10:1166–1173
14. Nielsen KF, Mansson M, Rank C et al (2011) Dereplication of microbial natural products by LC-DAD-TOFMS. J Nat Prod 74:2338–2348
15. Markham KR (1982) Techniques of flavonoid identification. Academic press, London
16. Molinski TF (2010) NMR of natural products at the 'nanomole-scale'. Nat Prod Rep 27:321–329
17. Wolfender JL, Queiroz EF, Hostettmann K (2005) Phytochemistry in the microgram domain—a LC-NMR perspective. Magn Reson Chem 43:697–709
18. Olson DL, Norcross JA, O'Neil-Johnson M et al (2004) Microflow NMR: concepts and capabilities. Anal Chem 76:2966–2974
19. Glauser G, Boccard J, Rudaz S et al (2010) Mass spectrometry-based metabolomics oriented by correlation analysis for wound-induced molecule discovery: identification of a novel jasmonate glucoside. Phytochem Anal 21:95–101
20. Guillarme D, Nguyen DTT, Rudaz S et al (2007) Method transfer for fast liquid chromatography in pharmaceutical analysis: application to short columns packed with small particle. Part I: isocratic separation. Eur J Pharm Biopharm 66:475–482
21. Guillarme D, Grata E, Glauser G et al (2009) Some solutions to obtain very efficient separations in isocratic and gradient modes using UHPLC technology. J Chromatogr A1216:3232–3243

22. Funari C, Eugster P, Martel S et al. (2012) High resolution Ultra High Pressure Liquid Chromatography Time-of-Flight Mass Spectrometry dereplication strategy for the metabolite profiling of Brazilian *Lippia* species. J Chromatogr A 1259:167–178
23. Waridel P, Wolfender JL, Ndjoko K et al (2001) Evaluation of quadrupole time-of-flight tandem mass spectrometry and ion-trap multiple-stage mass spectrometry for the differentiation of C-glycosidic flavonoid isomers. J Chromatogr A 926:29–41
24. van der Hooft JJJ, Vervoort J, Bino RJ et al (2011) Polyphenol identification based on systematic and robust high-resolution accurate mass spectrometry fragmentation. Anal Chem 83:409–416

Chapter 20

Statistical Analysis of Metabolomics Data

Alysha M.De Livera, Moshe Olshansky, and Terence P. Speed

Abstract

Statistical matters form an integral part of a metabolomics experiment. In this chapter we describe several important aspects in the analysis of metabolomics data such as the removal of unwanted variation and the identification of differentially abundant metabolites, along with a number of other essential statistical considerations.

1 Introduction

Statistical considerations inevitably play a vital role in a metabolomics experiment from the beginning through to the end. The aims in the statistical analysis of a typical metabolomics data matrix can be the identification and quantification of metabolites, the discovery of differentially abundant metabolites under different conditions (*groups* or *factors of interest*), classification where the factors of interest are predefined and the basis for classification needs to be identified, cluster analysis where the factors of interest are unknown a priori and need to be discovered, or correlation analysis where the correlations between metabolites or certain variables within or across certain factors of interest need to be determined.

A metabolomics experiment consists of several key stages and starts with abiological question of interest. A suitable metabolomics approach, suchas the *target analysis*, *metabolite profiling*, *metabolomics*, or *metabolic finger/footprinting* [1, 2], needs to be chosen. The nature of the metabolomics data, hence, the statistical considerations depend on this chosen metabolomics approach. For example, in *target analysis* a small and well-defined set of known metabolites (*targets*) which belong to certain classes (e.g., sugars, aminoacids, nucleotides, flavonoids, alkaloids, or fatty acids) are measured usingone particular analytical platform [3]. In *metabolomics*, on the

Ute Roessner and Daniel Anthony Dias (eds.), *Metabolomics Tools for Natural Product Discovery: Methods and Protocols*, Methods in Molecular Biology, vol. 1055, DOI 10.1007/978-1-62703-577-4_20, © Springer Science+Business Media, LLC 2013

otherhand, a large number of metabolites either identified or unidentified are measured using several analytical platforms [2] such as the nuclear magneticresonance (NMR) spectroscopy, gas chromatography–mass spectrometry(GC–MS), and liquid chromatography–mass spectrometry (LC–MS). Inaddition to detecting as many metabolites as possible, the use of severalanalytical platforms allows the option of integrating the measurements on thesame metabolites, taking into account the between and within platformvariation [4].

In the design of the experiment, the researcher has to consider a balance between practical and statistical matters [5, 6]. For example, *biological replication* where samples are taken from different subjects within factors of interest is required in order to extrapolate the conclusions from the experiment to a wider population of subjects, and *technical replication* where samples are taken from the same subject and split onto different subsamples is required to provide estimates of technical variability. However, it may not be possible to obtain the desired number of replicates due to restrictions involved in collecting the samples or the cost involved in running the samples. Careful attention should also be paid to *randomization* and other experimental factors which can be confounded with the biological factors of interest. For instance, the collection, preparation, or extraction of biological samples belonging to different factors of interest on different days may introduce a component of day variation which can interfere with the interesting biological variation. In addition, the use of *controls* [5], which is often overlooked in the metabolomics literature, should play an important role. The controls provide the researcher with a gold standard relative to which their statistical methods can be assessed. For example, the performance of a statistical method in identifying differentially abundant metabolites can be assessed by the use of *positive control* metabolites which are known to be changing and the use of *negative control* metabolites which are known not to be changing between factors of interest.

Data preprocessing then starts from a set of raw data files acquired from the instruments, each corresponding to a single biological sample. Mass spectrometry data is acquired as a time series of mass spectral scans, where each scan consists of a series of *m/z* intensity pairs [7], and NMR data contains a set of sine/cosine waves measured as a function of time decaying toward zero intensity at an exponential rate [8]. The goal of the preprocessing step is to obtain a data matrix from these raw data files that is suitable for downstream statistical analysis [7, 9] which consists of metabolite abundances of the samples, often under various factors of interest. Although most of the preprocessing software contain a simple normalization step, we recommend omitting this step and handling normalization as a part of the statistical analysis as described in Subheading 5.

2 Missing Values

Depending on the chosen metabolomics approach, a typical metabolomics data matrix contains a substantial amount of missing values (around 10–40%) [10], which can affect up to 80% of all variables [11]. These missing values can arise due to both biological and/or technical reasons. For example, the missing metabolite abundances may be lower than the threshold used in the detection algorithm, or may not be present in the samples due to genuine biological reasons, or may have occurred due to analytical/technical errors. We describe below some of the commonly used methods for handling missing values in metabolomics data matrices. A combination of these approaches is often used, depending on the nature of the data, possible reasons for the missing values, and the purpose of the statistical analysis. Prior to using these approaches, it is important to reduce the number of missing values as much as possible by using an effective preprocessing procedure. For example, a secondary peak picking method can be used for LC–MS data to fill in missing peaks which are not detected and aligned [12].

1. Excluding the variables containing missing values from the analysis

   Discarding the variables with missing values clearly leads to loss of valuable information [13]. However, if values are missing in most of the samples of a variable in all factors of interest, it may be sensible to exclude it from the analysis. For example, [14] employs a "80% rule" in which a variable is kept if it has non-missing values for at least 80% of all samples. Care must be taken in applying this rule, as the missing values appearing only in some of the biological factors of interest may indicate differentially abundant metabolites. After this treatment, there will still be a substantial amount of missing values in the data [14], which needs to be handled using one or more approaches described below.

2. Replace by a small predetermined value

   Here, a small predetermined value is used to replace the missing abundances. For instance, GeneSpring MS software-Agilent Technologies uses a value of 0.01 [11], and many others [10, 14, 15] simply use one or half of the minimum value in the entire data matrix. This method generally assumes that the missing abundances are lower than the threshold used in the detection algorithm, and a small value that is greater than zero is used to mimic this effect and to allow for any transformations to be carried out. A better alternative to this strategy is to replace these missing values by a random number between zero and the minimum value of the entire matrix. If the values are missing in all samples within a specific factor of interest for example, this will allow for essential within group variability that is required for some statistical approaches (e.g., *t* tests) to be carried out.

3. Leave as missing

   Some statistical approaches such as *t*-tests, fold changes, and more generally linear modeling can accommodate missing values [16]. Leaving the missing values as they are without replacing them should only be done for those values which are not assumed to be below the detection limit.

4. Replace by the mean or the median across different samples

   Here, the missing values are simply replaced by the mean or median of the metabolite abundances across different samples, often within a factor of interest [13, 17], assuming that the metabolite's abundances are similar across the samples.

5. Estimating the missing data by various algorithms

   Some of these algorithms include *weighted k-nearest neighbor algorithm* (KNN) [18], *Bayesian principal component analysis* (BPCA) [19], and *multivariate imputation by chained equations* (MI) [20]. These methods are often used prior to statistical approaches which require a complete data matrix (e.g., principal component analysis). A discussion on the performance of these methods for metabolomics data is given by [11].

3 Transformations

Abundances of metabolites in a data matrix are usually heteroscedastic and have a right skewed distribution. Therefore, an appropriate transformation is needed to obtain a more symmetric distribution and achieve a realistic variation component that is independent of the magnitude.

The metabolomics literature have discussed various transformations such as log, cubic, and square root as ways of handling these [15, 21], all of which belong to the family of power transformations [22]. It is often found that a log transformation is adequate for statistical purposes [16]. Although further gain in homoscedasticity may be obtained by using other power transformations, this comes at the cost of losing the simplicity of using a log transformation. For example, addition or subtraction on log scale corresponds to multiplication and division on original scale, making normalization and interpretation simple and straightforward.

The log transformed data does not need to be further scaled for the methods that we present in Subheading 6 for the identification of differentially abundant metabolites. However, for certain classification and clustering methods mentioned in Subheading 7, further scaling may be needed.

4 Visualizing the Data

The log transformed data matrix can then be visualized using various plots, in order to observe possible variation, clustering tendencies, trends, and outliers. One way of doing this is the use of *across group* or *within group* relative log abundance (RLA) plots [4].

Across group RLA plots can be obtained by standardizing the metabolites by removing the median from each metabolite across all factors of interest. The boxplots of these scaled metabolites then provide a way of comparing the behavior of metabolites between groups, hence can give an indication of variation in the data across groups. For example, the top panel of Fig. 1 shows the across group RLA plots for the GC–MS data set described by [23], which consists of 48 leaf samples obtained from 9 different plant species with 5–7 biological replicates for each. Refer to [23] for more details. Comparisons of the boxplots show that the metabolite abundances in the *G. pruinosa* and *H. kanaliense* plant species tend to be higher while those of *P. douarrei* tend to be lower compared to those in the rest of the plant species. These differences may be indicative of the variation in the weight of the leaf samples.

For within group RLA plots, each metabolite is scaled by removing the median within each factor of interest. Boxplots of these can be used to visualize the tightness of the replicates within groups, and should have a median close to zero and low variation around the median. The bottom panel of Fig. 1 shows the within group RLA plots for the leaf data set. The replicates in the *H. austrocaledonicus* and *G. hirsuta* plant species seem to be slightly more variable than those in the other plant species.

In addition to the above figures, the use of figures related to multivariate statistical analyses (e.g., principal component analysis, hierarchical cluster analysis, multidimensional scaling) mentioned in Subheading 7 can reveal any outlying samples, metabolites, and clustering tendencies in the data matrix.

5 Normalization

The purpose of normalization is to identify and remove unwanted variation seen in the raw metabolomics data, in order to focus on the biological variation of interest. It is an important part of metabolomics data analysis, as every metabolomics experiment is exposed to various sources of unwanted variation and the results obtained in subsequent data analysis can vary depending on the normalization method used to remove the unwanted variation [16].

5.1 Sources of Unwanted Variation

The sources of unwanted variation seen in metabolomics data can occur due to both experimental and biological reasons. Unwanted experimental variation can arise from human error

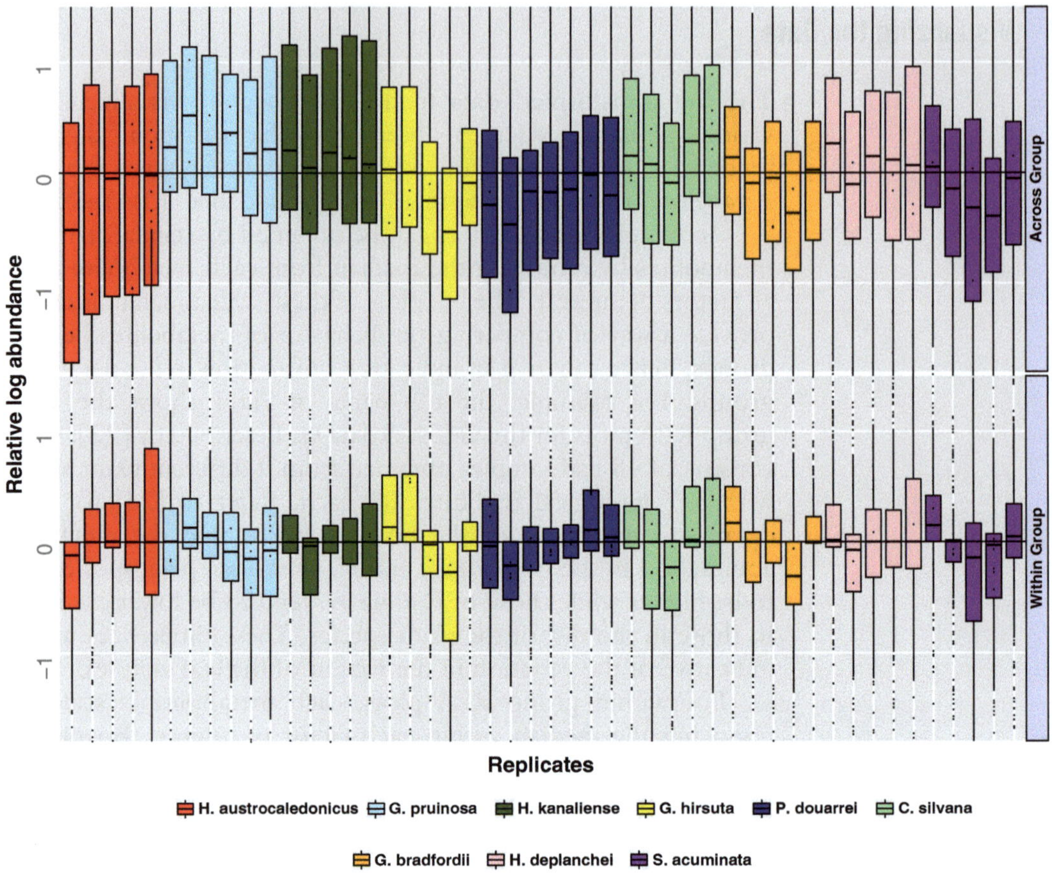

Fig. 1 Figures showing across group RLA and within group RLA plots of the normalized log transformed GC–MS data described by [23]

(e.g., sample extraction and preparation), within instrument variation (e.g., temperature changes within the instrument, sample degradation, or loss of performance of the instrument during long run of samples), between instrument variation, different batches, different laboratories, and different analytical platforms. Unwanted biological variation (e.g., the number of cells, concentration of biofluid) is also commonly encountered in metabolomics data, and is sometimes confounded with the factors of interest, making it difficult to extract only the unwanted variation component. Both the biological and experimental unwanted variation can be unobserved as well as observed. For example, due to long run of a large number of samples, it may not be possible to allocate a batch number for the samples, or the batch that each sample belongs to may not be known. Similarly, the unwanted biological variation may be observed (for example as weight or volume of the samples or the number of cells), and also may be unobserved (for instance in the form of cell sizes or concentration of biofluid).

5.2 Normalization Methods

Normalization methods available for metabolomics data can be divided into several main categories: (1) the use of internal or external standards, (2) the use of scaling factors, (3) the use of quality control (QC) samples which are composed of identical amounts of a representative pool of all the samples, and (4) the use of *non-changing* metabolites which may also include internal or external standards. In what follows, we describe normalization approaches available for each category.

1. The use of internal or external standards

   Internal or external standards are compounds which are added to each sample before and after extraction, respectively. The methods which use internal or external standards are limited in their ability to handle only some of the unwanted variation. For example, the external standards are not able to capture the unwanted variation occurred in the process of sample preparation. In addition, the unwanted biological variation, which is sometimes confounded with the factors of interest, is also ignored.

   Single Internal Standard Method

   The simplest of the normalization methods is the use of a Single Internal Standard (SIS) [24], and the normalized log abundances are obtained by subtracting the log abundance of the single internal standard from the log metabolite abundances in each sample. The SIS method implicitly assumes that every metabolite in a sample is subject to the same amount of unwanted variation, simply measured by a single internal standard, and is often found to be inadequate in removing unwanted variation. The variation in the abundance of an internal standard may not arise from the unwanted variation alone, but can be dependent on its chemical properties [25]. Other sources of variation can arise from inadequate chromatographical separation and ion suppression [26, 27]. As a result, it is sometimes seen in practice that multiple internal standards do not correlate strongly with each other. Hence, depending on which compound is used as the internal standard, the results obtained from subsequent statistical analysis for the identification of differentially abundant metabolites can vary. Moreover, even if the single internal standard perfectly describes the unwanted variation associated with the experiment, subtracting the log abundances of the single internal standard from the log abundances of each metabolite will lead to normalized values with higher variability, although the bias of the estimates may be reduced.

   The use of multiple standards helps to overcome some of the issues explained above, and in particular leads to lower variability of the normalized abundances [27]. The importance of using more than one standard for measuring unwanted instrumental variation has been noted by several authors in the recent literature [4, 25–27], introducing several normalization methods using multiple standards as described below.

Retention Index Method

One approach is to choose different internal standards according to the proximity of retention times to certain metabolite classes [25] and then use the SIS method to normalize within classes. Retention time, however, does not necessarily describe all chemical properties leading to unwanted variation [27].

Normalization Using Optimal Selection of Multiple Internal Standards

Another normalization approach was presented by Sysi-Aho et al. [27], in which they select the optimal combination of multiple internal standards using multiple linear regression. In this, they fit a linear model to log transformed abundances of each metabolite with a design matrix consisting of columns corresponding to mean centered internal standard log abundances and a column of ones representing an intercept term. In estimating the coefficients for the unwanted variation term, the model implicitly assumes that the mean centered multiple internal standard abundances are orthogonal to the experimental factors of interest. If this assumption is violated, it is possible that the biological variation of vital interest may also be removed along with the removal of the unwanted variation term.

Cross Contribution Compensating Multiple Standard Normalization

Redestig et al. [26] argue that the unwanted variation implied by the internal standard is influenced by contamination from the rest of the metabolites, and hence the direct use of internal standards for normalizing can also remove the true signal of interest. The authors refer to this concept as *cross contribution* and present Cross Contribution Compensating Multiple Standard Normalization (CCMN). Their model consists of several steps. First, each metabolite is transformed to have mean zero and variance one to give *z*-transformed log metabolite abundances. Then any direct dependence of the experimental factors on the abundances of the internal standards is removed using multiple linear regression with a design matrix consisting of observed factors of interest. This step assumes that the experimental factors of interest have direct influence on the internal standards and are independent of any unwanted variation implied by the internal standards. Any cross contribution occurred via other metabolites in this step is also ignored [26].

PCA is then performed on the residuals of the above regression model to obtain the row scores vectors for each sample, denoting the unwanted variation. Under the assumption that these score vectors are orthogonal to the factors of interest, the unwanted variation component is removed, using multiple linear regression with the scores vectors forming the design matrix. The normalized metabolite abundances are then obtained by reversing the preprocessing steps.

2. The use of a scaling factor

   Two of the commonly used methods which do not use any internal standards are the normalization to a total sum [28] and normalization by the median [29] of each sample. The total sum method scales each sample so that the sum of squares of all abundances in that sample equals one. The median method scales each sample so that the median of the metabolite abundances in the sample equals one. The use of the median method is found to be more practical than the sum method, especially in situations where several saturated abundances may be associated with some of the factors of interest. Other similar scaling methods include normalization by unit norm [30] and normalization to the total ion current [31]. These methods rely on the self-averaging property [27]. For instance, it is assumed that an increase in the abundances of a group of metabolites is balanced by the decrease in abundances of metabolites in another group. However, in many practical applications this property does not hold [26, 27].

3. The use of quality control samples

   Several metabolomics literature have presented ways of using QC samples [32–36] for removing unwanted variation occurred during large-scale metabolomics experiments which span over a long period of time. The use of QC samples for normalizing can only adjust for certain forms of unwanted variation. For example, it can be used to correct for the drift in signal over time, but cannot accommodate unwanted variation occurred during sample extraction or preparation.

4. The use of *non-changing* metabolites

   Remove Unwanted Variation: RUV-2

   RUV-2 method has been introduced for the purpose of identifying differentially abundant metabolites [4, 37] by accommodating both observed and unobserved unwanted technical/biological variation (refer to Subheading 6 for details). Instead of restricting only to internal or external standards, several *non-changing* metabolites are used in addition to measure unwanted variation. These non-changing metabolites should be present in the biological samples, exposed to the unwanted variation, and unassociated with the factors of interest. In metabolomics data, such non-changing metabolites can be discovered in various ways: Biological background of an experiment can provide insight into the metabolites which are known beforehand not to change with the experimental factors. The metabolites which correlate highly with the internal standards have been found to be good candidates for non-changing metabolites. In addition, the contaminant metabolites obtained from running *blank* samples and internal and external standards are also possibilities. As these non-changing metabolites are unaffected by the biology of the experiment, it can be safely assumed that any variation seen in these metabolites are arising from unwanted variation alone. The unobserved unwanted variation component for each sample is then estimated using these non-changing metabolites in classical methods of factor analysis. Therefore, this normalization approach attempts to ensure that the biological factors of interest are not removed along with the unwanted variation, and it can be applied to situations where multiple internal standards are not available. The experimental and biological unwanted variation is accommodated in the model, and it deals with both the observed and unobserved unwanted variations [4, 37].

In some experiments, it may happen that the technical variation is more prominent than the unwanted biological variation. In such situations, the use of metabolites which correlate highly with the internal standards as our only non-changing metabolites may not provide sufficient information to correct for the unwanted biological variation. We may also come across experiments where background information on non-changing metabolites is not available. A strategy for dealing with such experimental data is to fit the RUV-2 model in an iterative procedure. For example, we can use metabolites which highly correlate with the internal standards as an initial set of non-changing metabolites to fit the RUV-2 model and rank the metabolites using adjusted *p*-values to identify a further set of non-changing metabolites. Then the refined set of non-changing metabolites can be used to refit the RUV-2 model.

5.3 Choosing an Appropriate Normalization Method

The choice of an appropriate normalization method should consider several aspects. First, it should depend on the data and the experimental design. For example, the internal or external standards-based methods should not be used for experiments where it is known that unwanted unobserved biological variation exists. Normalization to a total sum and normalization by the median should not be used for experiments where the self-averaging property of the metabolome is unlikely to hold, for example in some time course experiments. Second, it should also depend on the purpose of the statistical analysis. For example, the normalized abundances obtained from the CCMN and RUV-2 methods which use the biological factors of interest in the normalization approach should not be used for the purpose of clustering or classification methods where the factors of interest must be treated as unknown. Third, it is equally important to consider both variability and bias reduction approaches [4]. For example, the assessment of the effectiveness of a normalization method should not be based solely on the reduction of variability, such as the tightness of the replicates or smaller coefficients of variation. Similarly, an increase in the number of differentially abundant metabolites found does not necessarily imply that a normalization approach improved the analysis. Depending on the association of the unwanted variation with the factors of interest, a good normalization approach can increase or decrease the number of differentially abundant metabolites [37]. Finally, the normalization method should not be carelessly chosen to best fit the expectations of the researcher (e.g., *method mining* as explained by [15]).

6 Identifying Differentially Abundant Metabolites

One of the main goals of the statistical analysis of metabolomics data is the identification of metabolites which are present in significantly differential abundances between biological factors of interest. The design of a typical metabolomics experiment can be represented in terms of a linear model fitted to each metabolite.

Let $\mathbf{Y}_{m\times n}$ be a matrix whose elements are the log transformed metabolite abundances obtained as described in the previous sections. The linear model can be described by

$$\mathbf{Y}_{m\times n} = \mathbf{X}_{m\times p}\,\gamma_{p\times n} + \mathbf{W}_{m\times q}\boldsymbol{\alpha}_{q\times n} + \boldsymbol{\varepsilon}_{m\times n}, \quad (1)$$

with $\text{Rank}[(X\,|\,W)] = p + q < m$, $\varepsilon \sim N(0,\sigma^2)$, and $\text{Cov}(\varepsilon_{ij}, \varepsilon_{i'j'}) = 0$ if $ij \neq i'j'$. Here, $\mathbf{X}$ is a matrix with columns indicating observed factors of interest, $\mathbf{W}$ is a matrix containing factors corresponding to unwanted variation, $\boldsymbol{\gamma}$ and $\boldsymbol{\alpha}$ are matrices with unobserved coefficients determining the influence of a particular factor on the metabolites, and $\boldsymbol{\varepsilon}$ is a matrix containing the unobserved error component.

The term $\mathbf{W}\boldsymbol{\alpha}$ can be optional. For example, if the unwanted variation has been adequately handled using a prior adjustment, this term can be excluded from Eq. 1. Otherwise, the columns of the matrix $\mathbf{W}$ can contain observed factors of unwanted variation such as the batch information or sample weights [4, 37–39]. In order to accommodate both the observed and unobserved variations, an alternative strategy is to estimate the matrix $\mathbf{W}$ using RUV-2 methodology [4, 37] as was explained in Subheading 5.

Some of the distributional assumptions of the linear model can be summarized as follows: for metabolite j, the linear model has residual variance σ_j^2 with sample value s_j^2 which is assumed to follow approximately a scaled chi square distribution with degrees of freedom d_j. The coefficient estimator $\widehat{\gamma_j}$ has the estimated covariance matrix given by $\text{Cov}(\widehat{\gamma_j}) = \mathbf{U}_j s_j^2$, where $\mathbf{U}_j$ is a positive definite matrix. Often certain contrasts of coefficients are of biological interest to the researcher, and these are defined by

$$\hat{\beta}_j = \mathbf{C}'\hat{\gamma}_j, \tag{2}$$

with the estimated covariance matrix $\text{Cov}(\widehat{\beta_j}) = \mathbf{C}'\mathbf{U}_j\mathbf{C}s_j^2$, where $\mathbf{C}$ is the relevant contrast matrix. These contrast estimators are assumed to be approximately normal with mean $\boldsymbol{\beta}_j$ and covariance matrix $\mathbf{C}'\mathbf{U}_j\mathbf{C}\sigma_j^2$.

For example, the researcher may be interested in obtaining *fold changes* for pairwise comparisons of some of the factors of interest. This can be extracted using an individual contrast, say $\hat{\beta}_j$, and the rank of the absolute value of these is often used to flag metabolites as differentially abundant. A disadvantage of using the fold changes for ranking metabolites is that it does not take the variability within each metabolite into account. If we do this, a highly variable metabolite with given average log fold change is treated no differently from a much less variable one exhibiting the average same fold change. A better choice for ranking metabolites is the use of *ordinary* t-statistic given by

$$t_j = \frac{\hat{\beta}_j}{\hat{\sigma}_j / \sqrt{v_j}}, \tag{3}$$

where v_j is the corresponding diagonal element of $\mathbf{C}'\mathbf{U}_j\mathbf{C}$. A drawback of using the ordinary t-statistic is that the observations with small standard errors produce large absolute t-statistics even when the fold change is substantially small. This issue has been discussed in the literature by several authors [40–43], and one of the popular means of accommodating this is the use of *moderated* t-statistic [43]

$$\tilde{t}_j = \frac{\hat{\beta}_j}{\tilde{\sigma}_j / \sqrt{v_j}}, \tag{4}$$

where the estimate $\hat{\sigma}_j = \dfrac{d_0\hat{\sigma}_0^2 + d_j\hat{\sigma}_j^2}{d_0 + d_j}$. Here, the standard deviation for each metabolite has been moderated across metabolites toward a common value using a simple Bayesian model, assuming an inverse chi square prior for the σ_j^2 with mean $\hat{\sigma}_0^2$ and degrees of freedom d_0. The moderated t-statistic has a t-distribution with

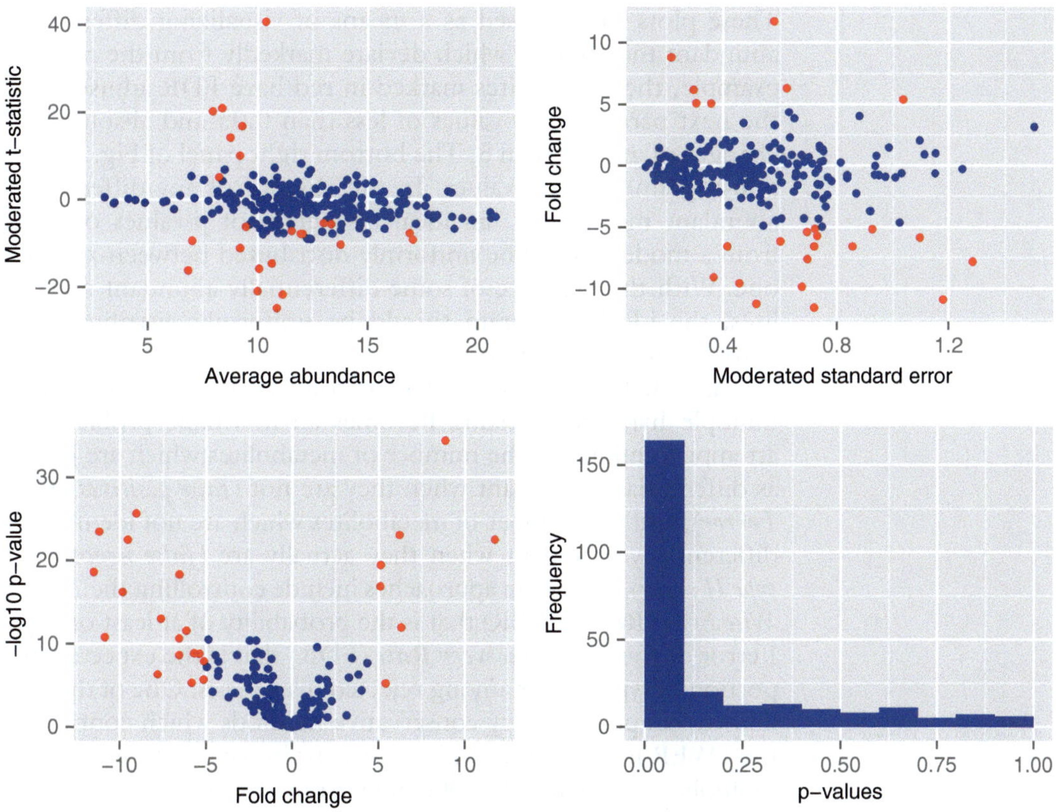

Fig. 2 Several graphical illustrations for the comparison between the *P. douarrei* and *G. pruinosa* plant species. The values were obtained using log transformed data

degrees of freedom $d_j + d_0$ under the null hypothesis $\beta_j = 0$, and is used in the same way as the ordinary *t*-statistic with the difference that the standard errors have been moderated across metabolites, taking into account the attributes of the whole ensemble of metabolites to support the inference about individual metabolites.

Several useful graphical illustrations for the comparison between the *P. douarrei* and *G. pruinosa* plant species of the leaf data set are shown in Fig. 2. Top left panel of Fig. 2 shows a plot of the moderated *t*-statistic versus the average log abundance of the metabolites. The metabolites with increased and decreased abundances are shown by the relatively large positive and negative *t*-statistics, respectively. Top right panel of Fig. 2 shows a plot of the log standard error against the absolute log fold change. The differentially abundant metabolites should have a high fold change and low standard error. The volcano plot is shown in the bottom left panel of Fig. 2, displaying the $-\log_{10}$ of the *p*-value against the log fold change. The differentially abundant metabolites have large absolute $-\log_{10}$ *p*-values and log fold changes.

These plots can be used as a means of visualizing differentially abundant metabolites which deviate markedly from the rest. For example, the metabolites marked in red have FDR adjusted (see the next paragraph) *p*-values of less than 0.05 and absolute fold changes of greater than 5. The bottom right panel of Fig. 2 shows a histogram of the *p*-values. In the absence of any differentially abundant metabolites, an ideal histogram of *p*-values obtained from a model should be uniformly distributed between zero and one. With the presence of some differentially abundant metabolites, the ideal histogram should be uniformly distributed but with a peak close to zero.

As a metabolomics experiment consists of many metabolites, multiple hypothesis testing becomes an inevitable problem. We attempt to minimize the number of metabolites which are labeled as differentially abundant when they are not (*false positives* or *type I error*), and the number of metabolites which are not identified as differentially abundant when they actually are (*false negatives* or *type II error*). Common approaches include controlling the *Family-Wise Error Rate* (FWER) that is the probability of at least one type I error or the *False Discovery Rate* (FDR) that is the expected proportion of type I error among rejected hypothesis. One of the well-known, simple, but very conservative methods which controls for the FWER is the Bonferroni adjustment in which the *p*-values are multiplied by the number of comparisons. Holm's method [44] is a similar, but less conservative adjustment, in which the unadjusted *p*-values are ordered from the smallest to the largest and are considered successively to make adjustments at each step. A more powerful and less stringent adjustment is the Benjamini–Hochberg adjustment [45], which controls for the FDR. Other methods include those given by [40, 46–48], some of which are yet to be explored in the context of metabolomics data acquired from different metabolomics approaches.

7 Concluding Remarks

In this chapter, we have presented an overview of some of the main steps involved in the statistical analysis of metabolomics data. In addition to the methods discussed in the preceding sections, classification and clustering methods are particularly important. For example, classification can be used to determine whether a patient has a certain disease, or to distinguish between different disease subtypes, or to identify the regions where the plants have been grown. Some of the multivariate statistical approaches which can be employed for such classifications include *Linear Discriminant Analysis* [49], *Partial Least Squares* [50, 51], *Support Vector Machines* [52], *Classification and Regression Trees* [53], and *Random Forests* [54]. Clustering, on the other hand, is used to discover previously unknown classes, for example, types of a disease or families of plant species. Most commonly applied methods are *Principal Component Analysis*, *Multi Dimensional Scaling* [55], *Hierarchical Clustering*, *k-means clustering* [56], and *Self Organizing Maps* [57]. The R codes for carrying out some of the methods described in this chapter will be available in the *metabolomics* package for R [58].

References

1. Fiehn O (2002) Metabolomics—the link between genotypes and phenotypes. Plant Mol Biol 48:155–171
2. Roessner U, Bowne J (2009) What is metabolomics all about? Biotechniques 46(5):363–365
3. Roessner U, Beckles DM (2009) Metabolite measurements. Springer, New York
4. De Livera AM, Dias DA, De Souza D, Rupasinghe T, Pyke J, Tull D, Roessner U, McConville M, Speed TP (2012) Normalising and integrating metabolomics data. Anal Chem 84(24):10768–10776. DOI:10.1021/ac302748b
5. Glass DJ (2007) Experimental design for biologists. Cold Spring Harbor Laboratory, New York
6. Montgomery DC (2008) Design and analysis of experiments. Wiley, Hoboken
7. O'Callaghan S, Desouza DP, Isaac A, Wang Q, Hodkinson L, Olshansky M, Erwin T, Appelbe B, Tull DL, Roessner U, Bacic A, McConville MJ, Likic VA (2012) PyMS: a Python toolkit for processing of gas chromatography–mass spectrometry (GC–MS) data. Application and comparative study of selected tools. BMC Bioinformatics 13(1):115
8. Schleif F-M (2007) Preprocessing of nuclear magnetic resonance spectrometry data. Technical report, August 2007
9. Katajamaa M, Orešič M (2007) Data processing for mass spectrometry-based metabolomics. J Chromatogr A 1158:318–328
10. Xia J, Psychogios N, Young N, Wishart DS (2009) MetaboAnalyst: a web server for metabolomic data analysis and interpretation. Nucleic Acids Res 37:W652–W660
11. Hrydziuszko O, Viant MR (2012) Missing values in mass spectrometry based metabolomics: an undervalued step in the data processing pipeline. Metabolomics 8(1):161–174
12. Katajamaa M, Oresic M (2005) Processing methods for differential analysis of LC/MS profile data. BMC Bioinformatics 6:179
13. Steuer R, Morgenthal K, Weckwerth W, Selbig J (2007) A gentle guide to the analysis of metabolomic data. Methods Mol Biol (Clifton, NJ) 358:105–126
14. Smilde AK, van der Werf MJ, Bijlsma S, van der Werff-van der Vat BJC, Jellema RH (2005) Fusion of mass spectrometry-based metabolomics data. Anal Chem 77(20):6729–6736
15. van den Berg RA, Hoefsloot HCJ, Westerhuis JA, Smilde AK, van der Werf MJ (2006) Centering, scaling, and transformations: improving the biological information content of metabolomics data. BMC Genomics 7:142
16. Temmerman L, De Livera AM, Bowne J, Sheedy RJ, Callahan DL, Nahid A, De Souza DP, Schoofs L, Tull DL, McConville JM, Roessner U, Wentworth JM (2012) Cross-platform urine metabolomics of experimental hyperglycemia in type 2 diabetes. Diab Metab vol S6:002. DOI:10.4172/2155-6156.S6-002
17. Roessner U, Nahid A, Chapman B, Hunter A, Bellgard M (2011) Metabolomics—the combination of analytical biochemistry, biology, and informatics, vol 1, 2nd edn. Elsevier B.V., New York
18. Troyanskaya O, Cantor M, Sherlock G, Brown P, Hastie T, Tibshirani R, Botstein D, Altman RB (2001) Missing value estimation methods for DNA microarrays. Bioinformatics (Oxford, England) 17(6):520–525
19. Oba S, Sato M, Takemasa I, Monden M, Matsubara K, Ishii S (2003) A Bayesian missing value estimation method for gene expression profile data. Bioinformatics 19(16):2088–2096
20. van Buuren S, Groothuis-Oudshoorn K (2011) MICE: multivariate imputation by chained equations in R. J Static Softw 45(3):1–67
21. Goodacre R, Broadhurst D, Smilde AK, Kristal BS, Baker JD, Beger R, Bessant C, Connor S, Capuani G, Craig A, Ebbels T, Kell DB, Manetti C, Newton J, Paternostro G, Somorjai R, Sjöström M, Trygg J, Wulfert F (2007) Proposed minimum reporting standards for data analysis in metabolomics. Metabolomics 3(3):231–241
22. Schlesselman J (1971) Power families: a note on the Box and Cox transformation. J R Stat Soc Ser B (Methodol) 307–311
23. Callahan DL, Roessner U, Dumontet V, De Livera AM, Doronila A, Baker AJM, Kolev SD (2012) Elemental and metabolite profiling of

nickel hyperaccumulators from New Caledonia. Phytochemistry 81:80–89
24. Gullberg J, Jonsson P, Nordström A, Sjöström M, Moritz T (2004) Design of experiments: an efficient strategy to identify factors influencing extraction and derivatization of Arabidopsis thaliana samples in metabolomic studies with gas chromatography/mass spectrometry. Anal Biochem 331(2):283–295
25. Bijlsma S, Bobeldijk I, Verheij ER, Ramaker R, Kochhar S, Macdonald I, Van Ommen B, Smilde AK (2006) Large-scale human metabolomics studies: a strategy for data (pre-) processing and validation. Anal Chem 78(2):567–574
26. Redestig H, Fukushima A, Stenlund H, Moritz T, Arita M, Saito K, Kusano M (2009) Compensation for systematic cross-contribution improves normalization of mass spectrometry based metabolomics data. Anal Chem 81(19):7974–7980
27. Sysi-Aho M, Katajamaa M, Laxman Y, Oresic M (2007) Normalization method for metabolomics data using optimal selection of multiple internal standards. BMC Bioinformatics 8:93
28. Crawford LR, Morrison JD (1968) Computer methods in analytical mass spectrometry. Identification of an unknown compound in a catalog. Anal Chem 40(4):1464–1469
29. Wang W, Zhou H, Lin H, Roy S, Shaler TA, Hill LR, Norton S, Kumar P, Anderle M, Becker CH (2003) Quantification of proteins and metabolites by mass spectrometry without isotopic labeling or spiked standards. Anal Chem 75(18):481848–26
30. Scholz M, Gatzek S, Sterling A, Fiehn O, Selbig J (2004) Metabolite fingerprinting: detecting biological features by independent component analysis. Bioinformatics (Oxford, England) 20(15):2447–2454
31. Cairns DA, Thompson D, Perkins DN, Stanley AJ, Selby PJ, Banks RE (2008) Proteomic profiling using mass spectrometry—does normalising by total ion current potentially mask some biological differences? Proteomics 8(1):21–27
32. Gika HG, Macpherson E, Theodoridis GA, Wilson ID (2008) Evaluation of the repeatability of ultra-performance liquid chromatography-TOF-MS for global metabolic profiling of human urine samples. J Chromatogr B Anal Technol Biomed Life Sci 871(2):299–305
33. Zelena E, Dunn WB, Broadhurst D, Francis-McIntyre S, Carroll KM, Begley P, O'Hagan S, Knowles JD, Halsall A, Wilson ID, Kell DB (2009) Development of a robust and repeatable UPLC-MS method for the long-term metabolomic study of human serum. Anal Chem 81(4):1357–1364
34. Lai L, Michopoulos F, Gika H, Theodoridis G, Wilkinson RW, Odedra R, Wingate J, Bonner R, Tate S, Wilson ID (2010) Methodological considerations in the development of HPLC-MS methods for the analysis of rodent plasma for metabolomic studies. Mol Biosyst 6(1):108–120
35. Dunn WB, Broadhurst D, Begley P, Zelena E, Francis-McIntyre S, Anderson N, Brown M, Knowles JD, Halsall A, Haselden JN, Nicholls AW, Wilson ID, Kell DB, Goodacre R (2011) Procedures for large-scale metabolic profiling of serum and plasma using gas chromatography and liquid chromatography coupled to mass spectrometry. Nat Protoc 6(7):1060–1083
36. Kamleh MA, Ebbels TMD, Spagou K, Masson P, Want EJ (2012) Optimizing the use of quality control samples for signal drift correction in large-scale urine metabolic profiling studies. Anal Chem 84(6):2670–2677
37. Gagnon-Bartsch JA, Speed TP (2011) Using control genes to correct for unwanted variation in microarray data. Biostatistics 13(3):539–552
38. Leek JT, Storey JD (2007) Capturing heterogeneity in gene expression studies by surrogate variable analysis. PLoS Genet 3(9):1724–1735
39. Leek JT, Storey JD (2008) A general framework for multiple testing dependence. Proc Natl Acad Sci USA 105(48):18718–18723
40. Tusher VG, Tibshirani R, Chu G (2001) Significance analysis of microarrays applied to the ionizing radiation response. Proc Natl Acad Sci USA 98(9):5116
41. Efron B (2007) Correlation and large-scale simultaneous significance testing. J Am Stat Assoc 102(477):93–103
42. Lonnstedt I, Speed TP (2002) Replicated microarray data. Stat Sin 12:31–46
43. Smyth GK (2004) Linear models and empirical Bayes methods for assessing differential expression in microarray experiments. Stat Appl Genet Mol Biol 3(1):1544–6115
44. Holm S (1979) A simple sequentially rejective multiple test procedure. Scand J Stat 6(2):65–70
45. Benjamini Y, Hochberg Y (1995) Controlling the false discovery rate: a practical and powerful approach to multiple testing. J R Stat Soc Ser B 57:289–300
46. Westfall PH, Young SS (1993) Resampling-based multiple testing: examples and methods for p-value adjustment. Wiley-Interscience, New York

47. Efron B, Tibshirani R, Storey JD, Tusher V (2001) Empirical Bayes analysis of a microarray experiment. J Am Stat Assoc 96(456):1151–1160
48. Storey JD, Tibshirani R (2001) Estimating false discovery rates under dependence, with applications to DNA microarrays. Technical report
49. Friedman J, Hastie T, Tibshirani R (2001) The elements of statistical learning, 2nd edn. Springer, New York
50. Frank IE, Friedman JH (1993) A statistical view of some chemometrics regression tools. Technometrics 35(2):109–135
51. Wold S, Sjostrom M (2001) PLS-regression: a basic tool of chemometrics. Chemom Intell Lab Syst 58(2):109–130
52. Vapnik V (1999) The nature of statistical learning theory. Springer, Berlin
53. Breiman L, Friedman JH, Olshen RA, Stone CJ (1984) Classification and regression trees. Wadsworth International Group, Belmont
54. Breiman L (2001) Random forests. Mach Learn 45(1):5–32
55. Cox TF, Cox MAA (2001) Multidimensional scaling. Chapman and Hall, Boca Raton
56. MacQueen J (1967) Some methods for classification and analysis of multivariate observations. In: Proceedings of the fifth Berkeley symposium on mathematical statistics and probability. University of California Press, Berkeley, pp 281–297
57. Kohonen T (1982) Self-organized formation of topologically correct feature maps. Biol Cybern 43(1):59–69
58. De Livera AM, Bowne J (2013) Metabolomics: A collection of functions for analysing metabolomics data. R Package Version 0.1.1

INDEX

Ute Roessner and Daniel Anthony Dias (eds.), *Metabolomics Tools for Natural Product Discovery: Methods and Protocols*, Methods in Molecular Biology, vol. 1055, DOI 10.1007/978-1-62703-577-4, © Springer Science+Business Media, LLC 2013

MIX
Papier aus verantwortungsvollen Quellen
Paper from responsible sources
FSC® C105338

If you have any concerns about our products, you can contact us on
ProductSafety@springernature.com

In case Publisher is established outside the EU, the EU authorized representative is:
Springer Nature Customer Service Center GmbH
Europaplatz 3, 69115 Heidelberg, Germany

Printed by Libri Plureos GmbH
in Hamburg, Germany